Hermann H. Hahn
Rudolf Klute (Eds.)

PRETREATMENT IN CHEMICAL WATER AND WASTEWATER TREATMENT

Proceedings of the
3rd Gothenburg Symposium 1988
1.-3. Juni 1988
Gothenburg

Springer-Verlag
Berlin Heidelberg New York
London Paris Tokyo

Prof. Dr. Hermann H. Hahn
Dr. Rudolf Klute
Institut für Siedlungswasserwirtschaft
der Universität Karlsruhe
Postfach 6380
D-7500 Karlsruhe 1

ISBN 3-540-19423-1 Springer-Verlag Berlin Heidelberg New York
ISBN 0-387-19423-1 Springer-Verlag New York Berlin Heidelberg

This work is subject to copyright. All rights are reserved, whether the whole or part of the material is concerned, specifically the rights of translation, reprinting, reuse of illustrations, recitation, broadcasting, reproduction on microfilms or in other ways, and storage in data banks. Duplication of this publication or parts thereof is only permitted under the provisions of the German Copyright Law of September 9, 1965, in its version of June 24, 1985, and a copyright fee must always be paid. Violations fall under the prosecution act of the German Copyright Law.

© Springer-Verlag Berlin Heidelberg 1988
Printed in Germany

The use of registered names, trademarks, etc. in this publication does not imply, even in the absence of a specific statement, that such names are exempt from the relevant protective laws and regulations and therefore free for general use.

The publisher cannot assume any legal responsibility for given data, especially as far as directions for the use and the handling of chemicals are concerned. This information can be obtained from the instructions on safe laboratory practice and from the manufactures of chemical and laboratory equipment.

Printing and Binding: Weihert-Druck GmbH, Darmstadt
2151/3140 - 543210

PREFACE

The International Gothenburg Symposia on Chemical Treatment have proven to be a unique platform for the exchange of ideas between theory and practice. They bring together administrators, engineers and scientists, who are concerned with water purification and wastewater treatment through precipitation, coagulation and subsequent solid/liquid separation.

This volume contains the proceedings of the 3rd Symposium, focussing on Pretreatment. Pretreatment is understood as the scene total of all measures taken at the pollutant source to protect water supply, the sewerage system, the central treatment plant, and the aqueous environment. It is, where applicable, the most efficient measure in ecological and economic respects.

The contributions of this third volume address questions of surveillance, automation and remote control of installations as well as the principles of legal, administrative and economic measures for regulations within the context of pretreatment. Special attention is given to the possibilities and limits of pretreatment of industrial discharges.

Again it is the editors' privilege to acknowledge the invaluable help from the authors of this book. It is the editors' hope that they might convey the significance and potential of pretreatment in water supply, in industrial waste management and in municipal wastewater treatment and sludge handling.

Again, Karin Knisely, made a coherent reading from a group of individual papers and Adam Leinz and Springer Publishing Company more than helped to make the book appear on time. Boliden Kemi of Helsingborg, Sweden not only initiated the International Gothenburg Symposia but also supported all meetings with their contributions. - The editors express their deep gratitude to all of them who made this book possible.

Karlsruhe, Germany
March 1988

H.H. Hahn
R. Klute

MEMBERS OF THE SCIENTIFIC COMMITTEE

Prof. E. Arvin, Denmark

Prof. T. Asano, Japan-USA

Prof. P. Balmér, Sweden

Dr. J. Bernard, France

Dr. M. Boller, Switzerland

Prof. A. Grohmann, FRG

Prof. H.H. Hahn, FRG

Dr. R. Klute, FRG

Prof. L. Lijklema, Netherlands

Prof. C. O'Melia, USA

Prof. R. Perry, England

Mr. K. Stendahl, Sweden

Prof. M. Viitasaari, Finland

Prof. H. Ödegaard, Norway

Contents

Section I: Water Supply

From Filters to Forests: Water Treatment and Supply 3
C. R. O'Melia

Pretreatment of Drinking Water to Control Organic Contaminants
and Taste and Odor 15
I.H. Suffet, R.J. Baker and T.L. Yohe

Water Quality Problems and Control Strategies for the Water Supply
of Tianjin City 41
X. Zhu

Humic Substances Removal by Alum Coagulation
- Direct Filtration at Low pH 55
J. Fettig, H. Ødegaard and B. Eikebrokk

Modeling the Effects of Adsorbed Hydrolyzed Al(III)-Ions
on Deep Bed Filtration 67
Z. Wang

Polyectrolytes for the Treatment of Tap and Filter Back Washing Water 91
J. M. Reuter and A. Landscheidt

New Coagulant Injection Process 103
C. Ventresque and G. Bablon

Odour Control by Artificial Groundwater Recharge 113
R. Sävenhed, B.V. Lundgren, H. Borén and A. Grimvall

Section II: Industrial Discharges

Pretreatment of Industrial Wastewater:
Legal and Planning Aspects - A Case Study 125
H.H. Hahn and K.-H. Hartmann

Clean Technology in the Netherlands: The Role of the Government 139
A.B. van Luin and W. van Starkenburg

Synergistic Approach to Physical-Chemical Wastewater Pretreatment
in the Food Industry 151
G. von Hagel

Separation of Heavy Metals from Effluents by Flotation 159
I. Zouboulis and K.A. Matis

Pretreatment of Wastewater from the Automobile Industry 167
R. Klute

Industrial Wastewater Pretreatment of a Dental-Pharmaceutical Company .. 179
F.W. Günthert and P.-M. Hajek

Membrane Separation Processes for Industrial Effluent Treatment 189
P.S. Cartwright

Alternative Treatment of De-icing Fluids from Airports 201
M. Boller

Separators and Emulsion Separation Systems for Petroleum, Oil,
and Lubricants 217
H. Nöh

Chemical Treatment of Flue Gas Washing Liquids 227
A.N. Grohmann, L. Bauch and H.-P. Scheerer

Section III: Wastewater and Sludge

Hydrogen Sulphide Control in Municipal Sewers 239
T. Hvitved-Jacobsen, B. Jütte, P.H. Nielsen and N.Aa. Jensen

Coagulation as the First Step in Wastewater Treatment 249
H. Ødegaard

Pre-precipitation for Improvement of Nitrogen Removal in Biological
Wastewater Treatment 261
I. Karlsson

Chemically Supported Oil and Grease Removal
in Municipal Wastewater Treatment Plants 273
H. Roggatz and R. Klute

Chemical-biological Treatment Versus Chemical Treatment
- A Case Study 281
A. Ilmavirta

Reuse of Chemical Sludge for Conditioning of Biological Sludges 291
Y. Watanabe and A. Toyoshima

Influence of Sludge from Chemical Biological Wastewater Treatment
on Nitrification and Digestion 307
C.F. Seyfried, H.-D. Kruse and F. Schmitt

Pretreatment of Sludge Liquors in Sewage Treatment Plants 319
B. Paulsrud, B. Rusten and R. Storhaug

Heavy Metal Removal from Sewage Sludge: Practical Experiences
with Acid Treatment 327
M. Ried

Treatment of Filter Effluents from Dewatering of Sludges by a New
High Performance Flocculation Reactor 335
U. Wiesmann, K. Oldenstein and L. Fechter

Pretreatment for Wastewater Reclamation and Reuse 347
T. Asano and R. Mujeriego

Section I

Water Supply

From Filters to Forests: Water Treatment and Supply

C. R. O'Melia

Abstract

Physical, chemical and biological processes in the atmospheric, terrestrial, and aquatic environments establish the major characteristics of water supplies and affect the concentrations of pollutants in these supplies. The design and operation of water treatment facilities should reflect characteristics of the water source. Important natural processes and water quality characteristics are identified and their effects on water treatment systems are indicated. Forests affect filters.

Introduction

In this paper the treatment of surface waters for the removal of particles and natural organic substances in the production of potable water supplies is traced from filters back to forests. The direction is "upstream", beginning with packed bed filters and moving backward through a conventional water treatment plant by considering in order sedimentation basins, flocculation facilities, rapid mixing tanks with coagulant additions, surface water supplies such as lakes and reservoirs, and ending in the drainage area tributary to the supply. Based on these discussions, conclusions are presented about the functioning of these components of a water supply system and about approaches to be used in the design and operation of these treatment systems.

Filters

Packed bed filters remove particles from the suspensions that flow through them. These particles may have been present in the raw water supply or have been formed in pretreatment. Particles in raw water supplies vary widely in origin, concentration, composition, and size. Some are constituents of land-based or atmospheric inputs (clays, microorganisms including pathogens, asbestos fibers, etc.) and others are produced by chemical or biological processes in the water source (algae, organic detritus, inorganic precipitates of $CaCO_3$, $FeOOH$, etc.). Soluble natural organic substances (e.g., humic substances) are present in all surface water supplies; these can originate on the land or in the water. Synthetic organic substances and many trace and toxic metals are particle reactive and are associated with particulate materials in the water source. The addition of chemical coagulants aggregates particles present in the water supply, adsorbs and precipitates humic substances, and forms new solid particles ($Al(OH)_3(s)$, etc.). The particles reaching a filter in a water treatment system reflect all of these inputs and processes.

Effective particle removal in filtration depends on physical and chemical factors. Particles must be transported to the surfaces of filter media and they must attach there for removal to occur. Both hydrodynamics and chemistry are important; filtration is a physicochemical process. Design criteria most often specify filtration rate, media size, and bed depth. These are pertinent to filter performance, but are not sufficient to assure filter effectiveness and usually have less impact on the process than several other properties of the system.

Pretreatment chemistry is the most important factor affecting particle removal in packed bed filters. If the solution chemistry of the water to be treated, the surface chemistry of the particles to be removed, and the surface chemistry of the media used to accomplish removal are not properly controlled, efficient particle removal just will not occur, regardless of the filtration rate, bed depth, and media size installed in a filter system. In practice, this is accomplished by proper selection of the type and dosage of pretreatment chemicals. These may include aluminium or iron(III) salts, polymeric inorganic coagulants such as polyaluminium chloride (PACl), polyiron chloride (PICl), and activated silica, synthetic organic polymers and polyelectrolytes, and even ozone.

The principal factors affecting head loss development in packed bed filters are the concentration and the *size* of the particles applied to the filters. Media size, filtration rate, and bed depth have important effects, but the physical properties of the particles applied to filters dominate head loss development in those filters. Small particles produce large head losses; submicron particles induce rapid head loss development when treated by conventional packed bed filters.

The size of the particles in the water being filtered also affects the removal efficiency of a clean filter bed. A critical suspended particle size exists, in the region of 1 μm, for which the filtration efficiency of clean beds is a minimum. Particles in this size region have the fewest opportunities for contact with the filter media and subsequent removal from suspension. Smaller particles are transported predominantly and effectively by convective diffusion and larger ones by fluid flow (interception) and gravity settling.

When filtration is successful in the early stages of a filter run, i.e., when the hydrodynamics of particle transport and the chemistry of particle attachment combine to produce effective particle removal in a clean filter bed, these retained particles can act as filter media during later stages of the run. This produces a ripening of the filter; the efficiency of removal of particles or turbidity improves with time and can be substantially greater than the "clean" bed in place when the filter run is begun. Particles removed by the filter become sites, collectors, or media for the deposition of additional material. The important result is that in many and perhaps most of the packed bed filters used in water and wastewater treatment, the actual filter media operative during most of a filter run are not the sand, coal or other media specified by the designer. Rather, they are the particles present in the water applied to the bed and removed in the filter. Stated another way, the actual filter media active during most of a filter run are formed from particles present in the water supply and altered during pretreatment processes and from new particles formed by precipitation during pretreatment. Hence the "design" of filter media really involves the selection of the raw water source and the design of the pretreatment facilities that are installed ahead of the filters. The design of filter media is established "upstream" of the filters by the concentration, size, and surface properties of the particles applied to filters and by the solution chemistry of the aqueous phase. With this established, the first step upstream from the filters is into the sedimentation basins.

Sedimentation Basins

Sedimentation basins remove particles by gravity, and conventional rectangular and circular basins are designed to do so on the basis of overflow rate. This overflow rate is the ratio of the volumetric flow rate applied to a sedimentation basin divided by the horizontal area available to collect particles in the basin. It is numerically equal to the gravitational settling velocity of those particles that should be completely removed by the basin.

In addition to gravity settling, coagulation occurs in all settling facilities receiving particles that can attach when they come into contact. Whenever sedimentation follows effective coagulant addition, some coagulation will occur. Contacts occur among particles in settling tanks by diffusion and by differential settling. These are analogous to contacts between suspended particles and filter media by convective diffusion and gravity forces in filtration. Both processes lead to the attachment of small particles to larger ones, so that settling basins aggregate and remove small particles effectively in many cases. Because this aggregation process takes time, conventional sedimentation basins are designed to provide a certain reaction or detention time.

Because sedimentation basins alter the concentration, size, and size distribution of particles in suspension, they have profound effects on subsequent filtration facilities. Small particles are aggregated to larger sizes. Large particles are removed. Both of these effects lengthen filter runs. Ripening periods are also lengthened; effects on filtrate quality are complex because of the one μm "window" in the effects of suspended particle size on filtration efficiency and the role of retained particles in filter ripening.

The separate effects of overflow rate and detention time on basin performance and, in turn, on filter performance, can be expected to differ. For example, the installation of tube or lamella settlers substantially decreases the overflow rate of a basin and leaves the detention time unchanged. The result may be an increased removal of small particles by gravity, and a decreased aggregation of small particles by diffusion and differential settling because of rapid removal of larger "targets" from suspension. The consequent effects on turbidity removal, ripening, and head loss development in filters are not clear or even established.

Sludge blanket clarifiers combine the processes of sedimentation and filtration. Particles entering these basins can collide with previously retained particles in the sludge blanket and can then be removed by gravity. Collisions among the particles in the blanket occur by diffusion, by bulk fluid flow, and by differential settling. Some aggregates may break up in the sludge blanket and its appurtenances. Aggregates can remain in the blanket for long periods of time, and their surface chemistry may change with aging. Particle concentrations, particle size distributions, and particle surface characteristics in the effluents from sludge blanket clarifiers can be expected to differ considerably from other types of settling units.

In summary, sedimentation facilities can combine characteristics of flocculation, pure sedimentation, and filtration. The particles in the effluents from these become the filter media in the packed bed filters to which these basins discharge. Because different sedimentation facilities utilize different physical processes to accomplish their function of suspended solids removal, the filterability of their effluents can be expected to differ accordingly. With this considered, the next and second step upstream from the filters is into the flocculation tanks.

Flocculation Tanks

Flocculation tanks change the size distribution of the particles in a suspension; a large number of small particles is transformed into a smaller number of larger aggregates. Traditionally, the objective of flocculation has been to produce aggregates that will settle well; the focus has been placed on forming particles that are larger than 100 μm or so in size and that may also be dense so that they can settle easily. More recently, as sedimentation basins are eliminated in the treatment of some water supplies (a process termed direct filtration), the emphasis in flocculation is on producing filterable particles. The meaning of filterable is not well established, but it is taken here to refer to a substantial reduction in the concentration of submicron particles that induce large head losses in filtration and also a reduction of particles in the 1 μm size region to minimize penetration of filter beds.

Flocculation facilities are designed to produce aggregation by fluid motion that induces fluid shear or velocity gradients within the fluid. Particles follow fluid elements and, since these travel at different velocities, the particles also move at different speeds. Collisions occur and, if the particles have been chemically pretreated and destabilized, aggregates are formed. Pertinent design criteria are generally considered to be the mean velocity gradient (G, in units of s^{-1}) and the hydraulic detention time to allow the flocculation reaction to proceed. As in filtration, these criteria are important but inadequate and incomplete. First, collisions occur by other physical processes; certainly diffusion causes collisions between submicron and larger particles, and differential settling may also occur. The size distribution of the particles in the suspension being treated is important in establishing the flocculation rates that actually occur in flocculation basins.

Second, flocculation is a second order reaction, so that flocculation rates are very dependent on particle concentration. In many cases, particles *must* be added to the water to provide any significant flocculation rate. This is accomplished by adding particles such as clays or, more often, precipitating metal hydroxides such as $Al(OH)_3(s)$ or $Fe(OH)_3(s)$. The formation of new particulate mass from the precipitation or adsorption of humic substances by aluminium or iron(III) coagulants can also enhance flocculation rates in this way. The use of sludge blanket clarifiers provides interparticle contacts and can accomplish flocculation. Finally, direct filtration can replace extensive flocculation as filter media provide contact opportunities for attachment (deposition or "aggregation") of suspended particles.

Third, fluid motion can be produced mechanically in well mixed tanks or reactors and also hydraulically in pipelines or long tanks. These two physical mixing processes should produce different effluents even when operated at identical detention times and velocity gradients. The contents of a stirred tank are comprised of fluid elements and particles that reside in the tank for a very wide range of times, ranging in fact from zero to infinity. The result is that some particles must escape with little or no flocculation or aggregation while others have an opportunity to grow to very large and perhaps dendritic, light, porous aggregates. This particle size distribution can be narrowed by using a number of such mixed tanks in series and by the breakup of larger aggregates by fluid shear, but it can be expected that the particle size distribution and density of aggregates leaving a mechanically mixed flocculation system will be different and more diverse than those achieved in a tubular flocculator. These effects may be unimportant when sedimentation tanks are used and in which additional flocculation continues to transform the particle size distribution, but it is plausible that they have real and important effects in direct filtration.

From flocculation, the next and third step upstream from the filters is into the rapid mixing tank, where chemical coagulants are added to the system.

Coagulant Addition and Mixing

Facilities for mixing coagulant chemicals with the water to be treated perform several functions and provide reactor volume for many diverse reactions. When aluminium or iron(III) salts are used as coagulants, an important and often overlooked function of rapid mixing facilities is the *formation* of the coagulant species that are operative in the coagulation process. Aluminium and iron(III) react rapidly and irreversibly with water under conditions typical of water and wastewater treatment. With aluminium, for example, monomers such as $AlOH^{2+}$, small polymers such as $Al_2(OH)_2^{4+}$, larger cationic polymers such as $Al_{13}O_4(OH)_{24}^{7+}$, and an amorphous precipitate, $Al(OH)_3(am)$, are among the species that can be formed. The monomeric species, Al^{3+}, is not present in sufficient concentration to serve as a coagulant. The actual coagulant species that are active in the process are produced primarily *within* the initial or rapid mixing facilities when alum or ferric chloride are used. These salts are, strictly speaking, not coagulants; they are the reactants or starting materials from which the actual coagulants are formed by hydrolysis reactions during and soon after mixing.

Some coagulants are produced externally to the treatment process and then added to the water to be treated in the rapid mixing facilities. Examples include synthetic organic polyelectrolytes and polymers and also partially neutralized inorganic coagulants such as activated silica, polyaluminium chloride (PACl), and polyiron chloride (PICl). The rapid mixing conditions necessary to disperse these preformed coagulants may differ from those necessary for effective formation of coagulant species from aluminium and ferric salts.

The second important function of rapid mixing facilities is dispersal of coagulant to promote the reactions of coagulant species with contaminants of concern. Much of the work on this subject has focussed on the reactions and removal of solid particles or turbidity. This emphasis is incomplete and can be seriously misleading. In addition to water itself, at least three different and distinct types of substances can react with coagulants in water and wastewater treatment. First, there are solid particles. These, as stated previously, usually receive the greatest attention; they are, however, often not important in establishing coagulant requirements. Second, there are soluble phosphorus species. These can be significant in contributing to chemical requirements when metal coagulants are used in wastewater treatment; they are generally not important in surface water treatment, even when phosphorus removal is the desired objective, as in the treatment of river inflow to Lake Tegel in Berlin. Third, there are humic substances and other natural organic materials. Humic substances are the major organic constituent of unpolluted waters. They are derived from soil and are also produced within natural waters and sediments by chemical and biological processes. Humic substances are anionic macromolecules of low to moderate molecular weight; their charge is due primarily to carboxyl and phenolic groups; they have both aromatic and aliphatic components and can be surface active; they are refractive and can persist for decades or longer. In natural waters, the concentration of soluble natural organic matter can range from less than one to more than 20 mg/l expressed as dissolved organic carbon (DOC).

Consider a water containing natural organic matter at a concentration of 10 mg DOC/l and also natural turbidity comprised of 10 mg/l of a kaolinite clay. The negative charge on humic substances at neutral pH is on the order of 10^{-2} eq/g of organic carbon and the cation exchange capacity of the clay would be about 10^{-2} eq/100 g of clay. If a coagulant were added to react stoichiometrically and neutralize these negative charges, the coagulant requirements would be 10^{-4} eq/l for the DOC and only 10^{-6} eq/l for the turbidity. The concentration of DOC establishes the coagulant requirements for this water; the charge and coagulant requirements for the turbidity are minor. While other raw water compositions will lead to other charge requirements and other coagulant dosages, this example is illustrative of many situations. In many cases in water and wastewater treatment (in the writer's opinion, in most cases in water and tertiary wastewater treatment), coagulant requirements are established by DOC concentrations, and not by turbidity or phosphorus levels. Often, and perhaps most often, DOC is not even measured in practice; emphasis is placed on constituents that have less or even negligible impact on chemical requirements.

Mixing is accomplished by creating a high intensity of turbulence in the water. Eddies are created that have a broad spectrum of sizes, with large eddies containing most of the energy and with small eddies dissipating most of the energy by inertial and viscous processes. The eddy size below which energy is dissipated by primarily viscous effects is termed the Kolmogoroff microscale, η. As power input and turbulence are increased, the size of the microscale is decreased. For the interaction between cationic hydroxoaluminium polymers and solid particles to achieve particle destabilization by neutralization of particle charge, Amirtharajah has proposed that mixing conditions that produce a Kolmogoroff microscale of the same order as the size of the particles to be destabilized are ineffective. Larger and smaller energy inputs (corresponding to smaller and larger microscales) are preferred. Velocity gradients of $3000 - 5000$ s^{-1} and $700 - 1000$ s^{-1} are suggested for rapid mixing facilities, with the intermediate region to be avoided. Similar analyses for other contaminants, other coagulants, other coagulation mechanisms, and for coagulant formation merit development.

Rapid mixing facilities include hydraulic jumps, static mixers, in-line blenders, diffusers, and stirred-tank reactors. A wide variety of hydraulic residence times and residence time distributions may be used, even at the same velocity gradient or energy input.

There are, therefore, many coagulant species and many contaminants; a wide variety of reactions can occur in parallel and in series. There are, in addition, many methods of mixing, and these affect product formation in such a diverse chemical system. It follows that the size, composition, and chemical reactivity of the particles and many of the soluble species leaving a rapid mixing facility to enter a flocculation basin can vary with raw water quality, type and dosage of coagulant, and type and duration of mixing.

From chemical addition and rapid mixing, the next and fourth step upstream from the filters is into the natural water supply.

The Natural Water Supply

Many and diverse physical, chemical, and biological reactions occur in natural waters. To narrow the scope of presentation somewhat, consideration is given here to natural lake and manmade reservoirs. These are viewed as sedimentation basins, receiving particulate inputs from surface runoff and atmospheric deposition, and within which particles are produced and destroyed by biological, chemical, and physicochemical reactions. Attention is given to coagulation as a physicochemical reaction that depends strongly on solution chemistry and that can increase the role of sedimentation in removing particles and particle-reactive pollutants from the water source. Coagulation and sedimentation in a natural water source are remarkably similar in function and in result to their analogues in a water treatment plant.

The extent of particle production in a natural water is related to the input of a limiting nutrient such as phosphorus. High phosphorus inputs result in extensive growth of algal biomass in a lake or reservoir, and the system may be termed eutrophic. When planktonic algae settle to the bottom waters, the process is reversed; organic matter is degraded, phosphorus is released, and oxygen is consumed. The formation of anoxic bottom waters is another characteristic of a eutrophic lake. Because of these effects, efforts have properly been directed at controlling phosphorus inputs to surface waters using such strategies as chemical and biological wastewater treatment, replacement of phosphorus in detergents, and various methods of fertilizer application and land management. Particulate organic matter may also reach the sediments and be deposited there, where anoxic decomposition can proceed.

Phosphorus can also have some beneficial effects on aquatic ecosystems. The uptake of phosphorus into biomass in the eutrophic zone of a lake is accompanied by an uptake of nitrogen. When this nitrogen is present in anionic form (NO_3^-), protons are consumed and the alkalinity of the water is increased. The acidifying effects of nitric acid in acid rain can therefore be at least partially offset by photosynthesis in lacustrine systems.

Phosphorus inputs produce particles in the photosynthetic zone of a lake, and these particles act as a conveyor belt for particle reactive pollutants, removing them from solution and transporting them to the sediments. Eutrophic lakes can thus be more efficient scavengers of pollutants than oligotrophic ones, if particles are buried in the sediments before they are mineralized in the water column. Metals such as iron and lead are particularly particle reactive; synthetic organic pollutants may partition into organic particles if they are sufficiently hydrophobic, a property described by their octanol-water partition coefficient (K_{ow}). It follows that efforts to reduce nutrient inputs to a lake in order to retard or reverse eutrophication can also result in changes in the processes that determine the transport and fate of particle reactive pollutants in the system.

The colloidal stability of the particles in a lake depends primarily on the solution composition of the lake. Major divalent cations such as Ca^{2+} and Mg^{2+} act to stabilize natural particles (i.e., they enhance interparticle attachment and accelerate coagulation rates). Natural organic substances (humic substances) act as dispersing or stabilizing agents (i.e., they retard interparticle attachment and slow coagulation rates).

The enhanced stability of natural particles in natural waters containing humic substances is a consistent observation without a clear cause. Humic substances adsorb on most surfaces, and it has been proposed that this adsorption produces electrostatic and perhaps steric repulsion. Divalent metal ions may enhance particle destabilization by chemical reactions with both the primary particles and the adsorbed humic substances. Conclusive evidence for such mechanisms has not yet been obtained.

Considering that particle stability in lakes is determined primarily by chemical characteristics of the lake water rather than by the nature of the particles themselves, the attachment probability or effectiveness of interparticle contacts can be expected to vary from lake to lake in a manner that is consistent with variation in lake solution chemistry. Low aggregation rates are expected in soft, colored waters and rapid coagulation can occur in hard waters that are low in DOC. Limited field observations on Swiss lakes support this view.

Coagulation, sedimentation, and deposition occur in all natural waters, although their significance can vary widely. The processes can have important effects on the fate of particles, pollutants, and nutrients in natural systems. Small particles collide with larger ones in lakes and reservoirs primarily by Brownian diffusion and differential settling; velocity gradient or shear-induced collisions may be important in rivers. Convective diffusion and gravity sedimentation are dominant in particle deposition in ground water aquifers. In this perspective, rivers resemble flocculation basins, lakes and reservoirs are analogous to sedimentation tanks, and ground water aquifers are similar to packed bed filters. Chemical coagulants and destabilizing agents originate in terrestrial systems from weathering reactions, natural vegetation, and man's activities. Some can also be formed in natural waters, as planktonic plants are produced by photosynthesis and degraded heterotrophically to yield, among other products, autochthonous aquatic humic substances.

The emphasis in this discussion of water quality in lakes is on phosphorus, hardness, and humic substances, and particularly on the latter. Phosphorus inputs from natural and manmade sources determine eutrophication in most cases, can lead to the formation of aquatic humic substances, and also establish particle production in the natural conveyor belt transporting pollutants to sediments and can generate alkalinity. Hardness destabilizes particles and accelerates the rate of pollutant removal by the particle conveyor belt. Dissolved natural organic substances stabilize particles in natural systems and retard pollutant removal to sediments by particle coagulation, sedimentation, and deposition. These substances are also the precursors of trihalomethanes when chlorine is used in water treatment, they establish design requirements when activated carbon is used, and they may also carry metals and synthetic organic substances through water treatment plants into potable water supplies. The dissolved organic carbon concentration in a water source is an important chemical characteristic in establishing source water quality and generally is the single most important parameter in water treatment plant design.

From the water supply, the last and fifth step upstream from the filters is into the tributary drainage area surrounding the water source.

Forests

The subtitle "Forests" is used here as a surrogate for the sources of inputs to the water source of substances of interest. It is selected in part because the emphasis here is on natural processes. Some persistent pollutants such as PCBs are no longer used commercially; consumption of others such as lead is declining rapidly; materials such as herbicides and pesticides can be managed by proper chemical selection and by land management; chemical and biological treatment can effectively remove most pollutants from wastewaters and, in the process, degrade many of them to inert materials. Without minimizing the difficulty or the importance of controlling the inputs of toxic substances and pathogens from terrestrial and atmospheric sources in the water source, the focus here is on phosphorus, hardness, and humic substances. The inputs of these substances to the water source have dominant impacts on source water quality and, in turn, on water treatment plant design and performance.

The use of phosphorus in agriculture, in detergents, and in other activities has lead to increased inputs of this element to natural waters. Schindler has demonstrated that physical and biological processes in lakes provide natural mechanisms for supplying sufficient carbon and nitrogen to utilize phosphorus inputs; phosphorus is normally the limiting nutrient in these systems. Autochthonous particle formation, productivity, biomass accumulation, and hypolimnetic oxygen consumption in lakes and reservoirs are ordinarily established by atmospheric and terrestrial phosphorus inputs to these systems. The fate of these particles in the water source can then be controlled by the hardness and humic substances in solution.

The inorganic chemistry of fresh water can be interpreted as the result of an acid-base titration between carbon dioxide in the atmosphere and that produced in the soil with bases of rocks. Atmospheric mineral acid deposition can also contribute to the reaction. The dissolution of carbonate rock can control the concentrations of Ca^{2+}, Mg^{2+}, HCO_3^-, and H^+ in fresh waters; the weathering of aluminosilicate minerals is the major source of Na^+, K^+, and $Si(OH)_4$, and also contributes Ca^{2+} and Mg^{2+}. The hardness of a water supply depends on the composition of the minerals in the area, and also the rate at which acids are introduced as reactants to the neutralization reaction. The presence of calcareous soils, aggrading forests, and acid deposition lead to rapid weathering and to hard water supplies. The resulting hard waters can be expected to assist particle destabilization and to enhance coagulation and sedimentation in surface water supplies and particle deposition (filtration) in ground water aquifers.

Allochthanous organic inputs to aquatic systems originate in two natural sources, soils and plants. Organic matter from soil has decomposed for a longer time and is less biodegradable than organic matter from plants. These two sources are the major inputs of dissolved organic carbon (DOC) to streams, small to moderate rivers, and small lakes. The effluents from well designed and operated biological wastewater treatment plants are an additional land-based source of recalcitrant dissolved organic matter in natural water. The effects of these effluents on the stability of natural particles in aquatic systems has not been investigated, but it is possible that they contribute to particle stabilization, thereby retarding natural coagulation and sedimentation and reducing pollutant removals in natural waters.

Algae are the principal source of autochthonous DOC in natural waters. Some DOC is released directly by algal cells and additional DOC is produced by heterotrophic organisms feeding on algae and organic detritus. Since phosphorus inputs establish algal productivity in lakes, the production of autochthonous DOC should be related to these inputs. The balance between natural allochthanous and autochthonous DOC inputs varies with lake size, with the proportion of DOC arising from autochthonous inputs increasing with lake size.

Summary

In this summary the perspective is returned to its normal direction, i.e., from forests to filters.

The chemical composition of a fresh water depends on the nature and the use of its tributary drainage area. The focus here is on phosphorus, humic substances, and hardness. Domestic and commercial production combine with natural processes to establish phosphorus inputs to natural waters. Terrestrial biological processes provide allochthanous inputs of humic substances to natural waters. Biological wastewater treatments may also be a contributor of recalcitrant dissolved organic matter. Chemical and biological weathering reactions in the titration of rocks with carbonic and mineral acids establish the hardness of aquatic systems.

Particle production in aquatic systems depends upon these phosphorus inputs. Algal production can result in the formation of autochthonous humic materials. The fate of particles in lakes, whether inorganic or organic, whether allochthanous or autochthonous, is affected by surface chemistry and gravity. Natural coagulation can increase sedimentation rates and contribute to the removal of particles and of particle reactive pollutants from fresh waters. Natural coagulation is accelerated by divalent metal ions and retarded by aquatic humic substances. The removal of particles and particle reactive pollutants is expected to be greater in hard waters that are low in DOC than in soft, colored waters. The quality of the water entering a water treatment plant depends in part on the particle production and the natural coagulation and sedimentation that occur in the water source.

The coagulant requirements in a water treatment plant are normally established by the DOC in the water. With conventional chemicals such as alum and ferric chloride, the actual coagulant species are formed within the water to be treated in rapid mixing facilities. There are many contaminants, many coagulants, and many mixing possibilities; the rapid mixing facilities may be the least understood component in a water treatment plant design.

Flocculation facilities alter particle sizes and particle size distributions. Particle size requirements for sedimentation and filtration differ, so that different flocculator designs are needed for these two different solid-liquid separation processes. Hydraulically mixed inline flocculators and mechanically mixed stirred tank flocculators have different residence time distributions, and this may have complex effects on the particle size distributions produced by these facilities.

Coagulation occurs in all sedimentation tanks when coagulants are used. Conventional tanks, lamella settlers, and upflow clarifiers differ in the interparticle contacts that they provide, and in the amount of coagulation that can occur within them. The particle size distributions in the effluents from these facilities should differ.

The design of filter media begins with the selection of the water supply source and continues through coagulant selection and pretreatment facilities design. The filter media operative during most of a filtration run are the particles that have been removed by the filter. These originate on the land, in the lake, and are produced by coagulants. Their concentration, size, and surface properties reflect all of these prior processes.

References

[1] Amirtharajah, A., Trusler, S.L. (1986) ASCE Jour. Envir. Engineering *112*, 1085
[2] Bernhardt, H., Hoyer, O., Schell, H., Lüsse, B. (1985) Z. Wasser-Abwasser-Forsch. *18*, 18
[3] Lawler, D.W., O'Melia, C.R., Tobiason, J.E. (1980) in: Kavanaugh, M.C., Leckie, J.O. (eds.): Particulates in Water: Characterization, Fate, Effects, and Removal. Adv. Chem. Ser. *189*, American Chemical Society, Washington, DC

[4] O'Melia, C.R. (1980) Envir. Sci. Tech. *14*, 1052
[5] O'Melia, C.R. (1985) ASCE Jour. Envir. Engineering *111*, 874
[6] Schindler, D.W. (1985) in: Stumm, W. (ed.): Aquatic Surface Chemistry. Wiley-Interscience, New York
[7] Stumm, W. (1986) Ambio *15*, 201
[8] Thurman, E.M. (1985): Organic Geochemistry of Natural Waters. Martinus Nijhoff/Dr. W. Junk, Dordrecht
[9] Weilenmann, U., O'Melia, C.R., Stumm, W. (1988) Limnol Ocean (in press)
[10] Westall, J., Stumm, W. (1980) in: Hunzinger, O. (ed.): The Handbook of Environmental Chemistry, Vol. 1/Part A. Springer, Berlin

C. R. O'Melia
The Johns Hopkins University
34th and Charles Streets
Baltimore, MD 21218
USA

Pretreatment of Drinking Water to Control Organic Contaminants and Taste and Odor

I.H. Suffet, R.J. Baker and T.L. Yohe

Abstract

This paper represents an evaluation of a laboratory study and related literature on pretreatment of drinking water for removal of organic contaminants and taste and odor. Pretreatment may include oxidation, aeration, biodegradation, or powdered activated carbon (PAC). The unit processes described are those used by the majority of water treatment plants in the USA, represented by the Philadelphia Suburban Water Company's (PSWC) Neshaminy Plant.

In a recent laboratory study, Neshaminy Creek water was spiked with a mixture of 15 low molecular weight, potentially organoleptic or hazardous compounds at low ppb concentrations. The solution was treated with oxidants or PAC, residuals of the test compounds were quantified and percent removal for each compound/treatment combination were calculated.

Laboratory results and the literature indicate that oxidants are ineffective for removing most low molecular weight organic compounds, although exceptions have been reported. All test compounds were removed to some degree by PAC, eight by greater than 90%. Aeration is effective for removing small volatile compounds.

Introduction

River water that is used for drinking can contain organic matter as high as 10 mg/l as dissolved organic carbon (DOC). It is estimated that 90% of the DOC in drinking water supplies is high molecular weight natural humic material which is apparently nontoxic [1]. The remaining 10% of the DOC is characterized as low molecular weight soluble organics. Many chemical identifications of the low molecular weight fraction that is non- polar, hydrophobic and volatile (b.p. < 400 x C) have been made. It is this fraction that contains the disinfection byproducts such as trihalomethanes, industrial compounds of health concern such as pesticides and chlorinated solvents, and compounds that cause taste and odor problems in drinking water such as geosmin. USA water treatment plants are designed to maintain drinking water standards below maximum contaminant levels (MCLs). Table 1A shows the primary organic chemical MCLs that must be met. Table 1B shows the unregulated synthetic organics to be monitored, and Table 1C shows candidates for future regulations [2].

Table 1. Current and proposed USEPA standards for organics in drinking water [2]

1A. USEPA standards for organic chemicals in drinking water		1B. Unregulated synthetic organics to be monitored	
Chemical	MCL [µg/l]	Monitoring required by all states	
Trichloroethylene	5	Bromobenzene	1,1-Dichloroethane
Carbon tetrachloride	5	1,1-Dichloropropene	Chloromethane
Vinyl chloride	2	1,3-Dichloropropene	1,1,2-Trichloroethane
1,2-Dichloroethane	5	Bromomethane	1,2-Dichloropropane
Benzene	5	Chlorobenzene	1,3-Dichloropropane
para-Dichlorobenzene	75	2,2-Dichloropropane	Dichloromethane
1,1-Dichloroethylene	7	Chloroethane	1,1,2,2-Tetrachloroethane
1,1,1-Trichloroethane	200	Styrene	p-Chlorotoluene
Trihalomethanes (total)	100	o-Chlorotoluene	1,1,1,2-Tetrachloroethane
Chloroform		Dibromomethane	Tetrachloroethylene
Chlorodibromomethane		o-Dichlorobenzene	Toluene
Bromodichloromethane		t-1,2-Dichloroethene	p-Xylene
Bromoform		c-1,2-Dichloroethene	o-Xylene
Pesticides		Ethylbenzene	m-Xylene
2,4-Dichlorophenol	100		
Endrin	0.2		
Lindane	0.4	Monitoring required for vulnerable systems	
Methoxychlor	100		
Toxaphene	5	Ethylene dibromide	
2,4,5-TP, Silvex	10	1,2-Dibromo-3-chloropropane (DBCP)	
1C. Candidates for future regulation		Monitoring required at states' discretion	
Zinc	Chloromethane	Bromochloromethane	Dichlorodifluoromethane
Silver	Bromobenzene	n-Butylbenzene	Fluorotrichloromethane
Sodium	Hypochlorite ion	Hexachlorobutadiene	Tert-butylbenzene
Aluminium	1,2,3-Trichloropropane	Isopropylbenzene	1,2,3-Trichlorobenzene
Molybdenum	Hexachloroethane	p-Isopropyl toluene	1,2,4-Trichlorobenzene
Vanadium	2,2-Dichloropropane	Naphthalene	1,2,3-Trichlorobenzene
Bromomethane	Dibromomethane	n-Propylbenzene	1,3,5-Trichlorobenzene
Chlorine	Chlorine dioxide	Sec-butylbenzene	
Chlorite	Cyanogen chloride		
Chloramine	2,4-Dinitrotoluene		
Ammonia	1,3-Dichloropropane		
Ozone	1,1,1,2-Tetrachloroethane		
Halonitriles	1,3-Dichloropropene		
Chloropicrin	o-Chorotoluene		
2,4,5-T	Acrylonitrile		
Isophorone	Ethylene thiourea		
Boron	Cryptosporidium		
Strontium	Ethylene thiourea		
Halogenated acids, alcohols			
Halogenated aldehydes, ketones			

A secondary drinking water standard for organic chemicals is the goal of all drinking water treatment, the absence of any tastes and odors. Figure 1 shows a classification of drinking water flavors [3]. Background odors and periodic episodes of malodorous water quality as shown on Figure 1 are endemic in the water supply industry. Organoleptic problems are associated with unsafe water by the public regardless of the actual quality of the water. This leads to customer dissatisfaction and in some cases to another source of water of unknown water quality. Thus, the water supply industry is very sensitive to taste and odor problems.

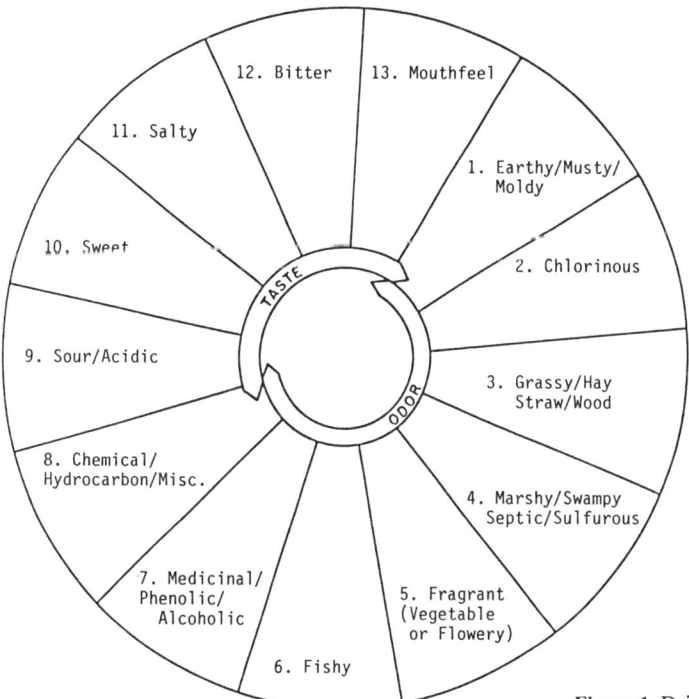

Figure 1. Drinking water flavour wheel

This paper will describe the pretreatment of drinking water in the USA for trace organic problems and the selection of treatment processes at a moderately sized drinking water plant. This paper does not address reservoir management practices used to minimize water quality degradation (e.g. algae control, withdrawal discharge depth, etc.). Pretreatment of drinking water is generally defined as those unit processes utilized before a coagulation system. These processes usually include oxidation (primarily chlorine-containing disinfectants, ozone and $KMnO_4$; aeration; biodegradation (storage or a form of soil filtration, e.g. river bank filtration); or the use of powdered activated carbon (PAC). The unit processes described are those used by the majority of water treatment plants in the USA, represented by the Philadelphia Suburban Water Company's (PSWC) Neshaminy Plant (Figure 2). The organoleptic quality of the water supply of this plant is shown in Table 2. The conventional coagulation system in U.S. water plants consists of a rapid mix of 0.5 minutes, flocculation of 30 – 60 minutes, and sedimentation of four hours with filtration of 30 – 60 minutes. Some of the general definition of a pretreatment process is fuzzy as shown at the Neshaminy Plant where the oxidants, PAC and even copper sulfate for algae control is added during the coagulation process. All U.S. primary drinking water standards are met at the Neshaminy Water Treatment Plant.

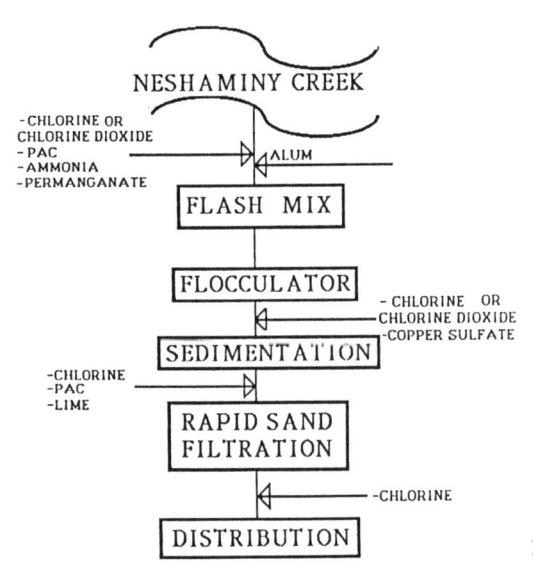

Figure 2. Neshaminy Water Treatment Plant

Tabel 2. Common flavours and aromas at the PSWC Neshaminy Plant

Sample location	Major response	Percent occurrence		Intensity
		inc. notes[*)]	ex. notes[*)]	(scale 0-12)
Effluent aroma	Chlorinous	47	34	3.6
	Musty	46	22	2.0
	Swimming pool	53	43	4.4
Effluent flavour	Chlorinous	84	72	2.2
	Metallic	56	38	2.1
	Astringent	56	28	2.0
	Bitter	47	22	2.7
Influent aroma	Sewage	84	83	4.3
	Creeky	47	20	3.5
	Musty	27	10	3.7

[*)] "Notes" occur when less than 50% of the panel detect a flavour or aroma. There is no intensity assigned to notes, thus the numerical average is based on the occurences excluding notes [ex. notes] rather than including notes (inc. notes).

A recent laboratory study was conducted to see how the basic pretreatment methods of oxidation or the use of powdered activated carbon (PAC) could affect the quality of drinking water at the Neshaminy plant. The basic laboratory approach taken was to spike Neshaminy Creek water with a mixture of 15 low molecular weight, potentially organoleptic or hazardous compounds. Many of these compounds have been identified in domestic and foreign water supplies [4]. Some have been detected in Neshaminy Creek water [5]. The compounds were chosen to provide a broad spectrum of functional groups representing types of structures known to be organoleptic and toxic. Table 3 shows the relationship of these probes to their taste and odor characteristics. The aqueous concentrations used were in the low ppb range where many of the compounds selected have threshold odor values. Compounds used are listed in Tables 3 and 4. The solution was then exposed to the highest dose and contact time of each oxidant or powdered activated carbon (PAC) that would be utilized in the Neshaminy water treatment plant for each unit operation. After treatment, residuals of the test compounds were quantified in the treated water. Percent removal for each compound/treatment combination were then calculated. Experimental and analytical details are published elsewhere [6].

Table 3. Test compounds and Treatment reagents used

Compound	Odor type	Aqueous conc.	
		Expts.#1&2	Expt.#3
Toluene	Model glue	10	20
Ethylbenzene	Floral	10	20
1,4-Dichlorobenzene	Penetrating	10	20
2-Ethyl-1-hexanol	Musty	50	100
2-Isopropyl-3-methoxypyrazine	Potato bin	15	30
Nonanal	Fruity/sweet	25	50
2-Methylisoborneol	Earthy-musty	50	50
Trans-2-nonen-1-al	Cucumber with skin	25	50
1,2,4-Trichlorobenzene		10	20
Napthalene	Camphorous	10	10
2-Isobutyl-3-methoxypyrazine	Green bell pepper	10	20
2,3,6-Trichloroanisole	Musty	10	20
Geosmin	Earthy	10	10
Diethylphthalate	Offensive	25	50
Diphenylamine	Fishy	10	20

Treatment Reagents

Alum	20 mg/l as $Al_2(SO_4)_3$
Chlorine	12 mg/l as Cl
Chloramines	12 mg/l as Cl
Chlorine dioxide	3 mg/l as Cl
Potassium permanganate	2 mg/l as $KMnO_4$
Powdered activated carbon	50 mg/l Aqua Nuchar

This paper represents a critical evaluation of the laboratory study and related literature on pretreatment for removal of organic contaminants and taste and odor. The pretreatment methods considered are oxidative processes, aeration and PAC. An evaluation of air stripping and ozonation for removal of five compounds used in the laboratory work was done by Lalezary and co-workers [7]. Additional air stripping work was done on a pilot scale by McCarty [8] and ozonation has been investigated by Anselme et al. [9] and Atashitari et al. [10] for taste and odor control

Table 4. Additional characteristics of compounds used in laboratory study

Compound	Treatment data available	Henry's law constant atm x m^3 x 10^3	GAC isotherm k, 1/n (a)	Toxicity LD50 mg/kg
Geosmin	Aer [7] GAC [44]	0.067 [7]	26.0, 0.59 [44]	
2-Methylisoborneol	Aer [7] GAC [44]	0.058 [7]	1.9, 1.65 [44]	
2-Isopropyl-3-methoxypyrazine	Aer [7] GAC [44]	0.067 [7]	1.54, 0.21 [44]	
2-Isobutyl-3-methoxypyrazine	Aer [7] GAC [44]	0.047 [7]	0.99, 0.52 [44]	
2,3,6-Trichloroanisole	Aer [7] GAC [44]	0.280 [7]	0.00, 5.50 [44]	
C2-Benzene (Ethylbenzene)	Ox (O$_3$) [20], Aer [46]	5.70 [46]	53, 0.79 [45]	3500 [49](b)
Naphthalene	Aer [46]	0.36 [46]	132, 0.42 [45]	1780 [49](b)
Toluene	Ox (O$_3$) [20]	6.60 [46]	26.1, 0.44 [45]	3000 [49](b)
Dichlorobenzene (1,4-isomer)	Ox (Cl,KMnO$_4$) [20] Aer [46, 47]	2.10 [46] 1.94 [48]	121, 0.47 [45]	500 [49](b)
Trichlorobenzene (1,2,4-isomer)	Aer [47]	1.42 [48]	157, 0.31 [45]	756 [49](b)
Diethylphthalate			110, 0.27 [45]	800 [49](b)
2-Ethyl-1-Hexanol				
Nonenal				
Nonanal				
Diphenylamine			120, 0.31 [45]	
Dichlorophenol				530 [50](b) 580 [50](b)

a Freundlich constants   b Toxicity on rat   Aer: aeration   Ox: oxidation

on a pilot scale. Therefore, air stripping and ozonation were not included as a laboratory procedure in this work, but these processes are evaluated as pretreatment processes for control of organic contaminants and tastes and odors.

Results and Discussion

Table 5 shows the percent removal of each organic for each treatment method in the laboratory study. Nonanal and trans-2-nonen-1-al were found to be unstable in the Neshaminy Creek water sample and degraded rapidly during the treatment contact time. Therefore, no treatment data is available for these compounds. This phenomenon was also noted in a nine day stability study where the test compounds were spiked into river water at concentrations similar to those listed in Table 3. All other compounds showed virtually no reduction during the nine day period.

Mixing and Aeration

The laboratory treatment experiments were run in closed systems to exclude air in order to minimize inadvertent air stripping. Table 5 shows that losses due to mixing were less than 10% for all compounds tested. Preliminary results obtained in a more open system suggested that air stripping from mixing and transferring samples was occuring, leading to false-high removal results. Some degree of aeration would

Table 5. Percent reductions of compounds from conventional water treatment unit processes

Compound	Treatment						
	Filtered blank	Coagulation	Chlorine	Chloramines	Chlorine Dioxide	Permanganate	PAC
Toluene	10.0 ± 8.5	13.7 (n=2)	16.7 ± 5.8	13.9 (n=1)	19.5 ± 14.9	10.3 ± 26.7	95.2 ± 8.0
Ethylbenzene	9.1 ± 2.4	8.6 ± 1.7	13.9 ± 11.5	3.6 ± 5.7	3.8 ± 6.0	7.4 ± 6.0	95.1 ± 4.4
1,4-Dichlorobenzene	7.9 ± 2.0	7.1 ± 6.0	11.8 ± 13.3	1.4 ± 9.0	-0.4 ± 4.5	6.9 ± 4.6	100.0
2-Ethyl-1-hexanol	-2.0 ± 9.1	-1.0 ± 9.1	4.3 ± 19.2	0.9 ± 7.0	-3.2 ± 10.7	1.5 ± 18.1	73.9 ± 26.8
2-Isopropyl-3-methoxypyrazine	1.0 ± 1.5	1.8 ± 3.7	3.5 ± 2.9	2.2 ± 8.4	3.5 ± 2.0	9.1 ± 2.7	86.1 ± 12.8
Nonanal[*]	---	---	---	---	---	---	---
2-Methylisoborneol	4.0 ± 0.6	4.5 ± 1.2	3.7 ± 2.0	4.2 ± 4.2	3.6 ± 1.6	1.8 ± 5.1	53.7 ± 1.8
Trans-2-nonen-1-al[*]	---	---	---	---	---	---	---
1,2,4-Trichlorobenzene	10.3 ± 3.6	-6.9 ± 10.0	15.6 ± 17.2	-2.7 ± 11.1	-3.0 ± 11.0	10.3 ± 8.6	100.0
Naphthalene	7.3 ± 2.2	4.7 ± 7.0	6.8 ± 10.0	-0.6 ± 6.6	1.3 ± 8.2	4.3 ± 8.6	100.0
2-Isobutyl-3-methoxypyrazine	5.0 ± 3.5	2.3 ± 6.5	8.5 ± 7.0	2.2 ± 3.3	8.9 ± 10.2	3.1 ± 3.4	93.1 ± 6.7
2,3,6-Trichloroanisole	-3.7 ± 1.1	1.2 ± 9.9	7.8 ± 8.8	-5.2 ± 7.1	-3.4 ± 8.6	3.7 ± 6.5	100.0
Geosmin	5.5 ± 3.8	0.6 ± 10.9	10.0 ± 13.2	-4.7 ± 11.3	-3.2 ± 10.3	1.1 ± 5.5	79.2 ± 5.2
Diethylphthalate	-3.7 ± 1.1	-2.5 (n=2)	-7.6 ± 5.9	-2.0 ± 3.2	1.3 ± 1.1	-14.7 ± 21.3	100.0
Diphenylamine	-5.2 (n=2)	1.6 ± 25.6	49.0 ± 10.5	-4.7 ± 37.1	36.3 ± 27.1	49.6 ± 28.3	37.5 ± 23.7

[*] Nonanal and nonenal were unstable in the water sampled, therefore no results are available for these compounds

always occur under treatment plant conditions since most operations take place in open tanks, and our preliminary experiments may have simulated this.

Air stripping is commonly used for volatile organic removal on a treatment plant scale [8]. In a treatment plant survey [11], only 10% of the plants responding experienced complete success in taste and odor removal by aeration. However, 50% reported partial success.

When an aqueous solution of organics is subjected to aeration, some organics will be transported to the vapor phase via air bubbles. The extent of this transport for an individual compound is defined by the compound's volatility, which is characterized by Henry's law [12]. The nondimensional Henry's law constant (H') of a compound is the partition coefficient of that compound between the vapor phase and liquid phase at equilibrium. Additionally, temperature must be specified, as it will effect the partitioning ratio.

McCarty et al. [8] conducted an extensive plant-scale study of volatile contaminant removal by air stripping as part of a wastewater reclamation project (Water Factory 21). Table 6 is a list of Henry's law constants compiled by McCarty and co-workers. Influent and effluent concentrations of 22 compounds from this list were reported for the air stripping process. They concluded that compounds with Henry's law constants greater than 10^{-3} m^{3}atm/mole can be effectively removed by aeration. Four of the model compounds in our study also appear in the study of McCarty et al. [8]. They are ethylbenzene, 1,4-dichlorobenzene, toluene and naphthalene. All of them have Henry's law constants greater than 10^{-3}. This suggests that aeration may be effective in removing these particular compounds from water. Additionally, aeration is very effective for VOCs including THMs with the possible exception of bromoform.

Lalezary et al. [7] determined Henry's law constants for five organoleptic microbiological metabolites: methylisoborneol (MIB), geosmin, 2-isopropyl-3-methoxypyrazine (IPMP), 2-isobutyl-3-methoxypyrazine (IBMP), and 2,3,6-trichloroanisole (TCA) (Table 4). The Henry's law constant for TCA was determined to be 2.88×10^{-4}, and the other four were considerably lower. Therefore, Lalezary and co-workers concluded that air stripping may be a viable treatment for removing TCA, but not for any of the other four compounds studied.

In summary, some toxic chemical problems and some taste and odor problems are treatable by aeration. Finally, the removal of organics will depend largely upon the Henry's law constants of the compounds involved, and this precludes the removal of common earthy-musty odors such as geosmin and MIB and many higher molecular weight toxic organics. A second consideration for any air stripping processes is added costs due to the control of air pollution problems as defined by different states in the U.S.

Filtering

In the laboratory study each treated sample was filtered through a 1.5 micron glass fiber filter (Whatman 934-AH, Whatman Ltd., Clifton, NJ) to remove suspended particulates before extraction. Table 5 shows that removal by filtration alone is not significant. This suggests that adsorption onto suspended sediment was minimal and that losses during the filtration process were not significant.

Organic removal during filtration at treatment plants has been observed. Rhine River water was analyzed for concentrations of several pollutants before and after the following unit operations: river bank filtration alone, ozonation with river bank filtration, and GAC filtration [13]. Biological activity was suggested as the removal mechanism accounting for substantial removals of dichlorobenzene, trichlorobenzene and dichlorotoluene. It was noted that removal efficiency decreased as the number of chlorines on the benzene molecule increased. Kobayashi and Rittmann [14] reported that biodegradability of chlorinated benzenes decreases as more

Table 6. Henry's law constant values from water factory 21 research

Compound	Henry's law constant atm m^3/mole	Compound	Henry's law constant atm m^3/mole
Vinyl chloride (a)	6.4	1,2-Dichloropropylene	2.0×10^{-3}
Dichlorofluoromethane	2.1	Alpha-BHC	2.0×10^{-3}
1,1-Dichloroethylene (a)	1.7×10^{-1}	1,2-Dichlorobenzene	1.7×10^{-3}
1,2-Dichlororthylene	1.7×10^{-1}	Anthracene	1.4×10^{-3}
Trichlorofluoromethane	1.1×10^{-1}	1,2-Dichloroethane	1.1×10^{-3}
Methyl bromide	9.3×10^{-2}	Hexachloroethane	1.1×10^{-3}
Toxaphene	6.3×10^{-2}	1,1,2-Trichloroethane	7.8×10^{-4}
Carbon tetrachloride (a)	2.5×10^{-2}	Bromoform (a)	6.3×10^{-4}
Tetrachloroethylene	2.3×10^{-2}	PCB (Arochlor 1242)	4.9×10^{-4}
Chloroethane	1.5×10^{-2}	1,1,2,2-Tetrachloro-ethane	4.2×10^{-4}
Beta-BHC	1.1×10^{-2}	Naphthalene	3.6×10^{-4}
Trichloroethylene (a)	1.0×10^{-2}	Fluorene	2.1×10^{-4}
Methyl chloride	8.0×10^{-3}	Acenaphthene	1.9×10^{-4}
PCB (Arochlor 1260)	6.1×10^{-3}	Phenanthrene	1.3×10^{-4}
1,2-Trans-dichloro-ethylene	5.7×10^{-3}	Bis(2-chloroisopropyl) ether	1.1×10^{-4}
Ethylbenzene	5.7×10^{-3}	Acrolein	9.7×10^{-5}
Toluene	5.7×10^{-3}	2-Nitrophenol	7.6×10^{-5}
1,1-Dichloroethane (a)	5.1×10^{-3}	Acrylonitrile	6.3×10^{-3}
Benzene (a)	4.6×10^{-3}	Di-n-butyl phthalate	6.3×10^{-5}
Chlorobenzene	4.0×10^{-3}	2,4-dichlorophenol	4.2×10^{-5}
1,1,1-Trichloroethane(a)	3.6×10^{-3}	4,4'-DDT	3.4×10^{-5}
Chloroform (a)	3.4×10^{-3}	2-Chlorophenol	2.1×10^{-5}
PCB (Alochlor 1248)	3.0×10^{-3}	Nitrobenzene	1.1×10^{-5}
1,3-Dichlorobenzene	2.7×10^{-3}	Isophorone	4.2×10^{-6}
Methylene chloride	2.5×10^{-3}	Pentachlorophenol	2.1×10^{-6}
Heptachlor	2.3×10^{-3}	Dimethyl phthalate	4.2×10^{-7}
PCB (Alachlor 1254)	2.3×10^{-3}	Lindane	3.2×10^{-7}
1,4-Dichlorobenzene	2.1×10^{-3}	Phenol	2.7×10^{-7}
Aldrin	2.1×10^{-3}	Dieldrin	1.7×10^{-7}
1,2-Dichloropropane	2.0×10^{-3}	4,6-Dinitro-o-cresol	1.7×10^{-7}

(From McCarty 1983 [8])   a) MCL drinking water standard has been established

chlorines are added. This further suggested that biological activity was responsible for reduction of chlorinated benzenes in the Rhine River water study [13].

Yagi et al. [15] found rapid sand filtration to be ineffective in removing geosmin and MIB, but slow sand filtration removals were greater than 98% for each compound. Removal mechanisms were not proposed, but apparently are related to biological activity.

In summary, therefore, in some instances slow sand filtration or river bank storage/filtration may remove some low molecular weight organics. The literature reviewed suggests that at least four of the compounds studied (dichlorobenzene, MIB, geosmin, and trichlorobenzene) might be removed to some degree during river bank storage/filtration or slow sand filtrations, and biological activity would be responsible for the removal. Such biodegradation is compound-specific. The filtration process itself does not appear to be an important process for the removal of trace organics.

Oxidation

Chlorine, chloramines, chlorine dioxide and potassium permanganate were used as oxidants in the laboratory study. Relative strengths of various oxidant reaction processes are listed in Table 7. The relative oxidative potential is not an indication of an oxidant's power to degrade a specific organic compound. Oxidation of organics is very complex. Reaction rates are functions of numerous factors, including strength of the oxidant (relative oxidation potential) [16], type of attack by the oxidant [17], functional groups on the organic molecule [17], and environmental factors [18]. Environmental factors include pH, temperature, concentrations of oxidants, reactions and product competition from other oxidants and/or organics (especially background humic materials), and the presence of interfering or catalyzing substances.

Table 7. Oxidation potential of treatment oxidants

OXIDANTS	E^o (VOLTS)	Ref.
$Cl_{2(aq)} + 2e^- \longrightarrow 2Cl^-$	+1.39	(19)
$ClO^- + H_2O + 2e^- \longrightarrow Cl^- + 2OH^-$	+0.89	(19)
$ClO_2 + H^+ + 2e^- \longrightarrow HClO_2$	+1.27	(16)
$ClO_2 + e^- \longrightarrow ClO_2^-$	+1.15	(19)
$O_3 + 2H^+ + 2e^- \longrightarrow O_2 + H_2O$	+2.07	(18)
$H_2O_2 + 2H^+ + 2e^- \longrightarrow 2H_2O$	+1.77	(16)
$O_3 + H_2O + 2e^- \longrightarrow O_2 + H_2O$	+1.24	(16)
$MnO_4^- + 4H^+ + 3e^- \longrightarrow MnO_{2(s)} + 2H_2O$	+1.70	(16)
$MnO_4^- + 8H^+ + 5e^- \longrightarrow Mn^{+2} + 4H_2O$	+1.51	(18)
$MnO_4^- + 2H_2O + 3e^- \longrightarrow MnO_{2(s)} + 4OH^-$	+0.59	(18)
$MnO_4^- + e^- \longrightarrow MnO_4^{-2}$	+0.56	(16)

The following discussion will concentrate on types of organic functional groups that can be attacked by oxidizing agents, and how these reactions relate to oxidation in water treatment. The discussion relating to organic oxidation is divided into the following sections:

1) Attacks on C=C double bonds
2) Attacks on functional groups containing oxygen
3) Other oxidation reactions of chlorine and chloramines
4) Other oxidation reactions of chlorine dioxide
5) Other oxidation reactions of permanganate
6) Other oxidation reactions of ozone
7) Oxidation under water treatment conditions

Information for sections one through six was extracted primarily from Morrison and Boyd [17], Weber [16], Masschelein [19], Hoigné and Bader [20, 21] and Katz [22]. References for section seven are cited individually.

Attacks on C=C Double Bonds. C=C double bonds are succeptible to attack by several oxidizing agents, but with different results. Ozonolysis (low pH) can result in cleavage of the molecule at the double bond:

$$R-\underset{R}{C}=\underset{R'}{C}-R' \xrightarrow{O_3} R-\underset{O}{\underset{|}{C}}-\underset{O}{\underset{|}{C}}-R' \xrightarrow{O_3} \quad (1)$$

$$R-\underset{O}{\underset{|}{C}}\overset{O}{-}\underset{O}{\underset{|}{C}}-R' \xrightarrow[H_2O]{Zn} R-\underset{R}{C}=O + R'-\underset{R'}{C}=O \quad (2)$$

(The "R" groups can be a hydrocarbon chain or a hydrogen atom, so the products can be either aldehydes or ketones.) Permanganate can attack a C=C double bond by hydroxylation. Instead of cleaving the carbon chain, hydroxyls are added, resulting in a glycol:

$$R-\underset{R}{C}=\underset{R'}{C}-R' \xrightarrow[> pH = 7]{MnO_4^-} R-\underset{OH}{\underset{|}{C}} - \underset{OH}{\underset{|}{C}}-R' \quad (3)$$

The glycol can be further oxidized to acids, ketones and CO_2. Carbon chain cleavage can also take place as follows:

$$CH_3-CH=\underset{CH_3}{\underset{|}{C}}-CH_3 \xrightarrow{MnO_4^-} CH_3COOH + CH_3\underset{CH_3}{\underset{|}{C}}=O \quad (4)$$

or:

$$CH_3-CH_2CH=CH_2 \xrightarrow{MnO_4^-} CH_3CH_2CH_2COOH + CO_2 \quad (5)$$

Chlorine has the capacity to add to C=C double bonds:

$$R-\underset{}{C}=\underset{}{C}-R' \xrightarrow{Cl_2} R-\underset{\underset{Cl}{|}}{\overset{\overset{R}{|}}{C}} - \underset{\underset{Cl}{|}}{\overset{\overset{R'}{|}}{C}}-R' \qquad (6)$$

This reaction mechanism is considered more important under normal treatment conditions than chain cleavage or oxygen addition [16]. This is part of the pathway leading to the formation of trihalomethanes, which will not be discussed here, but is well described by others (e.g. Rook [23] und Rice [24]).

Attack on Functional Groups Containing Oxygen. A definition of organic oxidation which is especially well suited to this type of attack was quoted by Weber [16]: "It appears that the early concept of oxidation as a reaction involving oxygen addition or hydrogen abstraction adequately describes oxidative degradation reactions of organic compounds in waters and waste waters". From this definition we can describe the following reactions as oxidation:

$$\text{alcohols} \xrightarrow{\text{oxidation}} \text{aldehydes, ketones} \xrightarrow{\text{oxidation}} \text{acids}$$

The behaviour of permanganate toward alcohols depends upon the placement of the hydroxyl group:

$$1° \text{ Alcohol: } R-CH_2OH \xrightarrow{MnO_4^-} R-COOH \quad (Acid) \qquad (7)$$

$$2° \text{ Alcohol: } R\underset{R'}{\overset{|}{C}}HOH \xrightarrow{MnO_4^-} R-\underset{R'}{\overset{|}{C}}=O \quad (Ketone) \qquad (8)$$

Permanganate can also oxidize aldehydes to acids:

$$\text{Aldehydes} \quad R-\underset{O}{\overset{\|}{C}}H \xrightarrow{MnO_4^-} R-COOH \quad (Acid) \qquad (9)$$

Chlorine dioxide reacts strongly with aldehydes and ketones to produce acids. For example, benzaldehyde is quickly oxidized to benzoic acid in water:

$$Ph-\overset{\overset{O}{\|}}{C}H \xrightarrow{ClO_2} Ph-\overset{\overset{O}{\|}}{C}OH \qquad (10)$$

Kraznov et al. [25] investigated the oxidation of primary aliphatic alcohols and aldehydes with ozone. Aqueous solutions of ethanol, butanol and octanol were exposed to 33 mg/l dosages of ozone. This produced aldehydes, which were in turn oxidized to acids, but not to CO_2. The rate of oxidation increased with increasing pH. Secondary alcohols were oxidized to ketones upon ozonation, which were further oxidized to acids:

$$\begin{array}{c} R' \\ \diagdown \\ CHOH \\ \diagup \\ R \end{array} \xrightarrow{O_3} \begin{array}{c} R' \\ \diagdown \\ C=O \\ \diagup \\ R \end{array} \xrightarrow{O_3} \begin{array}{c} R'-COOH \\ + \\ R-COOH \\ + \\ H_2O_2 \end{array} \quad (11)$$

Other Oxidation Reactions of Chlorine and Chloramines. Under water treatment conditions, chlorine probably plays a minor role in organic oxidation, and substitution reactions as previously described are favored [16]. Therefore, in addition to producing some oxidative degradation products, chlorination generates chlorinated organics. In addition to adding to the carbon chain, chlorine can add to the nitrogen of organic amines in aqueous solution [18].

Chlorination of methylamine is one example of this:

$$CH_3NH_2 \xrightarrow{Cl_2} CH_3NHCl \quad (12)$$

As the strength of the base increases, the rate of reaction with HOCl increases. Chlorine can add to the alkyl groups of alkyl benzenes:

$$\phi\text{-}CH_3 \xrightarrow{Cl_2} \phi\text{-}CH_2Cl \xrightarrow{Cl_2} \phi\text{-}CHCl_2 \xrightarrow{Cl_2} \phi\text{-}CCl_3 \quad (13)$$

Chloramines are weaker oxidants than chlorine, and therefore function even less effectively as organic oxidants. Thus, they react slower than chlorine.

Other Oxidation Reactions of Chlorine Dioxide. Primary and secondary aliphatic amines in aqueous solution react slowly or not at all with chlorine dioxide. However, tertiary amines react significantly [12]. For example:

$$(C_2H_5)_3N \xrightarrow{ClO_2} CH_3CHO + (C_2H_5)_2NH \quad (14)$$

Chlorine dioxide reacts strongly with phenols in neutral aqueous solution. Products are oxidized phenols which can be further oxidized to aliphatic acids and chlorinated phenols. The oxidation scheme in Figure 3 is from Masschelein [19]. Phenolic secondary amines also react with chlorine dioxide:

$$\phi\text{-}CH_2NHC(CH_3)_3 \xrightarrow{ClO_2} \phi\text{-}CHO + (CH_3)_3CNH_2 \quad (15)$$

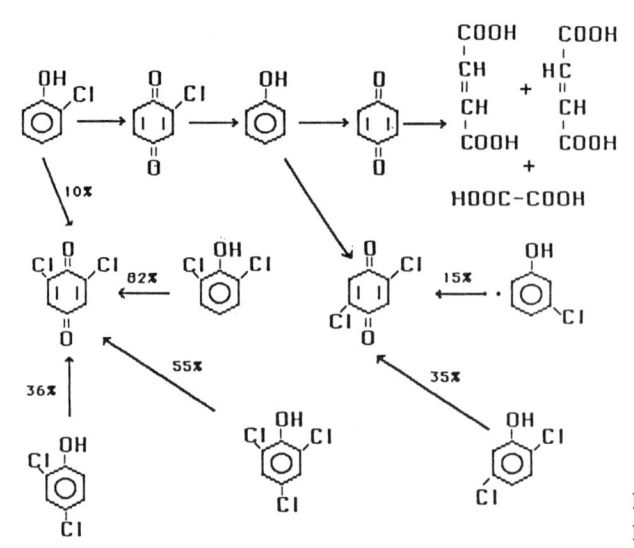

Figure 3. Oxidation of chlorinated phenols by chlorine dioxide

Table 8. Oxidation of organic functional groups by permanganate (from Weber [16])

FUNCTIONAL GROUP	OXIDIZED FUNCTIONAL GROUP
$-CH=CH-$	$-\underset{OH}{CH}-\underset{OH}{CH}-$
R-CHO	R-COOH
HCHO, HCOOH	$CO_2 + H_2O$
$R-CH_2OH$	R-COOH
R_2CHOH	$R_2C=O$
$R-NH_2, R_2NH, R_3N$	$R-CH_2COOH + NH_3$ + OTHER COM'DS
R-SH	$R-SO_3H$
R-S-R'	$R-SO_2R'$
R-S-S-R	$2R-SO_3H$
R-S-O-R'	$R-SO_2R'$
Ar-NH_2, Ar-NRH, Ar-OH	R-COOH, CO_2, NH_3 & OTHER COM'DS
Ar-CH_3, Ar-CH_2OH, Ar-CHO	Ar-COOH
FURANS	SO_4^{-2}, CO_2, H_2O

Other Oxidation Reactions of Permanganate. Table 8 shows reactions of permanganate with organic species. Some of these reactions could be significant for taste and odor removal. Additionally, the alkyl group of alkyl benzenes can be attacked by permanganate:

$$\text{C}_6\text{H}_5\text{-CH}_2\text{CH}_3 \xrightarrow{\text{MnO}_4} \text{C}_6\text{H}_5\text{-CHOH} + CO_2 \quad (16)$$

Other Oxidation Reactions of Ozone. Dore [26], based upon the work of Hoigné and co-workers [27–31], indicates that ozone can react directly to form carbonyls by reacting as a dipole on C=C double bonds as discussed above (equations 1 – 4). Ozone can also react directly as an electrophilic agent on aromatics by ring hydroxylation or a nucleophilic agent on C=N double bonds.

Ozone can react indirectly as a free radical to form carbonyls. Scavengers such as carbonate can stop the reaction. During water treatment, intermediate reaction products can form. The type and amount of products are a function of ozone dose, reaction time, scavengers present, and pH.

Direct ozone reaction rate constants have been determined for a wide range of organics at pH 2 – 3 [21]. The rate laws are assumed to be first order with respect to ozone (O_3) and solute (M) concentration. Therefore, the reaction can be expressed as a second order reaction:

$$M + nO_3 \xrightarrow{k_{O_3}} \text{Products} \quad (17)$$

$$\frac{-d[M]}{dt} = \frac{k_{O_3}}{n} [O_3]^{1.0}[M]^{1.0} \quad (18)$$

Hoigné and Bader [21] proposed the following relationship for compounds treated with ozone under water treatment plant conditions:

$$t_{1/2} = \frac{0.69}{[O_3] k_{O_3}/n} = \frac{0.69}{10^{-5} k_{O_3}/n} \quad (19)$$

This follows mathematically from the second order rate equation. The right-hand approximation can be made assuming 10^{-5} M ozone (0.5 mg/l, typical in treatment) and that $n = 1$.

Using this approximation, compounds having rate constants greater than 100 M^{-1} s^{-1} are significantly oxidized in 10 minutes by the direct attack mechanism. Hoigné and Bader suggest that those compounds would be oxidized by ozone under water treatment plant conditions at low pH. At higher pH's, where the hydroxyl free radical predominates, equations 18 and 19 hold with $O^\bullet$ (representing any oxygen radical like the hydroxy radical) replacing O_3.

$$M + O^\bullet \xrightarrow{k'} \text{Products} \quad (20)$$

$$-\frac{d[M]}{dt} = k^{\prime} [O^{\bullet}]^{1.0}[M]^{1.0} \tag{21}$$

Staehelin and Hoigné [32] conclude that in the case of impure water (e.g., water treatment), the overall kinetics due to radical-type chain reactions may still be first order in ozone concentration if the concentration of solutes stays constant. However, as a result of a sum of chain reactions of each solute with the free radicals, each solute concentration is affected differently. For these reasons, prediction of products at present has not been accomplished.

Removal of Organics from Water Using Oxidation. Surveys of treatment plants indicated that oxidation does not have a high percentage of success in treating taste and odor problems [11]. About 20% of taste and odor problems were completely solved by using either chlorine, chlorine dioxide or superchlorination. Chlorine dioxide was reported less effective than the other two oxidants. Over all, more than half of the plants reported no success in taste and odor reduction by oxidation. A later study by Sigworth [33] indicated that permanganate is not superior to the other chemical oxidants in reducing tastes and odors.

Table 9. Oxidation of organics by chlorine, permanganate and ozone

Compound	Organic compound reaction half-lives		
	Chlorine	Permanganate	Ozone
Acetophenone (a)	2.6 days	43 days	25 min
Benzaldehyde	>3.2 days	0.60 hours	28 min
Benzothiazole	8.2 min	>5.8 days	22 min
1,2-Bis(2-chloroethoxy)ethane	>20 days	67 days	50 min
Bis(2-chloroethyl)ether (a)	>20 days	15 days	21 min
Borneol	1.4 days	7 days	53 min
Camphor	>3.2 days	>5.8 days	>12 min
p-Dichlorobenzene (a)	>4.2 days	>22 days	Not run
p-Nitrophenol	2.1 hours	1.1 days	2.0 min
Methyl-*m*-toluate	>20 days	22 days	5.5 min
p-Tolunitrile	>20 days	28 days	6.4 min
Diacetone-*L*-sorbose	100 days	>14 days	2.8 hours
Diacetone-*L*-xylose	>15 days	>14 days	2.3 hours
Toluene (a)	Not run	Not run	2.8 min
Ethylbenzene (a?)	Not run	Not run	2.8 min
1,2,3-Trimethylbenzene (a?)	Not run	Not run	1.9 min

Conditions: (a) Priority pollutant (From McGuire et al. 1978 [34])
pH: 7
Concentration of test compounds: 0.1-10 mg/l
Chlorine dosage: 45-140 mg/l
Permanganate dosage: 95 mg/l as potassium permanganate
Ozone: 14.2-14.6 mg/min O_3 delivered to a 48 x 25 inch tube reactor containing 2 litres test solution

McGuire et al. [34] conducted a laboratory scale oxidation study of 19 trace organic compounds of health concern by treating them with chlorine, permanganate and ozone (Table 9). Test compounds were present in the low mg/l range, chlorine was added at 45 – 140 mg/l, potassium permanganate at 95 mg/l and ozone was fed at a rate of 14.2 and 14.6 mg/min in a 2.5 x 48 inch tubular or semibatch reactor. Ozone concentration varied depending upon utilization rate of the compound being tested. The results of this study showed that performance by chlorine and permanganate were poor, as shown by the long half lives in Table 9, indicating that these oxidants are not efficient treatment techniques for removing the organic compounds tested. However, ozone appears to be a promising treatment method for all compounds tested, based on these data.

Lin and Carlson [35] measured the time to 95% removal of heteroarenes by chlorine, chloramines and chlorine dioxide. Heteroarene compound levels of 0.1 to 1.0 mg/l and oxidant levels of about 1 to 10 mg/l were used. Contact times to achieve >95% removal ranged from 0.25 to >50 hours (Table 10). Again, different oxidants exhibited different removal efficiencies for the same compound with chloramine reaction rates being significantly less than those of chlorine or chlorine

Table 10. Chlorine Oxidation of Heteroarenes

Compound	Oxidant	pH	Results
Indoles a) Indole b) 3-Methylindole c) 2-Phenylindole d) N-Phenyl-propyrrole	NaOCl, ClO$_2$	5.3, 7	>95% reduction in 15 min. exept 2-phenyl (ClO$_2$, pH=7 is 1 hr.) and N-phenyl (ClO$_2$, pH=5.3, 45 min.)
Indoles (a-d)	NaOCl/NH$_3$	5.3, 7	Slower oxidation rates. (pH=5.3 > rate pH=7 by 3-14x) N-phenyl < 3-methyl < indole < 2-phenyl
Carbazole	ClO$_2$	5.3, 7	>95% reduction in 15 min.
	NaOCl	5.3, 7	>95% reduction in 1 and 2 hours, resp.
Quinoline	NaOCl	5.3, 7	>95% reduction in 6 hrs., 80% in 50 hrs., resp.
Quinaldine	NaOCl	5.3, 7	50% reduction in 32 hrs., >95% in 50 hrs., resp.
Isoquinoline	NaOCl	5.3, 7	>95% reduction in 1 hr., >95% in 4 hrs., resp.
Phenanthridine	NaOCl	5.3, 7	>95% reduction in 0.25 hrs., >95% in 1hr., resp.
	ClO$_2$	5.3, 7	58% reduction in 50 hrs., 47% in 50 hrs., resp.
Acridene	NaOCl	5.3, 7	>95% reduction in 0,25 hrs., >95% in 0.75 hrs., resp.
	ClO$_2$	5.3, 7	>95% reduction in 0.25 hrs., >95% in 46 hrs., resp.
4-Azafluorene	NaOCl	5.3, 7	8% reduction in 50 hrs.
5,6-Benzoquinoline	NaOcl	5.3, 7	33% reduction in 50 hrs., 19% in 50 hrs., resp.
7,8-Benzoquinoline	NaOCl	5.3, 7	34% reduction in 50 hrs., 27% in 50 hrs., resp.
Benzothiazole	NaOCl	5.3, 7	>90% reduction in 24 min., >90% in 48 min., resp.

(From Lin and Carlson, 1984 [35])

dioxide (3 − 14x) for a series of indoles. At pH 5.3 (100% HOCl) these reactions have faster rates than at pH 7 (OCl⁻ and HOCl).

In summary, the literature reviewed describes oxidation of a wide range of compounds and functional groups under many different conditions. Although it is difficult to compare data between these different studies, they all point toward the same general conclusion: that oxidation is not generally an effective way of removing low molecular weight organics, with the exception of ozone, which shows promise for removal of many types of organic compounds. Exceptions exist, of course, as some treatment plants completely eliminate taste and odor problems by chemical oxidation on a routine basis. However, reasons for the removal efficiency are not known.

Of the compounds tested in the laboratory (Table 5), only diphenylamine was significantly reduced by oxidation. Aromatic secondary amines are susceptible to oxidation, as described by Masschelein [19]. In contrast, Hoigné and Bader showed significant removals of unprotonated tertiary amines by chlorine dioxide, but poor removal of other amines.

Results of laboratory studies were consistent with most of the literature reviewed. The algal and actinomycete metabolites MIB and geosmin were not affected by any of the oxidants used. The chlorinated benzenes were not significantly reduced, in contrast with Hoigné and Bader's work [21], but consistent with McGuire et al. [34]. In pilot plant work at Osaka, Japan, [10] ozone was observed to remove MIB and geosmin. Apparently the presence of humics in the system helps oxidation.

At this time, when odor events occur, identification of specific organoleptic compounds in water is difficult. What is needed is a concerted effort of routine sampling, with chemical analysis for known odorants and taste and odor panel analysis. Even when the odor agent is known, predicting the compound's response to oxidants is at best an educated guess. Therefore, treating these problems by trial dosing in the laboratory or at the plant and evaluating odor reduction by a sensory based procedure [36 − 38] would appear to be the most viable approach [3]. Future research is ongoing to help define causes of taste and odor [39].

Powdered Activated Carbon

Powdered activated carbon (PAC) is used extensively to remove toxic chemicals and tastes and odors from water supplies. PAC is normally added directly to water during pretreatment sometime before sedimentation. The ability of a particular carbon to remove organic compounds is often characterized by the amount of the compound adsorbed by a specified amount of carbon under controlled, specified conditions. This adsorptive capacity is often expressed as a plot of the equilibrium liquid phase concentration versus the amount adsorbed by the carbon (isotherm).

Table 4 shows constants taken from the literature for compounds evaluated in this study. McGuire and Suffet [40] have presented a general isotherm for adsorption of organic compounds from water. This is shown as a dotted band on Figure 4. Also shown are the isotherms for compounds from Table 4. These compounds appear to follow the general isotherm. However, there are several compounds for which more than one isotherm was available, and differences were often encountered in literature values. These differences could arise from the use of different carbons, different equilibration times, or differences in experimental conditions. Regardless, it is important to understand the effects of these differences when considering carbon. This can be seen in Figure 4, compounds numbered 8, 9, and 10, where the three literature isotherms for toluene vary over two orders of magnitude in surface concentration for the same liquid phase concentration.

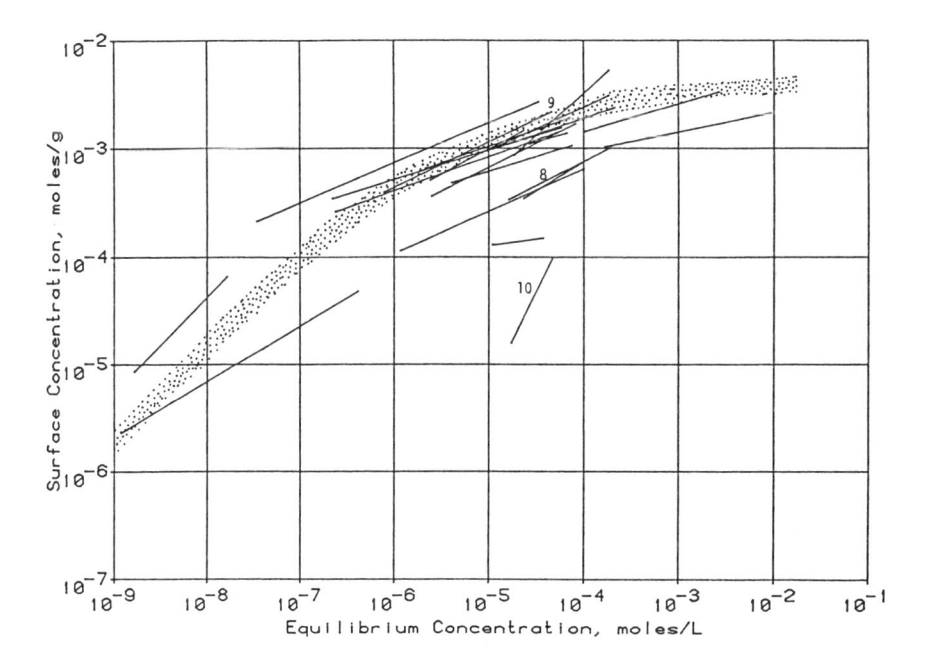

Figure 4. GAC Isotherms for study compounds from literature sources

In the laboratory work all compounds were simultaneously removed by 50 mg/l PAC. Eight of the 15 test compounds were removed by greater than 90% and only one (diphenylamine) by less than 50%. Carbon was by far the most effective treatment method for the test compounds used in this work. Most of the literature reviewed indicated that activated carbon is effective for removing nonpolar organic compounds and/or solving taste and odor problems. Many of the test compounds used in our work were also evaluated in previous studies. Water plant surveys [11, 33] indicated that treatment plants have higher success rates with carbon (PAC or GAC) than with any other traditional treatment for taste and odor removal. Adsorption was discussed by Montiel [41] as a taste and odor removal treatment in his review article. He concluded that activated carbon adsorption is usually the most effective means available for removing odorous compounds from drinking water, and that the literature reviewed indicated that GAC is more effective than PAC for geosmin and MIB removal.

Cherry [42] described steps taken during episodes of musty-moldy taste and odor at the water treatment plant in Cedar Rapids, Iowa. The plant uses surface water which had heavy algae growth and an overloaded sewage treatment plant upstream. PAC was found to be effective when combined with permanganate treatment. In order to reduce the threshold odor number (TON) from 200 – 400 to 5 – 7, the following treatment scheme was required:
1) Move chlorination and ammoniation from the coagulation slow mix tank to after filtration.
2) Add up to 6 mg/l $KMnO_4$ at the raw water intake.
3) Add 35 to 70 mg/l PAC to the coagulation slow mix tank.

Order of addition may be a key element in this treatment scheme. Cherry found that chlorination increased odors when chlorine was added before permanganate and PAC treatment.

Herzing et al. [43] studied the adsorption of MIB and geosmin on a laboratory scale, using natural and distilled water. Isotherms and breakthrough studies were done for solutions with and without humics and copper ions present, and at various pH values. Both compounds were strongly adsorbed under all experimental conditions. However, the presence of humics reduced the carbon's adsorptive capacity for both MIB and geosmin. Copper ions had no such effect. The conclusion of Herzing et al. is that the two compounds are strongly adsorbed and efficiently removed by GAC, and that the reduction in adsorption caused by humics is not significant enough to render the process ineffective for treating taste and odor problems.

Yagi et al. [15] evaluated the efficiency of traditional unit operations for removing MIB and geosmin at several treatment plants on Japan's Lake Biwa. Table 11 shows some of their data. PAC was found to be somewhat effective in reducing MIB and geosmin from Lake Biwa water. 25 mg/l of PAC reduced MIB from 116 to 45 ng/l, and 20 ng/l of geosmin was reduced to 10 ng/l by 10 mg/l of PAC. GAC was found to be more effective than PAC. A GAC filter that had been in service for one year and was saturated with respect to trihalomethanes could still effectively remove MIB and geosmin. At a 40 cm bed depth geosmin was reduced from 60 ng/l to below detectable levels. MIB was reduced from 201 to 2 ng/l with 70 cm bed depth of GAC. Empty bed contact times and fluxes were not specified.

The literature indicates that PAC is the treatment most likely to succeed in removing toxic and organoleptic compounds from drinking water. PAC would appear to be a viable first course of action when an odor problem arises in a plant. GAC is more costly in terms of capital and space, but in some instances it is the right choice for a long-term solution.

Plant-Scale Experience to Control Taste and Odor

Removal of tastes and odors was monitored for a one year period in a plant-scale evaluation at Morsang-sur-Seine near Paris, France [9]. Figure 5 shows the frequency of occurrence of six odors as the water moves through the treatment

Table 11. Removal of Geosmin and MIB from Lake Biwa

Location	Date	Compound	Unit process	Concentration Raw water ng/l	Concentration Treated water ng/l
Ohtsu City	8/25/81	Geosmin	Slow sand	360	6
Ohtsu City	8/25/81	Geosmin	PAC, rapid sand	360	160
Kyoto City	8/26/81	Geosmin	Rapid sand	270	*
Murano	8/26/81	Geosmin	Rapid sand	30	30
Osaka City	9/07/81	Geosmin	Rapid sand	122	94
Osaka City	9/12/81	Geosmin	PAC, rapid sand	130	52
Osaka City	9/24/81	MIB	Rapid sand	37	27
Ohtsu City	6/21/82	MIB	PAC, rapid sand	69	27
Ohtsu City	6/21/82	MIB	Slow sand	69	0
Koyoto City	6/15/82	MIB	GAC, rapid sand	36	13
Neyagawa	6/24/82	MIB	Slow sand	25	3
Otsu City	6/19/82	MIB	PAC (5 mg/l)	116	84
Otsu City	6/19/82	MIB	PAC (25 mg/l)	116	45
Otsu City	9/08/81	Geosmin	PAC (10 mg/l)	20	10
Otsu City	9/08/81	Geosmin	PAC (20 mg/l)	20	0
Kyoto City	8/20/81	Geosmin	GAC (5 cm bed)	60	30
Kyoto City	8/20/81	Geosmin	GAC (40 cm bed)	60	0
Kyoto City	9/29/82	MIB	GAC (5 cm bed)	201	146
Kyoto City	9/29/82	MIB	GAC (70 cm bed)	201	2

* Not measured (From Yagi et al. 1983 (15))

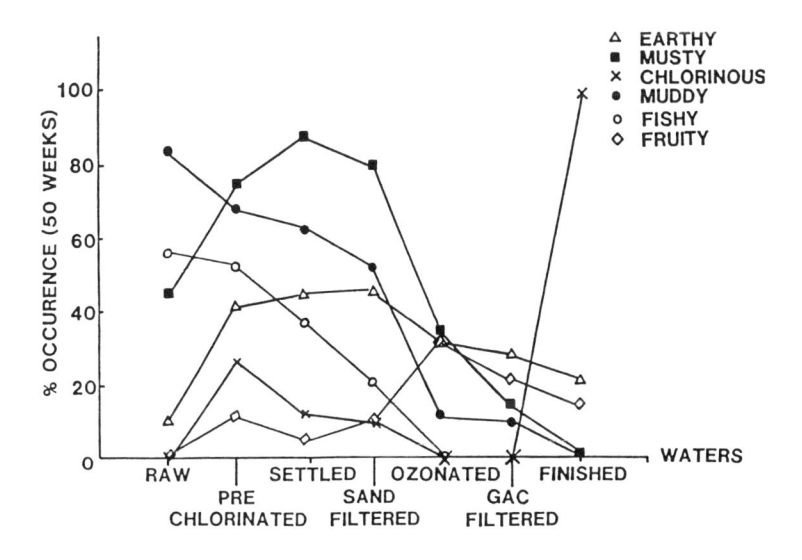

Figure 5. Occurance frequency of odors, Morsang line 2

plant. It was determined that coagulation did not effectively remove odors. Chlorination removed fishy odors, probably by oxidizing amines, but increased the occurrence of earthy/musty odors. Ozone removed most odors but sometimes added a sweet or fruity residual, and PAC removed muddy tastes and odors. GAC removed some tastes and odors, but the GAC influent was relatively taste and odor free due to the effectiveness of ozone.

These results were consistent with the laboratory study and the literature reviewed. GAC, PAC and ozone would generally be expected to remove tastes and odors more effectively than chlorine-based oxidants, and coagulation should have little effect, as was observed. The production of sweet or fruity compounds (probably aldehydes and ketones) may pose problems in areas where customers are not accustomed to such flavors in their water.

The Philadelphia Suburban Water Company utilizes many sources for the production of drinking water. Of these sources, the Neshaminy watershed is the most highly industrialized and the most prone to intermittent upset due to anthropogenic activities. The Neshaminy creek has a drainage area of 544 km^2 and an average discharge of 8.326 m^3/s which is augmented by releases from five upstream reservoirs. There are over 25 private and public domestic wastewater treatment plants located upstream of the Neshaminy Falls Drinking Water Treatment Plant. The combined discharge of these wastewater treatment plants is approximately 1.1 m^3/s. In addition, over 300 factories and businesses are located within the watershed.

Three tiers of protection are utilized to assure that the drinking water produced at the Neshaminy facility meets all applicable drinking water standards (see Table 1). The watershed is monitored on a daily basis upstream of the treatment plant, raw water quality is monitored routinely at the plant and finished water is monitored prior to distribution.

The watershed is patrolled daily upstream of the treatment plant to evaluate the conditions of the stream. The watershed patrolmen routinely sample above and below major discharges and are equipped to do field monitoring for pH, ammonia, cyanide and other parameters.

If a problem is detected, the patrolman will

1) notify the treatment plant and estimate the time of travel,
2) attempt to isolate the source of the problem, and
3) try to work with the polluter to abate the problem.

In spite of the above precautions, contamination is often not detected until it is observed at the treatment plant. Raw waters are monitored routinely at the plant and treatment schemes are modified to optimize finished water quality (see Figure 2). A few examples of how treatment flexibility allows rapid correction of degraded water quality are given below.

On several occasions, medicinal-type odors have been created upon chlorination of untreated Neshaminy water. It was found that these odors were eliminated by using chlorine dioxide in place of chlorine during pretreatment. Further study showed that the source of the odors was iodoform. The iodoform was created during chlorination of the raw water in the presence of iodide. It is hypothesized that the chlorine oxidized the iodide to iodine, which then entered into the haloform reaction.

The intermittent discharge of industrial wastes can often cause a rapid degradation of water quality. Routine jar testing allows for optimized treatment to rectify these problems. PAC is routinely fed at a dosage of about 10 mg/l for organic removal and taste and odor control. However, during pollution events this dosage may be increased to over 200 mg/l. Cyanide, if found in the raw water is removed by superchlorination and elevated pH throughout the 12 hour sedimentation process.

Finally, in the event of catastrophic pollution events, the raw water intakes can be shut down and water from other sources can be delivered through a highly interconnected distribution system.

Summary and Conclusions

A laboratory study and a literature review were conducted to determine the most effective conventional pretreatment processes to remove low molecular weight organics, including priority chemicals and organoleptic compounds. The processes evaluated were primarily aeration, oxidation and powdered activated carbon. Removal efficiencies of fifteen organic compounds representing a range of chemical structures were studied.

Chlorine, chlorine dioxide, chloramines, and potassium permanganate were the oxidants tested. This laboratory experiment and the literature suggest that oxidants are not an effective way of removing most low molecular weight organic compounds from water. Nevertheless, some water treatment plants have had good results in solving specific taste and odor problems using oxidants. Oxidants, therefore may be good for specific contaminants.

This laboratory experiment found that all test compounds were removed to some degree by a high dose of PAC (50 mg/l). Only one of the 15 compounds tested was reduced by less than 50% and eight compounds were reduced by greater than 90%. PAC appears to be the most effective pretreatment for removing taste and odor causing organic compounds and many priority chemicals. PAC should be considered as a first course of action when taste and odor problems arise.

In all cases, background water quality and the mixture of organic compounds in the water will effect the oxidation and adsorption processes. Therefore, routine sensory analyses and chemical analysis for known odorants should be completed. At the time of an odor event, laboratory analyses and treatment test evaluations are always advisable before taste and odor treatment procedures are implemented in the plant. The testing procedure should be rapid to enable a quick response to the problem.

At this time, when odor events occur, identification of specific organoleptic compounds in water is available for known odorants, such as geosmin and MIB. There are many odorants that have not been determined. What is needed is a concerted effort of routine sampling, with chemical analysis for known odorants and taste and odor panel analysis to define organoleptic problems. Even when the odor agent is known, predicting the compound's response to oxidants is at best an educated guess. Therefore, treating these problems by trial dosing in the laboratory at the plant and evaluating odor reduction by a sensory based procedure are the most viable approach. Future research is needed to help define causes of taste and odor.

Acknowledgement

This work is based partially on the final report for the AWWA Research Foundation project "Taste and Odor in Drinking Water Supplies", Project Officer Jon DeBoer.

References

[1] National Academy of Science, Safe Drinking Water Committee (1977): Drinking Water and Health, Vol. 1. National Academy Press, Washington
[2] Code of Federal Regulations (1987) Parts 141, 142, July 8, pp. 25690-25717
[3] Malevialle, J., Suffet, I.H. (1987): Identification and Treatment of Odors in Drinking Water. AWWA Research Foundation, Denver
[4] Kopfler, F.C., Melton, R.G., Mullaney, J.L., Tardiff, R.G. (1977) in: Suffet, I.H. (ed.), Fate of Pollutants in the Air and Water Environment, Vol. 2. Wiley, New York
[5] Khiari, D., Brenner, L., Suffet, I.H., Gittelman, T. (1987): Development of a Data Base of Trace Organic Chemicals for Control of Tastes and Odors at the Neshaminy Creek Water Supply. Submitted to the Philadelphia Suburban Water Co, Bryn Mawr, PA
[6] Baker, R.J., Suffet, I.H., Anselme, C., Mallevialle, J. (1986) in: Water Quality Technology Conference Proceedings. AWWA, Nov. 16-20, Portland, OR
[7] Lalezary, S. et al. (1984) J. AWWA *67*, 3, 83
[8] McCarty, P.L. (1983) in: Control of Organic Substances in Water and Waste Water, EPA-600/8-83-011
[9] Anselme, C., Mallevialle, J., Suffet, I.H. (1988): Removal of Tastes and Odors by the Ozone-Granular Activated Carbon Water Treatment Process. AWWA (in press)
[10] Atashitani, K., Hishida, Y. (1988): Water Science and Technology. J. of the International Association of Water Pollution Research and Control (in press)
[11] Sigworth, E.A. (1957) J. AWWA *42*, 12, 1507
[12] Lyman, W.J., Rehl, W.F., Rosenblatt, D.H. (1982): Handbook of Chemical Property Estimation Methods. McGraw-Hill, New York
[13] Kuhn, W. et al. (1978) J. AWWA *70*, 4, 326
[14] Kobayashi, H. Rittmann, B.E. (1982) ES&T *16*, 3, 170A
[15] Yagi, M. et al. (1983) Wat. Sci. Tech. *15*, 311
[16] Weber, W.J. (1972): Physiochemical Processes for Water Quality Control. Wiley, New York
[17] Morrison, R.T., Boyd, R.N. (1971): Organic Chemistry. Allyn and Bacon Inc., Boston
[18] Snoeyink, V.L., Jenkins, D. (1980): Water Chemistry. Wiley, New York
[19] Masschelein, W.J. (1979): Chlorine Dioxide. Ann Arbor Science Pub. Inc., Ann Arbor
[20] Hoigné, J., Bader, H. (1982) Vom Wasser *59*, 253
[21] Hoigné, J., Bader, H. (1983, 1983a) Water Research *17*, 2, 173 and 185
[22] Katz, J. (1980): Ozone and Chlorine Dioxide Technology for Disinfection of Drinking Water. Noyes Data Corp., Park Ridge, NJ
[23] Rook, J.J. (1977) ES&T *11*, 5, 478

[24] Rice, R.G. (1980): Ozone. Science and Eng. 2, 75
[25] Kraznov, B.P., Pakul, D.L., Kirillova, T.V. (1974) Khim Prom 1, 28 (English trans. in: Intl. Chem. Eng. 14, 4, 745)
[26] Dore, M. (1985) in: Perry, R., McIntyre, A.E. (eds.) Proceedings of the International Conference, The Role of Ozone in Water and Wastewater Treatment. SP Press, London
[27] Hoigné, J., Bader, H. (1976) Water Research 10, 377
[28] Hoigné, J., Bader, H. (1977) 3rd International Ozone Symposium (IOA), Paris
[29] Hoigné, J., Bader, H. (1977) Symposium on Advanced Ozone Technology (IOA), Toronto
[30] Hoigné, J., Bader, H. (1978) Prog. Wat. Techn. 10, 657
[31] Hoigné, J. (1984) in: Rice, R.G., Netzer, A. (eds.) Ozone Technique and its Practical Application, Vol. 1. Ann Arbor Science, Ann Arbor, MI
[32] Staehelin, J., Hoigné, J. (1983) Vom Wasser 61, 337
[33] Sigworth, E.A. (1968) Taste and Odor Control Journal 34, 34
[34] McGuire, M.J. et al. (1978) in: Ozone Technology Symposium. Int. Ozone Inst., Los Angeles
[35] Lin, S., Carlson, R.M. (1984) ES&T 18, 743
[36] Standard Methods for the Examination of Water and Wastewater (1976) APHA, AWWA and WPCF, 14th Ed., Washington
[37] Krasner, S.W. et al. (1985) J. AWWA 77, 3, 34
[38] Bartels, J.H.M. et al. (1986) J. AWWA 78, 3, 50
[39] Bartels, J.H.M., Brady, B.M., Suffet, I.H. (eds.) (1987): Taste and Odor in Drinking Water Supplies. (Prepared for AWWA Research Foundation, Denver)
[40] McGuire, M.J., Suffet, I.H. (1980) in: Activated Carbon Adsorption of Organics from the Aqueous Phase. Ann Arbor Pub. Corp., Ann Arbor, MI
[41] Montiel, A.J. (1983) Wat. Sci. Tech. 15, 279
[42] Cherry, A.K. (1962) J. AWWA f54, 419
[43] Herzing, D.R., Snoeyink, V.L., Wood, N.F. (1977) J. AWWA 69, 4, 223
[44] Lalezary, S., Pirbazari, M., McGuire, M.J. (1986) AWWA 78, 76
[45] Dobbs, R.A., Cohen, J.M. (1980): Carbon Adsorption Isotherms for Toxic Organics. EPA-600/8-80-023
[46] McKay, D., Wolkoff, A.W. (1973) ES&T 7, 611
[47] Love, T.O. Jr. et al. (1981): Treatment of Volatile Organic Compounds in Drinking Water. EPA-600/8-83-019
[48] Warner, P.H., Cohen, J.M., Ireland, J.C. (1980): Determination of Henry's Law Constants of Selected Priority Pollutants. USEPA
[49] Zoeteman, B.C.J. (1980): Sensory Assessment of Water Quality. Pergamon Press, Oxford
[50] Patty, F.A. (1967): Industrial Hygiene and Toxicology, Vol. 2. Interscience Publishers, New York

T. L. Yohe
Head of Research
Philadelphia Suburban Water Co.
Bryn Mawr, PA 19010
USA

I. H. Suffet, R. J. Baker
Environmental Studies Institute
Drexel University
Philadelphia, PA 19104
USA

Water Quality Problems and Control Strategies for the Water Supply of Tianjin City

X. Zhu

Abstract

Water supply development must be concerned with both the quantity and quality of water required to meet the needs of man in an efficient and economical manner. The diversion of water from the Luan River into Tianjin is a vast urban water supply project. There are some obvious and potential pollution hazards for the drinking water supply, especially eutrophication is a serious problem and results in treating difficulties at the water supply plant. This paper attempts to provide information on this issue and to discuss the program of water quality conservation and management for the water supply of Tianjin based on various treatment techniques used in the basin, the reservoir and the water treatment plant.

Introduction

Surface water sources are being used increasingly to supply a growing population, industry, and agriculture. In some areas the water supply falls far short of today's demand. This results in the necessity for these continuously growing communities to look elsewhere for a water supply to meet people's continuously growing demand. At the present time many cities in the world transport water over a long distance at great costs. Obviously, the diverted water would provide good conditions for city development and spur on the economy. Often some problems of water quality arise due to the environmental impact of diversion or new pollution problems appear. Sometimes the water resource may be in danger of being lost if the water quality is unsatisfactory. We must remember that water supply development must be concerned with both the quantity and quality of water required to meet the needs of man in an efficient and economical manner. Neither factor can be neglected. The usefulness of the water supply is determined to a large extent by its quality.

Tianjin is the third largest city in China. After the 1970's, the inadequacy of the water supply posed great difficulties to the development of the national economy and the life of the people. With the approval of the Chinese government, the project of diverting the water of the Luan River into Tianjin was started in May 1982 and completed in Sept. 1983. The diversion system has already been used for four years and it has provided a steady water supply for developing Tianjin. Many people have been pleased and satisfied with the sufficient water quantity. But experiences with water quality in the Tianjin water supply during the four years after completion of the diversion project showed that the water supply is not safe. Water quality problems related to the diversion project itself include the excessive growth of algae and aquatic plants in storage reservoirs, rivers and open channels; excessive coliform groups and other bacteria; periodical high concentrations of iron and manganese; organic wastes and so on. Eutrophication of rivers and reservoirs due to the diversion project was serious and resulted in treating difficulties at the water supply plants. This paper attempts to provide information on this issue and discusses the conservation and control strategies for the Tianjin water supply.

Diversion Project of Tianjin

The diversion of water from the Luan River into Tianjin was a vast urban water supply project. It extended across provinces, cities and river basins, with a complete system of channeling, conveying, storing, treating and distributing water. Crossing six counties, the system has a total length of 234 km. A tunnel 11 km long had to be dug, and as many as 215 construction projects had to be completed such as pump stations, reservoirs, waterworks, underground and above ground channels, piping systems, reverse siphons, sluice dams and power stations (see Fig. 1). The people of Tianjin completed the entire diversion project within a year and four months. The water source of the project is the Pan Reservoir with a capacity of 29×10^8 m^3, which is an impoundment of the Luan River. The reservoir dam is over 107.5 m high. As shown in Figure 1, the water from the Luan River released from the Pan Reservoir flows down to the Daheiting Reservoir, which is shallow and has a storage capacity of 3×10^8 m^3. From there, water reaches the beginning of the diversion system. From that point, water is transported by means of a tunnel, which is 11.2 km long and 3.5 m in diameter, across the mountains at a flow rate of 80 m^3/s, then released into the original channel of the Li River in the Tianjin area. The water follows the 57.4 km channel at a rate of 65 m^3/s to the Yuqiao Reservoir which is shallow with an average depth of about 3.8 m and a total capacity of about 4.2×10^8 m^3. The water leaves the reservoir at 40 m^3/s and turns south to flow in a 56 km river and a 64.2 km diversion channel. There are three pump stations for transporting water along the open channel. After the third pump station, the water enters two underground pipelines which are 2.5 m in diameter and which carry the water to the Tianjin water supply treatment plant at a flow rate of 10 m^3/s. A small reservoir for pre-storage of the water, with a capacity of 0.45×10^3 m^3, was constructed near the last pumping station.

This is the largest diversion project in China. In normal years, an annual supply of a billion cubic meters is available. This greatly relieves the city's water demand and promotes its functions.

Water Quality Problem

A comprehensive water quality program was undertaken for monitoring water from the Luan River to the Tianjin water supply plant by some environmental and

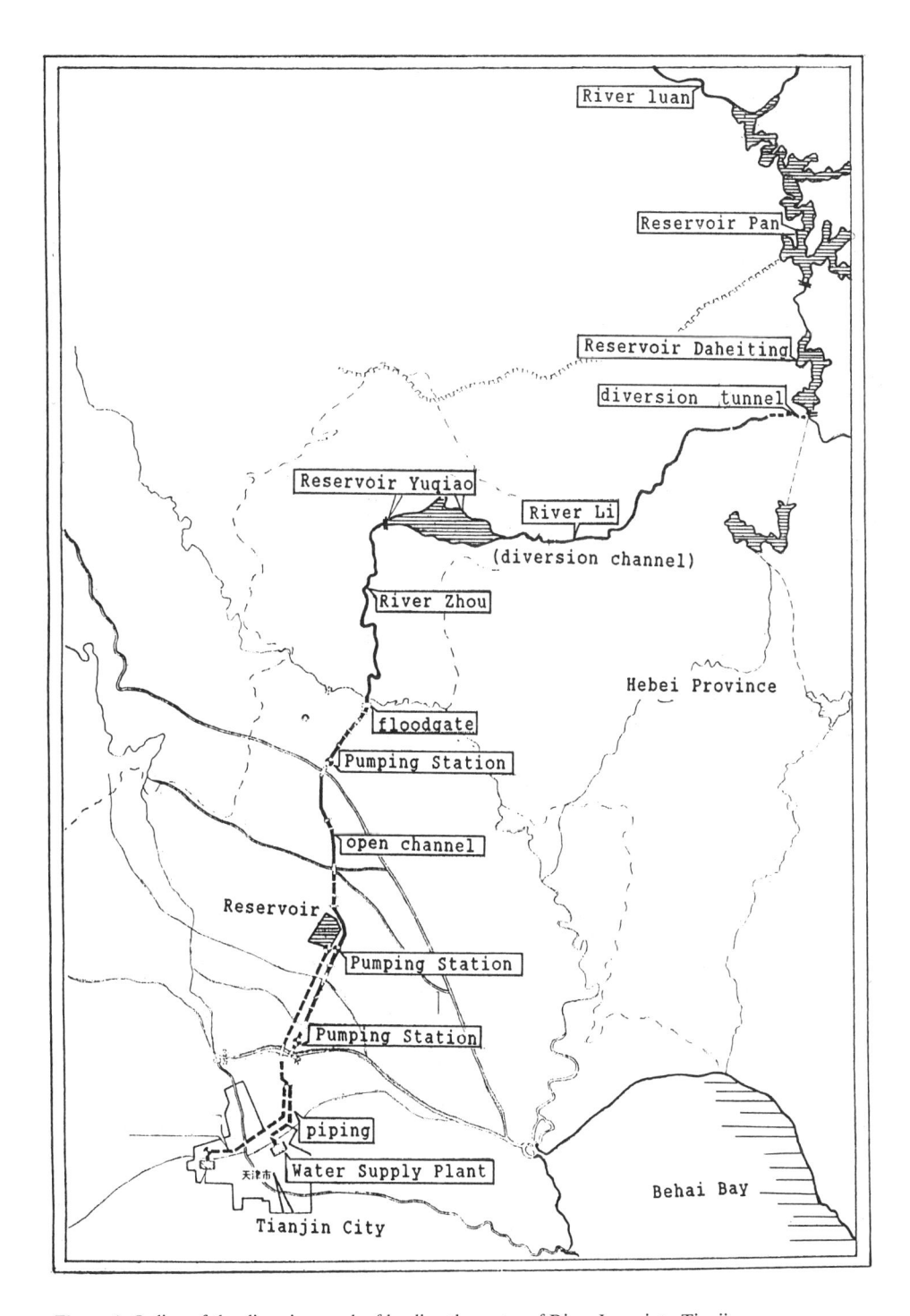

Figure 1. Ouline of the diversion work of leading the water of River Luan into Tianjin

hydrological stations and institutes. Based on the average monthly assessments of water quality, it was found that most of the water quality standards for surface water were met.

As the water was transported through this complex system of storage areas and transport pathways over a long distance, water quality was modified from upstream to downstream due to various nutrients, as shown in Figure 2.

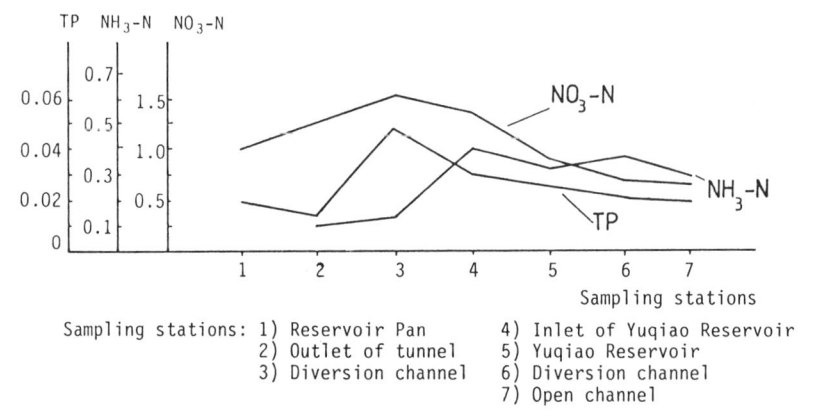

```
TP    NH_3-N  NO_3-N
      0.7
0.06        1.5                              NO_3-N
      0.5
0.04        1.0
      0.3
0.02        0.5                                      NH_3-N
      0.1                                      TP
   0
              1    2    3    4    5    6    7
                                      Sampling stations

Sampling stations: 1) Reservoir Pan        4) Inlet of Yuqiao Reservoir
                   2) Outlet of tunnel     5) Yuqiao Reservoir
                   3) Diversion channel    6) Diversion channel
                                           7) Open channel
```

Figure 2. Nutrient concentrations variance in the water along the diversion work

It was clear that the concentrations of some nutrients in the water were higher downstream than in the Pan Reservoir, and this may be related to the effect of the diversion channels which carry the water, the internal nutrient cycle in the reservoir, chemical, physical and biological processes, and finally an adjustment which occurred in the water body itself. The main water quality problems of the diversion project are analyzed subsequently.

Eutrophication of the Yuqiao Reservoir

The Yuqiao Reservoir was impounded in 1960 for purposes of irrigation. After completion of the diversion project it became a very important storage and regulating reservoir as well as the drinking water source for Tianjin. The area of the basin is about 2060 km² with three tributaries: Sha River (largest in the basin), Lin River and Li River (which also forms part of the diversion channel). The annual input from the three tributaries is about 4×10^8 m³. About 1 billion cubic meters of water from the Luan River enter the Yuqiao Reservoir via the Li River after completion of the diversion project. Thus some changes in the hydrology and loading conditions occurred in the Yuqiao Reservoir, as shown in Table 1.

	Before diversion work	After diversion work
influent flows	4×10^8 m³	14×10^8 m³
effluent flows	3×10^8 m³	10×10^8 m³
detention time	1.4 Years	0.42 Years

Table 1. Changes of limnology conditions of Yuqiao Reservoir

Before the diversion work in 1976 and in 1982 – 83, a chemical and biological examination of the Yuqiao Reservoir was undertaken. It was shown that the concentrations of nitrogen and phosphorus were higher than usual and thick algal growth was present in the Yuqiao Reservoir in spring and autumn. Most of the algal species were blue-green algae. A large number of macrophytes was found in the middle and upstream areas. Temperature and dissolved oxygen stratification was particularly pronounced in the summer and severe anoxic conditions were present near the reservoir bottom in the summer months. The Secchi Disk reading was only 0.5 m in summer. It was obvious that the Yuqiao Reservoir was on its way to becoming a eutrophic reservoir.

Some people expected the trophic state of the reservoir to improve due to the higher flushing rate and shorter detention time in the reservoir after completion of the diversion project. In fact, transparency was increased to 1.2 m in summer, 3 to 5 m in winter; the concentration of total phosphorus in the water had decreased to 0.02 to 0.04 mg/l. Some algal species and some macrophytes decreased, but some also increased because of the higher light intensity in the water due to increased water transparency. The stratification of temperature and oxygen in the reservoir changed slightly, but anoxic conditions near the bottom remained. See Figure 3.

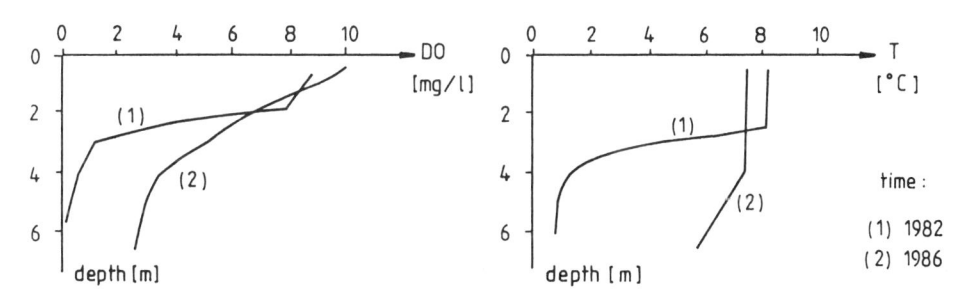

Figure 3. DO stratification of Yuqiao reservoir

Excessive Growth of Macrophytes

There were many species of macrophytes growing in the Yuqiao Reservoir, including emergent, floating-leaf and submersed varieties. Biomass was very high and in some parts of the reservoir the plants formed a thick forest which hindered movement of the boat. In addition to the macrophytes found in the reservoir, there were also plants downstream in the old and new channels. Some had been flushed from the reservoir, others grew there. The biomass of the macrophytes was surprisingly high in the growing season, causing decreased water flow and raising the water level significantly so that strong pressure affected the channel banks. A surprising event took place in the spring of 1985: macrophytes in the open channel had clogged the screen of the Chaobei pumping station. About 100 people stood on these aquatic plants, and their density was so high that no one fell into the water! It was necessary to harvest these macrophytes day and night during the growing season.

Other Water Quality Problems

Other water quality problems associated with the diversion project were as follows:

1) Periodically, high concentrations of iron and manganese were present in the reservoir and channel.
2) The populations of coliform groups and other bacteria were very large, above the national standard, upstream as well as downstream of the diversion system.
3) High concentrations of chloride were present in some parts of the open channel for short periods, and were due to the effect of salt land near the open channel.
4) The COD in the water in the diversion channel ranged from 3 to 6 mg/l, sometimes exceeding the national standard for surface water. Phosphate concentrations were often higher than 0.5 mg/l.

Water Treatment in the Tianjin Water Supply Plants

The Tianjin water supply plants have the obligation to provide potable water which is chemically and bacteriologically safe for human consumption and which is of an adequate quality for industrial users. Three plants provide 106 m^3 per day for Tianjin City. Before the diversion project, the plants received water from the Hai River which had a high chloride concentration due to the sea water. The diversion project ended the unpleasant history of people drinking salt water and provided sufficient quantity and better water quality for Tianjin.

The largest and most important of the three plants is called the Jieyan water supply plant, and was built in the 1930's but has been expanded since then. The design capacity of the plant is 5 x 10^5 m^3 per day. The plant consists of 2 horizontal flow basins, 2 tube settlers and 8 rapid sand filters. The treatment process consists of pre-chlorination, addition of chemicals, rapid mixing, coagulation, precipitation, filtration and chlorination. The water conditions and dosage of chemicals were as follows: water temperature 20 to 32°C, turbidity 20 to 40 mg/l, FeSO$_4$ dosage 5 to 6 mg/l (in summer) and 3 mg/l (in winter), chlorination 3 to 5 mg/l.

The treated water discharged from the plant met most of the national standards for water quality. But there were many problems with treatment due to the eutrophication of the water source, however, after completion of the diversion project. The problems encountered in the Tianjin water supply plant included: rapid clogging of filters by algae, unpleasant tastes and odors, risk of increased bacterial growth in the drinking water due to the fouling of the distribution network and the nutrient content. In order to counteract these problems, twice the amount of chemicals was added, which increased the frequency of filter backwashing. This led to a reduction in the water production. It is clear that the eutrophication problem at the water source has caused expensive and time-consuming treatment problems.

Control Strategies

The diversion project has played the most important role for improving people's lives in Tianjin and for economic development. We must take effective protective measures and use conservation strategies to ensure safe water transport and a developing economy for a long time.

Eutrophication is one of the most important water quality problems worldwide. Solving this problem will entail more advanced technology, high costs and detailed management planning. In spite of the economic conditions in our developing country, we cannot ignore the economic and political realities in favor of a technical approach.

The final decision on an appropriate control strategy must be based on the relevant social, technical, economic and ecological aspects.

We have proposed a research program for selecting effective strategies for controlling eutrophication of the Tianjin water supply. This program is based on the analysis of the following problems:

1) The water quality of the Pan Reservoir (Luan River impoundment and the water source of the diversion project) is better than that of the Yuqiao Reservoir. There are no obvious problems except the high concentrations of nitrogen produced by soil erosion.
2) All the areas through which the diverted water passes are agricultural areas without developed industry. The emphasis on pollution control is therefore in controlling nonpoint sources.
3) Since the water is transported over a long distance, it is affected by different storage areas and channels which are in turn affected by ecology, limnology and land use. A system of control strategies is needed for water quality conservation, which must be regarded either as a special problem or one for which the whole state carries a responsibility.

The research program is primarily concerned with control measures in five major categories:
1) Control of nutrient sources in the watershed
2) Treatment of the tributaries which feed into the Yuqiao Reservoir
3) Eutrophication control in the reservoir itself (biological methods)
4) Intensifying the treatment process in the Tianjin water supply plant
5) Legislation for and management of the watershed

Nutrient Source Control in the Watershed

Experience has shown that the most effective approach is to treat the most readily-controllable cause of the problem. In this case it is the input of excessive quantities of phosphorous and nitrogen from the drainage basin into the water. The control program must be directed towards the major sources of these nutrients in the drainage basin.

According to the examination of pollutant load inputs along the route of the diverted water, approximately 80% of the loads were contributed by the tributaries of the Yuqiao Reservoir.

A detailed investigation of nutrient loading into the reservoir was carried out in 1982 to 1984 simultaneously with investigations of water quality and quantity. The amount of phosphorus and nitrogen was estimated for each storm event and for each month of the year.

The results of total nutrient flux including runoff, precipitation and ground water into the Yuqiao Reservoir showed that:

1) Runoff from nonpoint agricultural sources was the main source of these nutrients
2) The nutrients input during the flood season were over 80% of the total annual load.
3) Approximately 50% of the annual nitrogen load comes from point sources. There are two fertilizer plants upstream of the Yuqiao Reservoir near the diversion channel. Another 50% comes from nonpoint sources. Almost all the phosphorus comes from nonpoint agricultural sources.

Two major approaches were undertaken in attempting to reduce the potential impact of agricultural activities on the eutrophication of the Yuqiao Reservoir:

1) Application of natural and mineral fertilizer in a manner that inhibits their transport into water bodies.
2) Maximum prevention of soil erosion.

The nutrient flux data indicated that the concentration of solids and suspended solids from agricultural runoff reached over 5000 mg/l in the Yuqiao Reservoir. The sediment mass flux was estimated at 24×10^4 tons per year, which played a major role in the eutrophication of the reservoir. Therefore, it is necessary to control the input of sediment in order to protect the reservoir and its water quality. Some typical management practices that should be adopted are as follows:

1) Contour farming to reduce erosion
2) Leaving buffer strips of trees and shrubs along streams to reduce soil erosion
3) Grass seeding or other vegetation to stabilize areas near the stream and reservoir banks particularly to reduce flushing of soil and erosion
4) Use of culverts and other engineering structures to control water flow and thereby reduce erosion
5) Nutrient management practices including the formulation of a fertilizer application rate, application technique and timing of the applications
6) Pesticide management practices including application methodology, timing and rate of application.

The total surface area of the Yuqiao Reservoir was 104 km² at normal water levels. The loads of total phosphorus and total inorganic nitrogen in the reservoir were 0.71 g/m³ yr and 4.31 g/m³ yr, respectively. It is clear that these loads are much higher than those recommended by Vollenweider. As noted above, the external nutrients came mainly from the agricultural nonpoint sources in the catchment area.

Treatment of Tributary Influent Waters

Pre-reservoir. The use of pre-reservoirs as bioreactors for phosphorus elimination has been investigated and applied successfully to some lakes and reservoirs around the world.

The elimination of phosphorus in the pre-reservoir is related to an enhancement of bioproductivity. The phosphorus is fixed in the increased algal and macrophyte biomass in the pre-reservoir and is thus largely retained there via sedimentation.

The Yuqiao Reservoir already seems to act as a pre-reservoir. Upstream of the reservoir, approximately 25% of the total surface area was covered with macrophytes which helped to absorb nutrients and increase the retention time of the water. The area is about 30 km² with depths of 0 to 3 m. These were good conditions under which to construct a pre-reservoir although it was provided by nature and had not required extensive planning.

In order to build an effective pre-reservoir, the following questions must be answered through research:

1) What retention time is required in order to ensure bioproductivity?
2) How can the pre-reservoir project be completed taking into account the natural conditions, and how efficient will it be for phosphorus elimination?

The research plan for the Yuqiao reservoir is shown in the form of a flow chart in Figure 4.

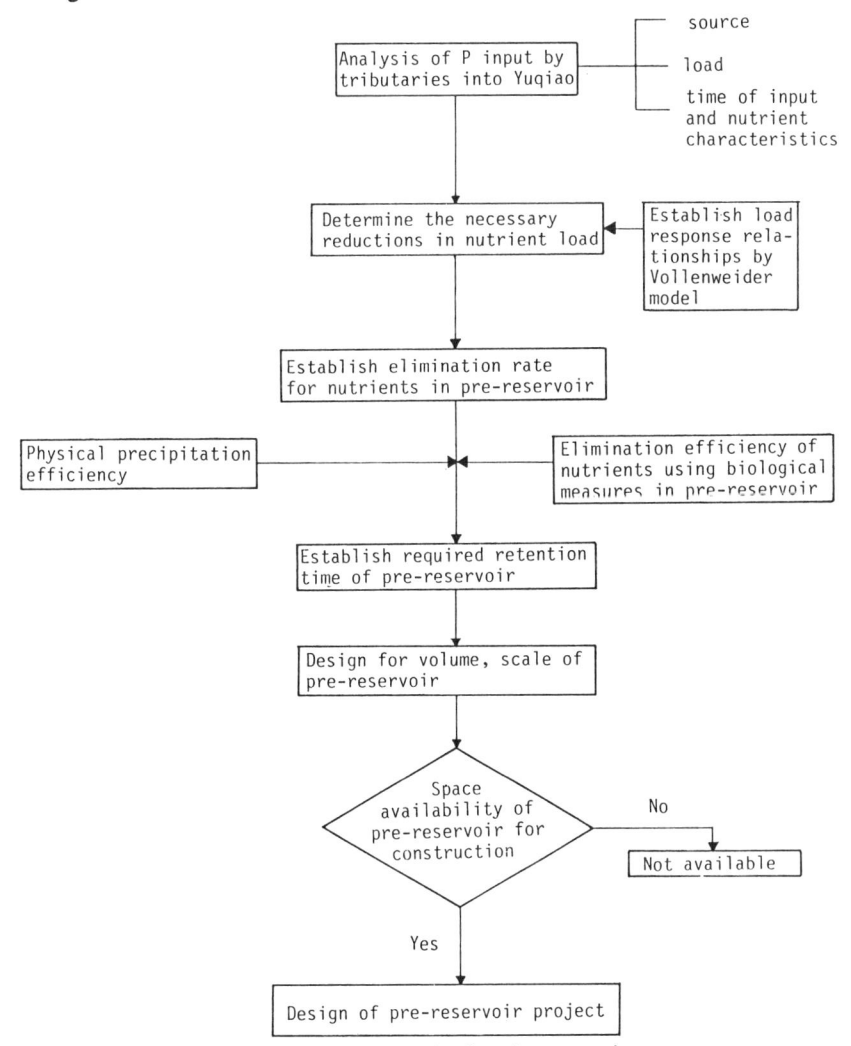

Figure 4. Research procedure for pre-reservoir of Yuqiao reservoir

Direct Addition of Phosphorus Precipitating Chemicals to the Influent Waters. The shallow Yuqiao Reservoir has three main tributaries. The large phosphorus load is contributed by the tributaries in a short period of 3 summer months. If the water is allowed to reach the treatment plant, pre-treatment is too expensive and this method cannot be used throughout the year. Thus, it seems reasonable to add the chemicals directly at the point where the water flows into the reservoir. The following questions must be answered through research:

1) Can addition of phosphorus-precipitating chemicals be correlated with the actual water flow and the concentration of phosphorus in the tributaries entering the reservoir?
2) What amount of chemicals must be used? What should be done about silting in the reservoir due to the precipitating phosphorus?

Eutrophication Control in the Reservoir Itself

Growth of the macrophytes in the Yuqiao Reservoir was prolific, covering the area at a depth of 0 to 3 m from April to November. If these plants are not removed from the water, they can function as a nutrient "pump" which transports nutrients from the sediments back into the water. This may result in an increase in nutrients in the reservoir and progressive eutrophication. We should determine the feasibility of harvesting macrophytes for the removal of nutrients from the reservoir and develop ways for reusing the biomass as raw materials and animal fodder.

Intensification of the Treatment Process at the Tianjin Water Supply Plant

Although significant progress has been made concerning the water supply in Tianjin, in many cases the water service is not nearly as good as it could and should be. Some aspects of water quality fell short of present daily standards. Monitoring, operator and manager training and water research do not stand up to today's requirements.

In order to counteract the eutrophication problem at the water source, management and conservation practices must be strengthened to achieve adequate surface water quality in the catchment area. On the other hand, in order to intensify the treatment processes at the water supply plant, research must be undertaken in the following areas:

Improvement of Water Treatment Processes. When water is supplied from rivers, this normally requires the most extensive treatment facilities and greatest operational flexibility. The water treatment plant must be capable of handling day-to-day variations and anticipate changes in quality. The Tianjin water treatment plant receives its water from rivers and reservoirs which are eutrophic. Rivers and reservoirs are subject to seasonal changes. Algal blooms frequently occur in early spring and late summer. Heavy algal growth, particularly that which includes certain species of blue-greens, produces tastes and odors which are difficult to remove. In order to deal with these changes in water quality at the source, the operator should have the means to change the point where certain chemicals are added. For example, chlorine feedlines are normally provided for pre-, intermediate and post-chlorination. Multiple chemical feeders and storage tanks should be supplied so that various chemicals can be employed in the treatment process.

Development of New Techniques. We should energetically pursue the development of new techniques in water treatment to improve treatment efficiency and to satisfy the increasing demands for high water quality. For example, proper instantaneous and thorough mixing of chlorine, and treatment which provides better disinfection; the use of granular activated carbon for removing trace organic compounds and for complete color removal and taste and odor control; the use of biological treatment to eliminate algae; the use of polymers to improve coagulation, flocculation and settling; the use of methods for continuously monitoring and controlling water treatment processes and other methods.

Legislation for and Management of the Watershed

Since Tianjin's water supply is transported over a long distance, across provinces and river basins, water quality is governed by processes in these basins, ecological characteristics and human activities. For example, deforestation, the excessive use of chemicals and fertilizer, the industrialization of agricultural regions, fish farming, irrigation practices, precipitation, and soil erosion, etc. may contribute to an increase in the nutrient and pollutant load which leaves the basin and promotes eutrophication of the water. Thus, in order to obtain a long-term reliable and good water supply, management and planning for efficient conservation, development and utilization of the water resource must be undertaken from the standpoint of the range of the watershed. The Tianjin Diversion Project Management Agency was established in 1984, and is responsible for managing and protecting the water supply. Although the transport of water and regulation of the quantity of the water are fairly simple tasks, implementation of the eutrophication control program is difficult due to its division among several administrative regions, range of responsibility and power. Control of eutrophication may contain some conflicting elements, for example, the Yuqiao Reservoir belongs to the Tianjin area but its basin to Hebei Province. Implementation of this control must involve limiting industrialization and excessive use of fertilizer and increasing forest cover in the upstream basin of Yuqiao Reservoir, among other things. The government of Hebei Province does not welcome this program, however, because it does not benefit them.

The first step, therefore, is to establish a single agency which has jurisdiction over the environmental management program in the national government, or through a cooperative hierarchical framework with the national government and local governmental entities.

Secondly, legislation and regulation should be examined as a necessary component of a pollution control program.

Research Plan for Decision Strategies

Summarizing the research plan, the basic approach for achieving the above-mentioned objectives consists of the following steps, shown in Figure 5:

– Identify water quality problems and establish management goals
– Assess how much information is available on the watershed
– Identify the options for managing water quality
– Analyze all costs and expected benefits of alternative management and control options
– Evaluate the adequacy of the existing institutional and regulatory framework for implementing alternative management strategies
– Select desired control strategies
– Submit progress reports on the control program to the government periodically.

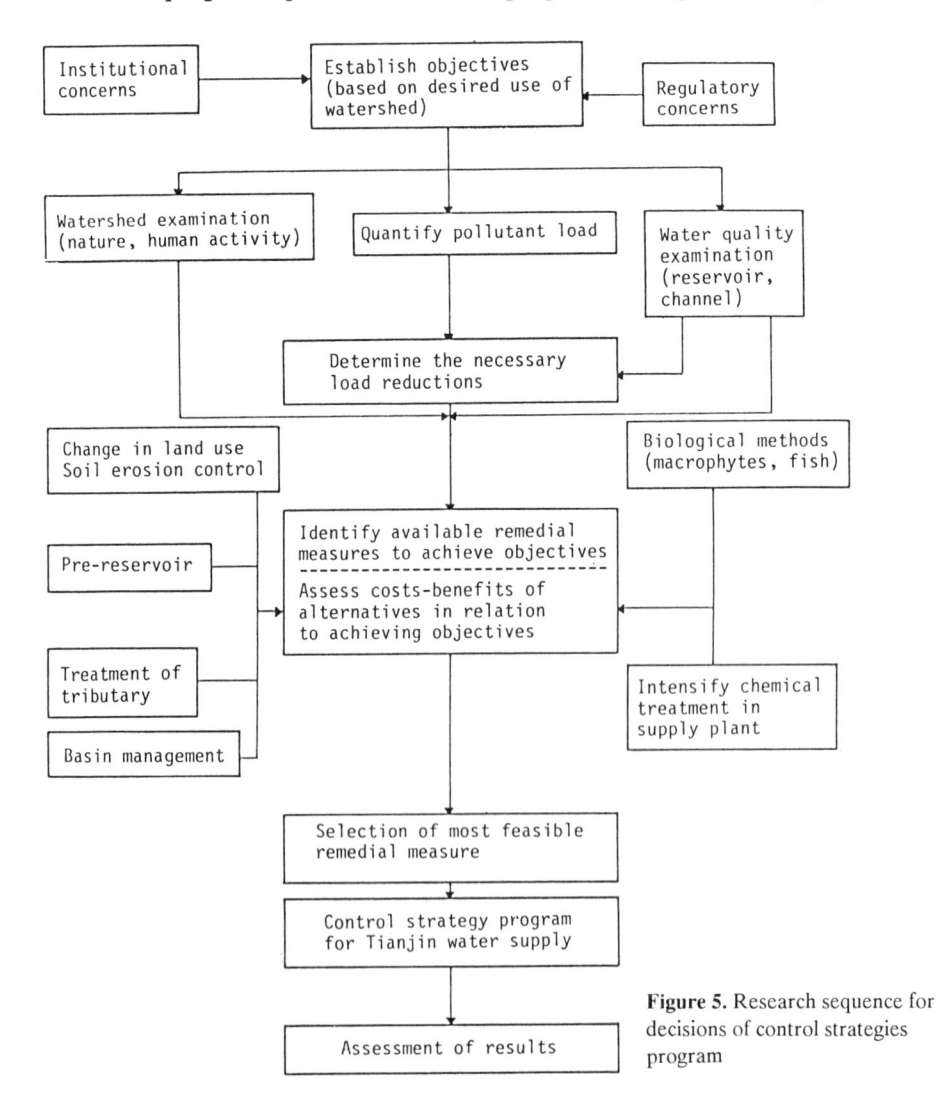

Figure 5. Research sequence for decisions of control strategies program

References

[1] Clark, J.W., Viessman, W. Jr., Hammer, M.J. (1977): Water Supply and Pollution Control. New York
[2] Ryding, S.O., Rast, W. (1987): Control of Eutrophication of Lakes and Reservoirs. IVL
[3] Lamb, B.L. (1980): Water Quality Administration. Ann Arbor Science, Michigan
[4] Gower, A.M. (1980): Water Quality in Catchment Ecosystems. John Wiley

X. Zhu
Tianjin Institute of Environmental Protection
and Sciences 17
Fu Kang Road
Tianjin
China

Humic Substances Removal by Alum Coagulation - Direct Filtration at Low pH

J. Fettig, H. Ødegaard and B. Eikebrokk

Abstract

The potential of using alum for the coagulation of humic substances was evaluated by treating a colored surface water. Jar testing revealed that only one major removal domain existed with pH = 5.5 being an optimum pH value. Sedimentation as well as filtration behaviour of the flocs was studied between pH = 4.7 and pH = 7. For the latter experiments, sand filters of different bed lengths were used. While sedimentation worked best at pH = 6, filtration efficiency was best for pH < 5. From the point of view of floc separation by direct filtration, therefore, conducting the process at pH < 5 offers advantages. However, the kinetics of floc formation were found to depend on pH, that is, they were much slower at pH = 4 compared to pH = 6. Data from preliminary technical-scale experiments conducted with the same raw water are presented.

Introduction

Aquatic humic material resulting from decay of plant residues is found in all natural waters. At higher concentrations it causes a distinct brownish-yellow color and sometimes noticeable taste and odor. It is also a precursor for trihalomethanes and other chlorinated compounds that are formed during water chlorination. Furthermore it has a certain binding potential for multivalent metal ions and for hydrophobic organic substances, for example, many pesticides. Therefore, the concentration of humic substances in drinking water ought to be kept low. The current Norwegian quality standard is met by a color value of 15 mg/l Pt which corresponds to raw water UV-absorbances (254 nm) of 6 – 8 l/m and to TOC values of about 1.5 – 2 mg/l. Since primarily surface water from lakes is taken for the drinking water supply in Norway, many raw waters exceed those levels, thus requiring pretreatment.

At NTH-SINTEF in Trondheim several studies have been conducted on different treatment techniques, e.g. membrane processes, ion exchange, activated carbon adsorption and coagulation/direct filtration [1]. The latter method was determined to be the least expensive alternative for raw water colors from 30 to 80 mg/l Pt [2]. Compared to the conventional coagulation process (rapid mix, slow mix, sedimentation, filtration), direct filtration offers economic advantages, and many Norwegian lake waters are well suited to this type of treatment because of their low turbidity. In preceding studies alum was primarily used as a coagulant. The pH that gave the best filtrate quality could be determined from jar tests, whereby optimum removal was achieved for pH = 6.3 – 6.7 for initial color values from 30 to 60 mg/l Pt [3]. No stoichiometry was observed at that pH range thus indicating sweep coagulation. Another study revealed, however, that filter performance was also excellent for pH < 5. The filtrate quality was found to be as good as at pH = 6.5 while the floc storage capacity of the filter was even larger [4]. Since most Norwegian waters have a low buffer capacity, it would be relatively inexpensive to shift pH to such low values. Therefore, the effect was investigated in more detail with respect to the phenomenon as well as to practical application.

The purpose of the study presented herein was twofold: The alum coagulation domains for organics removal from untreated lake water were to be established using jar tests paying particular attention to low pH. Then, test methods proposed for estimating floc separation efficiency by rapid sand filtration were to be evaluated.

Ultimate Removal Efficiency of Alum Coagulation

Background. While in earlier years much research had been focused on coagulation for removal of particulate or colloidal material [5, 6], more recently attention has also been paid to the removal of dissolved organics. In some of the studies surface or ground water was taken directly, whereas other investigations were conducted on organic substances isolated from natural waters by means of extraction, adsorption, etc. (humic and fulvic acids). Ultimate removal efficiencies of alum coagulation are summarized in Table 1. All of the waters were low in turbidity, the initial TOC varied from 2 to 20 mg/l. The specific aluminium dosages are optimum values, representing the lowest amounts of Al needed in order to achieve good removal at a given pH. We will operationally define ultimate removal by the concentrations of filtered samples (0.45 μm membrane filters) before and after coagulation/flocculation/sedimentation. However, in some of the studies listed in Table 1, other floc separation methods were used, e.g. glass fiber filtration or centrifugation. Although there is much variation among the results that might also be due to floc separation, some general conclusions can be drawn. The removal efficiencies for humic and fulvic acids seem to be higher than for natural waters. In some studies two major removal domains have even been observed with humic and fulvic acids [7, 8]. The specific Al dosage is lowest for pH = 5 – 6 with natural waters and almost as low for pH = 5 – 6 and pH = 6 – 7 with humic and fulvic acids. Removal efficiencies decrease for pH > 7 and pH < 5, although fairly good removal has been found at low pH.

Materials and methods. The raw water was taken from Lake Hyllvannet in Trmndelag, about 15 km east of Trondheim. It is characterized by the following parameters: turbidity = 0.45 ntu, pH = 6.8, specific conductivity = 6.5 mS/m, alkalinity = 0.42 mmol/l, color = 55 mg/l Pt (pH = 6.8) and 40 mg/l Pt (pH = 3), UV-absorbance (254 nm) = 32 1/m (pH = 3), TOC = 6.5 mg/l. Reagent-grade

Table 1. Ultimate removal of organics by alum coagulation

pH	Humic and fulvic acid			Natural water		
	mg Al/mg TOC	% removal	Ref.	mg Al/mg TOC	% removal	Ref.
<5	0.77	85 (UV254)	[7]	0.94	55 (Toc)	[11]
	0.12 - 1.10	>80 (Color)	[8]	≈ 0.70	>90 (Color)	[14]
5-6	0.77	>90 (UV254)	[7]	0.44	45 (UV254)	[14]
	0.97	>90 (UV254)	[9]	0.52	81 (UV254)	[14]
	0.41	>85 (TOC)	[10]	0.63	60 (TOC)	[15]
				0.58	53 (TOC)	[16]
				0.24	57 (TOC)	[17]
				0.24	90 (Color)	[2]
6-7	0.77	>90 (UV254)	[7]	0.94	50 (TOC)	[11]
	0.60	85 (UV254)	[11]	2.05	57 (TOC)	[12]
	0.42	>90 (TOC)	[12]	2.25	41 (TOC)	[13]
	0.50	48 (TOC)	[13]	0.31	59 (TOC)	[18]
	0.24	>90 (Color)	[8]			
>7	0.77	80 (UV254)	[7]	2.20	40 (TOC)	[18]

alum was used, whereby stock solutions (1 g/l Al) were prepared weekly, and reagent-grade HCl and $NaHCO_3$ were used for pH-adjustment. Batch experiments were conducted in 1-l jars with 600 ml water samples. Alkalinity was increased to 5 mmol/l by adding $NaHCO_3$, then pH was lowered by adding HCl. The solution was then rapidly stirred with a magnetic stirrer, and alum was added into the vortex. After one minute the jar was set on a paddle mix unit for 30 minutes of slow mixing. The pH values reported were measured at the end of that period.

After one hour of settling, samples were withdrawn from the supernatant and analyzed for turbidity (HACH 2100 turbidimeter) and for UV-absorbance (254 nm) after acidification to pH = 3 (HITACHI 100-20 spectrophotometer). Supernatant samples were also filtered through 0.45μm membranes (Schleicher & Schüll BA 85) and analyzed for color, UV-absorbance and TOC (Bybron-Barnstead OC-analyzer). The temperature was always 22 ± 2°C.

Results and Discussion. Ultimate removal with respect to UV-absorbance is shown in Figure 1. The percentiles are based on some 80 data points. Only one major removal domain was found, with the optimum pH being 5.5 – 5.9. There, more than 80% removal could be achieved using a dosage of 0.37 mg Al/mg TOC. This value compares well with the results listed in Table 1. In order to relate UV-absorbance data to color and TOC removal, respectively, the following correlations may be used:

1) Color [mg/l Pt] = 1.25(UV254 [l/m]) - 5.0,   UV254 ≤ 30 l/m
2) TOC [mg/l]   = 0.16(UV254 [l/m]) + 1.6,   UV254 ≤ 30 l/m

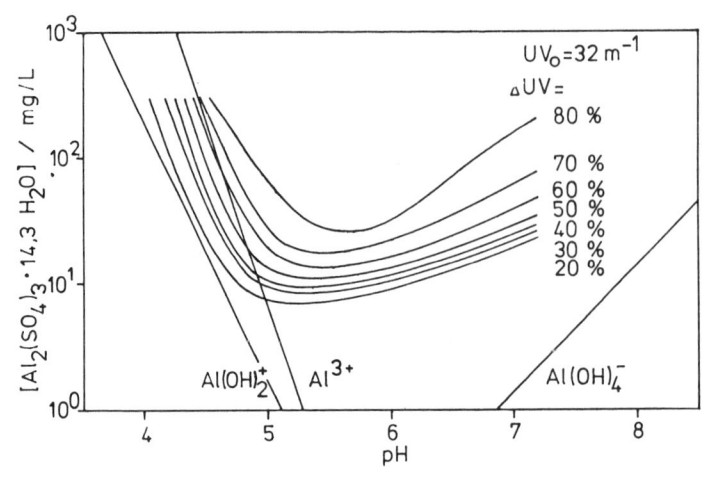

Figure 1. Removal of organics by coagulation, flocculation, sedimentation and 0.45 μm membrane filtration

Thus, 80% removal based on UV-absorbance corresponds to about 90% color removal and 60% TOC removal. Compared to the data of Edwards and Amirtharajah [8] who used a commercial humic acid, the pH range for 90% color removal is broader here while there is no particularly good removal for pH < 5, as can be seen in Figure 2. Additional experiments were conducted where the raw water was diluted with organic-free water prior to coagulation. At pH = 5.5 and pH = 5.0 the ratio between alum dosage and initial concentration of organics was stoichiometric for a given removal efficiency. At pH = 4.7 the residual Al concentration was too high to allow for this examination.

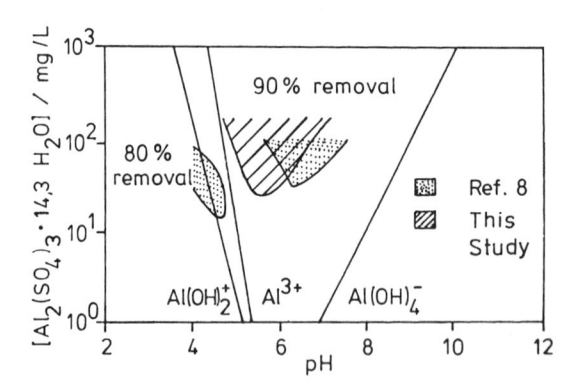

Figure 2. Comparison of color removal data to results given in (8) for initial humic acid concentrations from 2 to 20 mg/l as TOC

According to Dempsey et al. [7] the major reaction mechanisms for humic substances and alum are precipitation by monomers of Al for pH = 4.5 – 5, precipitation by polymers of Al for pH = 5 – 6 and adsorption onto precipitated $Al(OH)_3$ for pH ≥ 6.2 – 7. While all of these reactions are supposed to be quite fast, precipitation of $Al(OH)_3$ may take from hours to years, depending on pH. In order to check on the kinetics of floc formation, a series of tests was conducted as described by Snodgrass et al. [19], where the humics-containing solution was stirred rapidly, alum was added in the beginning and pH was controlled. At certain intervals, samples were drawn and analyzed with respect to turbidity and to number of particles > 1 μm (HIAC PC 320). Although this instrument is not well suited for floc size measurements [20, 21], it should allow for a qualitative evaluation of coagulation/flocculation under certain conditions.

Results obtained with a constant Al dosage at four different pH values are shown in Figures 3a and 3b. While turbidity and particle concentration reach a steady-state after 5 – 15 minutes at pH ≥ 5.5, it takes almost one hour at pH = 5.0 and more than two hours at pH = 4.7. Based on the conceptual model of Snodgrass et al. [19], this could be due either to the slow aggregation rate of small Al-humate nuclei (perikinetic flocculation) or to the slow formation of $Al(OH)_{3,S}$ submicron particles which then may serve as building material for larger flocs and as a sorbent for humic substances which are destabilized after complexation by Al monomers. The first hypothesis is supported by the fact that Al-humate particles carry an increasing

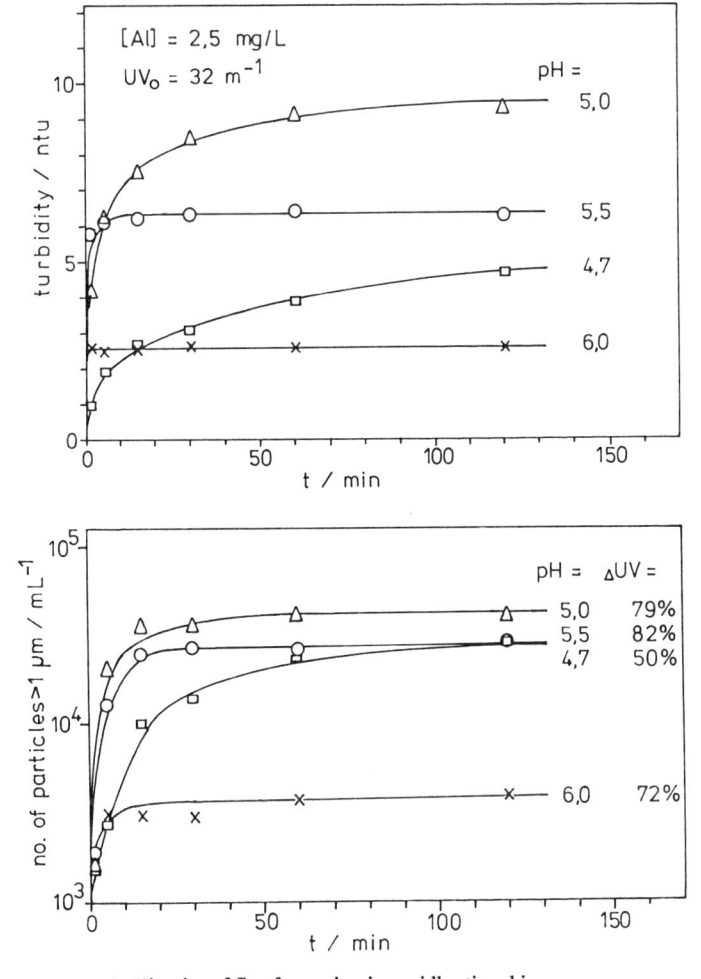

Figures 3a,b. Kinetics of floc formation in rapidly stirred jars

positive charge with decreasing pH for pH < 6, see, for example, [14], thus creating electrostatic repulsion. The second hypothesis is related to the fact that alum solutions age. For instance, Al solubility is about 5 mg/l at pH = 4.7 and 0.7 mg/l at pH = 5.0 in freshly prepared solutions, while it is much lower in old solutions. Furthermore the amount of Al monomers being complexed with humic substances for pH = 4.4 – 4.8 was found to be less than 0.04 mg Al/mg TOC, with 80 mg/l TOC and 5.4 mg/l Al [22]. Accordingly there may be a certain amount of Al in solution that is not taking part in immediate complexation reactions with humic substances but will precipitate more slowly as $Al(OH)_{3,S}$.

The effect of different slow mixing conditions on ultimate removal at pH = 4.8 is shown in Figure 4. Mixing intensity as well as flocculation time were varied. Mixing intensity seems to have no systematic impact, and differences among the results for 30 minutes of flocculation could be due to temperature effects. Meanwhile there is an increasing removal efficiency with increasing flocculation time. This points in the same direction as the results shown in Figure 3, indicating that there is an additional potential for organics removal for pH < 5 when flocculation time increases. With respect to direct filtration, however, flocculation time should be kept short. Therefore, it was reduced to 15 minutes, except for pH < 5 where it was 30 minutes when direct filtration experiments were conducted on a lab scale.

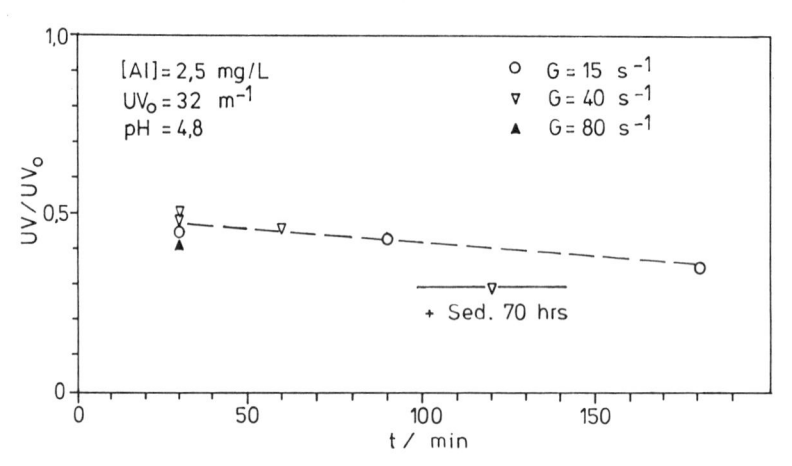

Figure 4. Effect of flocculation on ultimate removal at low pH

Floc Separation Evaluation

Lab Test Procedures. The sedimentation behaviour of flocs can be studied by measuring turbidity and other parameters in the supernatant of jars after settling. Typical results obtained in this study are shown in Figure 5 with respect to residual turbidity. While the settleability of flocs is poor for pH < 5.8 and pH = 7, it is quite good at pH values close to 6. According to results obtained by Vik et al. [14], this region corresponds to conditions where the net charge of the flocs is very small. Thus, separation of flocs formed by alum coagulation of humic substances by means of sedimentation seems to be strongly dependent on charge effects.

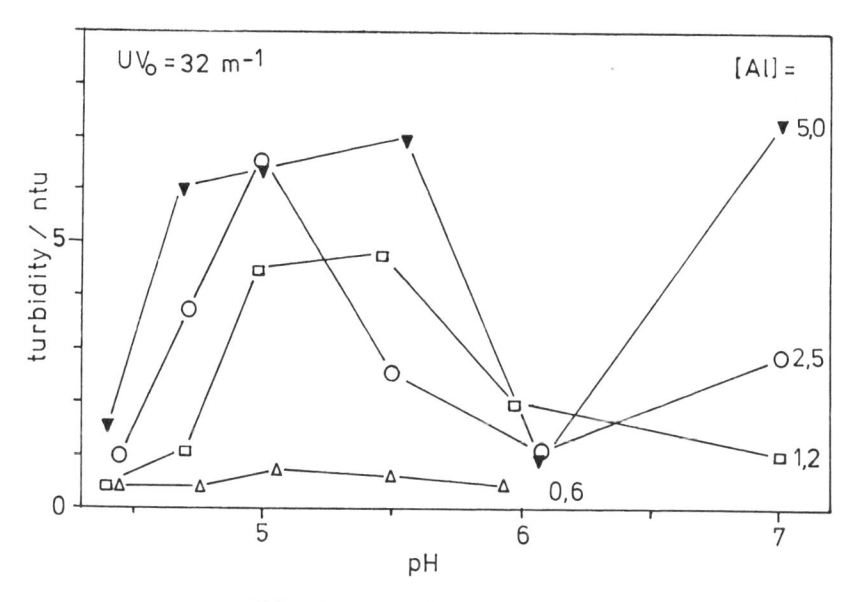

Figure 5. Supernatant turbidity after coagulation, flocculation and sedimentation

In order to estimate the filterability of a suspension, Wagner and Hudson [23] determined the quality of samples that were first coagulated and flocculated in jars and then without sedimentation filtered through Whatman 40 filter paper. This method was applied here, but the results did not show whether there was a pH range of particularly good filterability. The filtrate quality was always poorest at pH = 5.5.

Another test method described by TeKippe and Ham [24] is the refiltration technique where 100 ml of flocculated sample are filtered through a 0.45 μm membrane. Then the filtrate is filtered through the same membrane again, and the time required for filtering 80 ml is recorded. The refiltration time is dependent on the cake formed during the first filtration step. A small value is supposed to correspond to good filterability. According to results obtained by this method, filtration behaviour should be good at low pH values while for pH > 6 filter clogging occured, thus indicating poor filterability. However, both methods are considered to be incompatible with deep bed filtration conditions, and are thus not helpful in optimizing direct filtration.

A test method which closely simulates direct filtration is the operation of a lab scale filter. Experimental set-ups of different studies are summarized in Table 2. Most often a one-medium sand filter was used with an average bed length of 15 cm, the average filter velocity being 12 m/h, ripening period 20 minutes and operation period 95 minutes. The experimental design of this study is also given in Table 2. Alum was added to rapidly stirred samples in 1-l jars which then were poured into a 6-l flocculation tank connected to a small filter. Thus coagulation was conducted batchwise, while flocculation was done quasi- continuously, the detention time being 15 or 30 minutes, respectively. Samples taken from filter influent and effluent were analyzed for turbidity, color, UV-absorbance, TOC and number of particles > 1 μm.

Table 2. Experimental conditions for direct filtration testing

Filter-M Ø in mm	Removal	Coag.	Flocc. min	v_F m/h	l cm	Ripening min	Test period min	Ref.
Sand 0.6	Particles	Alum	20	51.0	8.2		0-1	[24]
Sand 0.8	Particles	$FeCl_3$	9-27	12.0	11.0	8	8-16	[25]
Sand 0.5	Particles	Alum + Polym.	0-45	11.5	30.0	≤20	20-64	[26]
Sand 0.4 + Polym.	Particles	Polym.	?	4.2	2.0	30	30-150	[27]
	Particles	Polym.		4.2	3.5	60	60-150	
	Particles	Polym.		4.2	14.0	50	50-150	
Sand 1.0	Particles	Polym.	0	5-8.5	15.0	5	5-80	[28]
Sand 0.6	Humics	Polym.	13	4.7	14.0	20	20-240	[29]
3M-Sand	Humics	Alum	15	12.0	37.6	≈5	5-6.5	[30]
Sand 0.9	Humics	Alum	15	10.0	10.0	20	20-50	This study
	Humics	Alum	30	10.0	30.0	20	20-50	

Results and Discussion. Typical breakthrough curves with respect to turbidity for 30 cm of bed length are shown in Figure 6. Effluent quality is almost constant with time at pH = 4.75 and pH = 6.2 while it deteriorates rapidly at pH = 5.25 and is very poor at pH = 5.7 from the beginning. In the last two experiments there is no

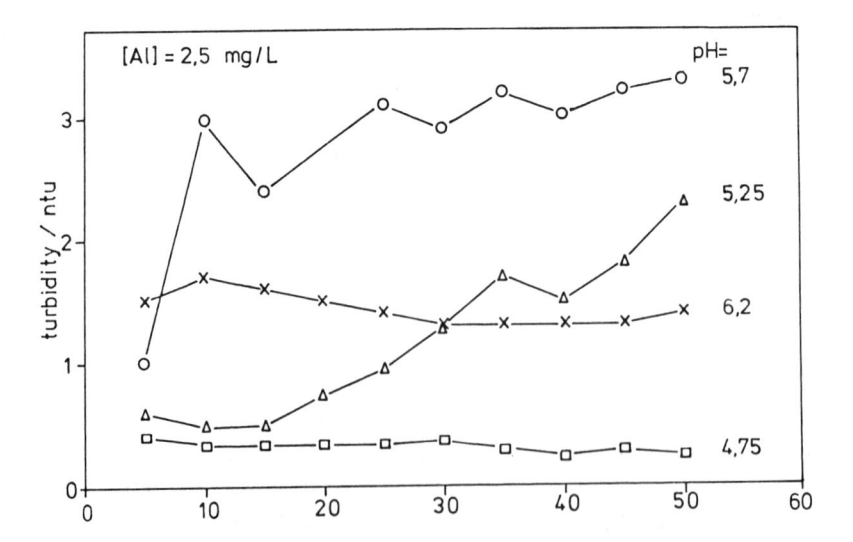

Figure 6. Breakthrough of turbidity in 30 cm filters at different pH values

evidence for filter ripening as can be seen at pH = 4.75 and pH = 6.2. These results were confirmed by particle size analysis data: at pH = 4.75 the number of particles > 1 µm was less than 100/ml during the entire filter run and at pH = 6.2 it was about 2000/ml, while at pH = 5.25 it increased from 100/ml after 10 minutes of operation to 7000/ml after 45 minutes. The average value at pH = 5.7 was 6000/ml. Based on the results presented in Figure 3, in-bed flocculation is not assumed to play an important role at pH values less than 5.5 because the packed-bed contact time was only 40 seconds with the 30 cm bed. Therefore the dramatic change in floc filterability from pH = 4.75 to pH = 5.25 is attributed to the flocs' chemical properties, e.g. flocs formed from humics and Al monomers are more effectively removed by sand filtration than flocs formed from humic and Al polymers.

A comparison of all of the experimental data is presented in Figures 7 and 8. Accordingly, filter efficiency is best for pH < 5 even with the 10 cm bed, it is poor at pH = 5.7 and it improves for pH > 6. Large differences between the results obtained with 10 cm and 30 cm beds indicate that deep bed filtration effects become more important, but floc storage capacity may then also be lower. An investigation of filter capacity, however, requires experiments with technical-scale filters. Filtrate quality with respect to humic substances as shown in Figure 8 is limited by the

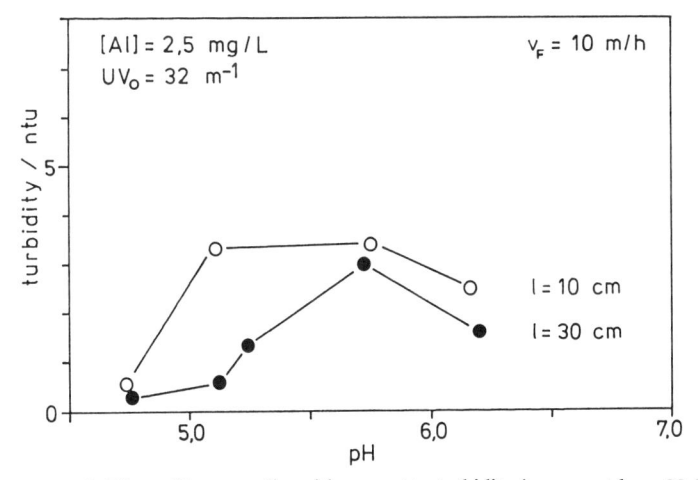

Figure 7. Filter effluent quality with respect to turbidity (average values, 20-50 minutes of operation)

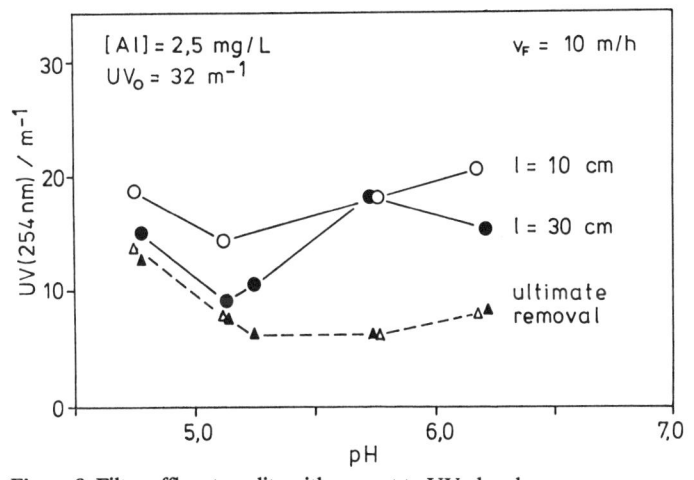

Figure 8. Filter effluent quality with respect to UV-absorbance

ultimate removal efficiency. A quality standard of UV- absorbance = 6 – 8 l/m can only be met for pH = 5.1 – 6.2. For pH < 5.1 a larger Al dosage would be required but even then the high residual Al concentration will create afterfloc formation. Technical application of low-pH direct filtration, therefore, must also involve stabilization of the water followed by a second filtration process. Thus the benefits of excellent floc filterability and probably high filter capacity for pH < 5 could be used for the removal of most of the organic material, and only a small amount of organics had to be removed at pH ≥ 6.

Technical Scale Data. The concept described above is currently being realized and investigated at Lake Hyllvannet. The set-up consists of an alum feed unit as described by Vråle [31], a two- media filter (70 cm anthracite, 30 cm sand), a feed unit for $CaCO_3$ slurry and another two-media filter. pH of the raw water is adjusted with HCl prior to alum addition; no separate flocculation tank is used. Figure 9 shows that the effluent turbidity of filter 1, although lower (≤ 0.1 ntu for pH = 4.7 – 4.9), follows the same pH-dependence as found with the 30 cm lab scale filter. Since the time prior to filtration is too short for complete floc growth to supramicron size (Fig. 3), it is concluded that filtration of submicron particles also works very efficiently for pH < 5. Further experiments are planned to show how the operation of both filters can be optimized.

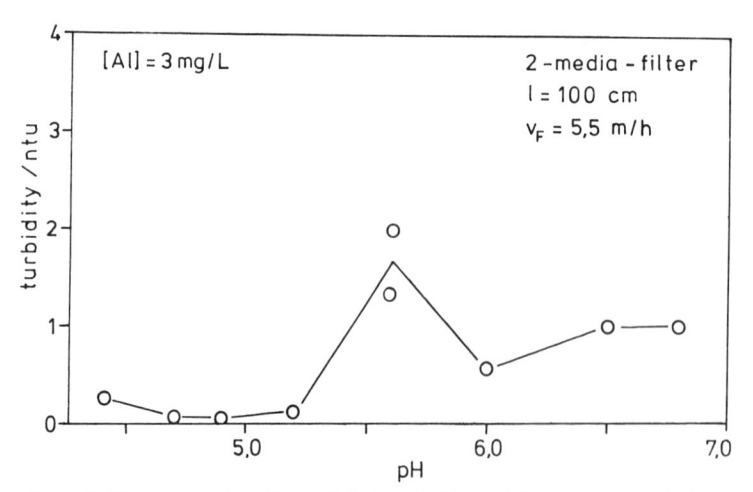

Figure 9. Effluent quality of a technical-scale filter with respect to turbidity, measured on-line after filter ripening

Summary

1) Alum coagulation is an efficient method for removing organic substances from natural lake water. At the optimum pH of 5.5, removal of 60% with respect to TOC, 80% with respect to UV- absorbance and more than 90% with respect to color can be achieved using a specific Al dosage of 0.37 mg Al/mg TOC.

2) Floc formation kinetics depend strongly on pH. For pH < 5 floc build-up takes on the order of hours.
3) Floc separation by sedimentation works best at pH = 6, while at lower and higher pH values floc settleability is poor.
4) Paper filtration as well as the refiltration method do not provide information about floc filterability under the conditions examined here. On the other hand, operation of a lab scale sand filter gives insight into direct filtration behaviour of flocs. Floc filterability deteriorates dramatically from pH < 5 to pH = 5.5 – 5.8 and improves for pH > 6. This is assumed to be related to the chemical structure of the flocs.
5) Results obtained with a 30 cm lab scale filter show good agreement with technical-scale data. A 10 cm bed seems to be too short to allow for scale-up conclusions, but the comparison of results obtained with 10 cm and 30 cm beds indicates the significance of deep-bed filtration effects for floc separation.

References

[1] Ødegaard, H., Brattebo, H., Eikebrokk, B., Thorsen, T. (1986) Wat Supp *4*, 129
[2] Eikebrokk, B. (1987) Report 55/86, NTNF program VAR-teknikk, Trondheim
[3] Eikebrokk, B. (1982) Ph.D. dissert., Norwegian Inst. Technol., Trondheim
[4] Eikebrokk, B. (1984) Report 6/84, NTNF program VAR-teknikk, Trondheim
[5] Committee report (1979) J. AWWA *71*, 588
[6] Committee report (1980) J. AWWA *72*, 405
[7] Dempsey B.A., Ganho, R.M., O'Melia, C.R. (1984) J. AWWA *76*, 141
[8] Edwards, G.A., Amirtharajah, A. (1985), J. AWWA *77*, 50
[9] Hall, E.S., Packham, R.F. (1965) J. AWWA *57*, 1149
[10] Babcock, D.B., Singer, P.C. (1979) J. AWWA *71*, 149
[11] Albert, G. (1975) in: Publ. Water Chemistry Div., Vol. 9. Engler-Bunte-Inst. Karlsruhe University
[12] Randtke, S.J., Jepsen, C.P. (1981) J. AWWA *73*, 411
[13] Weber, W.J., Jodellah, A.M. (1985) J. AWWA *77*, 132
[14] Vik, E.A., Carlson, D.A., Eikum, A.S., Gjessing, E.T. (1985) J. AWWA *77*, 58
[15] Knocke, W.R., West, S., Hoehn, R.C. (1986) J. AWWA *78*, 189
[16] Semmens, M.J., Field, T.K. (1980) J. AWWA *72*, 476
[17] Edzwald, J.K., Becker, W.C., Tambini, S.J. (1987) J. Environm. Engrg. 113, 167
[18] Chadik, P.A., Amy, G.L. (1983) J. AWWA *75*, 532
[19] Snodgrass, W.J., Clark, M.M., O'Melia, C.R. (1984) Water Res. *18*, 479
[20] Gibbs, R.J. (1982) Environm. Sci. Technol. *16*, 298
[21] Reed, G.D., Mery, P.C. (1986) J. AWWA *78*, 75
[22] Bakkes, C.A., Tipping, E. (1987) Water Res. *21*, 211
[23] Wagner, E.G., Hudson, H.E. (1982) J. AWWA *74*, 256
[24] TeKippe, R.J., Ham, R.K. (1970) J. AWWA *62*, 594
[25] Klute, R., Bernhardt, H., Hahn, H.H., Schell, H. (1979): Zeitschr. Wasser Abwasser Forsch. *11*, 193
[26] Treweek, G.P. (1979) J. AWWA *71*, 96
[27] Habibian, M.T., O'Melia, C.R. (1975) J. Environm. Engrg. *101*, 567
[28] Yeh, H.H., Gosh, M.M. (1981) J. AWWA *73*, 211
[29] Scheuch, L.E., Edzwald, J.K. (1981) J. AWWA *73*, 497
[30] Collins, M.R., Amy, G.L., Bryant, C.W. (1987) J. Environm. Engrg. *113*, 330

[31] Vråle, L. (1985) In: Proc. 1st Gothenburg Symp., Grohmann, A., Hahn, H.H., Klute, R. (eds.), Fischer, Stuttgart, New York

J. Fettig, H. Ødegaard
Div. of Hydraulic and Sanit. Engineering

B. Eikebrokk
NHL-SINTEF
Norwegian Institute of Technology
7034 Trondheim
Norway

Modeling the Effects of Adsorbed Hydrolyzed Al(III)-Ions on Deep Bed Filtration

Z. Wang

Abstract

A model for determining the effects of hydrolyzed aluminium ions on filtration of dilute suspensions by deep bed filters has been developed and experimentally tested in laboratory scale. The model relates solution conditions, especially pH and concentrations of Al(III), to the performance of clean filter beds. A surface precipitation model is used to describe surface properties of the suspended particles in terms of solution chemistry. This chemical model is combined with flow models and particle capture formulations within the filter bed to predict the filter performance. The complete model has been tested with experiments using suspensions of polystyrene latex particles with a diameter of 0.2 μm applied to filter media comprised of glass beads with a diameter of 0.4 mm. In the presence of Al(III), a favourable filtration region appeared in the middle of the pH range due to charge reversal of the particles. The width of the favourable pH region depended upon the concentrations of Al(III) and particle surface area. The present model can qualitatively explain the above observation.

Introduction

Deep bed filtration is a basic unit operation in potable water treatment. The initial removal und deposition of suspended particles in a filter bed involve physical processes such as flow through porous media of the filter bed and mass transport of suspended particles towards each single grain of media. These processes are dependent on physical parameters, e.g., size of particles and grains, filtration rate, temperature, and porosity and depth of the filter bed. Transport models which stress these physical aspects of filtration have been developed, based on mass transfer of suspended particles in a creeping flow field around a single collector (grain) [1, 2]. These basic aspects have been experimentally validated in defined filtration systems [1, 2, 3]. The most important contribution of these physical models to design and operation of deep bed filtration is to identify the mechanisms of mass transport and to establish a quantitative relationship between the efficiency of collection of suspended particles by a single collector (i.e., the single collector efficiency) and the dimensionless particle size, i.e., the ratio of the diameter of a suspended particle to that of the stationary collector.

In recent years, effects of colloidal interactions between suspended particles and collectors on particle deposition in filter beds have received more attention and experimental studies have revealed a sensitivity of particle deposition to these colloidal chemical factors such as zeta potentials of particles and collectors [4, 5] and solution ionic strength [6]. For example, when zeta potentials of particles and collectors have the same signs, a dramatic decrease in the deposition rate has been observed [4]. For ionic strength lower than a critical value, large changes of measured particle removal occur for a small change in the ionic strength. High sensitivity to the colloidal factors is a well known phenomenon in the stability of lyophobic colloids and can be explain by the DLVO-theory of Derjagium and Landau [7] and Verwey and Overbeek [8]. A surface force boundary layer model has been presented to predict the effects of colloidal chemical factors on the deposition of submicron particles [9, 10]. In this model, particle deposition is considered to occur in two sequential steps, i.e., particle transport to the collector by convective diffusion, and subsequent attachment onto the collector, whereby transported particles overcome the colloidal energy barrier and make contact. Available experimental data show that the model predictions underestimate measured particle depositions in the presence of high energy barriers [6, 11].

The author experimentally studied effects of the solution chemical factors, such as solution pH, simple cations and anions (e.g., Ca^{2+}, PO_4^{3-}) specifically adsorbed on the particle surfaces, on deep bed filtration. A chemical model for the effects on particle deposition on a single collector was presented by combining surface and solution chemistry with the surface force boundary layer model and the particle transport models [12]. Both theoretical analysis and experiments observation have shown a significant influence of these chemical factors upon the performance of clean filter beds, particularly, the solution pH. The objectives of this research are (1) to include effects of adsorbed Al(III) species on suspended particles in the fundamental theories that describe the deposition of submicron particles in clean granular media filters, (2) to test the theories experimentally, and (3) to use these results in assessing requirements for effective filtration and the capabilities of the process. The desire to gain an improved understanding of the role of aluminium coagulant chemistry in deep bed filtration is in part motivated by a practical requirement of design and operation of contact and direct filtration of surface waters using inorganic coagulants, e.g., aluminium sulfate.

The scope of this work is limited to the initial deposition of submicron particles in clean filter beds. Filter repining and particle detachment that may lead to a deterioration in filtrate quality are not considered.

Theoretical Framework

In order to model the effects of solution chemistry on particle deposition, three types of theories are considered jointly, i.e., particle transport, surface interaction forces, and surface chemistry, including a surface precipitation model.

Particle Transport. The fundamental basis of deposition theory is an examination of the mechanisms governing particle transport in a well defined flow field around (or through) a specified collector such as a sphere (or a tube). Fluid flow, gravity, diffusion and the finite size of the suspended particle all affect its transport towards the collector surface. The filter bed is modelled as an assemblage of idealized single collectors on which particles may deposit. Some of the models for the filter bed that have been used in establishing fundamental filtration theory include an isolated sphere in Stokes flow [1], a constricted tube [13] with internal laminar flow, and several porous medium models, e.g., Happel's sphere-in-cell model [14, 15] (see Figure 1).

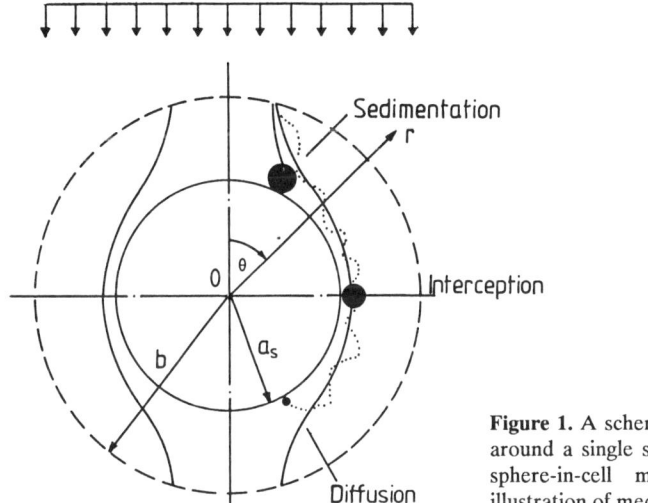

Figure 1. A schematic diagram of the flow field around a single spherical collector in Happels's sphere-in-cell model, and the schematical illustration of mechanisms of particle transport

Surface Interaction Forces. As particles and collector surfaces come close together, the colloid chemical interaction force or potential energy that have been widely studied and quantified can be considered. These include the double layer electrostatic force and the London-van der Waals or dispersion force which is usually attractive in aqueous systems and is a result of interactions between electronic dipoles of the molecules comprising the surface and solution. The magnitude of this force depends on material properties of the particles, filter media and intervening solution which are represented by the Hamaker's constant [16].

The double layer force between a particle and the collector is due to overlapping of their diffuse double layers. Its sign and magnitude are determined by signs and magnitudes of the diffuse layer potentials, i.e., the potential at the beginning of the diffuse layer of ions adjacent to the charged surface.

The total potential energy of the colloid chemical interaction between the particle and collector can be calculated by

$$V_T = V_D + V_L \tag{1}$$

V_D and V_L are the dimensionsless potential of the double layer and of the London-van der Waal attraction [16], respectively.

The colloidal potential energy ($V_T(H)$) is a function of the distance of separation (H). When the zeta potentials of the particle and collector have the same sign, an energy barrier or peak (V_m) appears in the colloid energy profile, i.e., $V_m = \max(V_T)$. The constant charge approach (no relaxation of the double layers) yields the largest energy barrier while the constant potential assumption (fully relaxed double layers) yields much lower values [18]. Intermediate to these approaches is regulated surface interaction [19] wherein surface chemical equilibria are maintained during the interaction. Another approach is to consider the dynamic aspects of surface interaction [20].

When the effect of hydrodynamic retardation is included in the deposition model, an attractive London-van der Waals force must be included so that deposition may occur [21].

Surface Chemistry. The nature of the interfacial region between a solid and the bulk solution, and thus the double layer, is determined by interactions between dissolved species and the solid surface. Theories of surface chemistry can be used to model predict measureable properties such as surface charge, zeta potential, and metal or ligand adsorption [22, 23]. For these particles with aqueous ionizable surfaces such as metal oxides (e.g., heamatite) and polystyrene latex particles and glass beads (i.e., SiO_2), their surface charges, and thus their zeta potentials, are produced by ionization of the surface function groups (e.g., hydroxyl and carboxyl groups), a site-binding reaction between the surface groups and an indifferent electrolyte, e.g., $NaNO_3$, and surface complex formation between the surface groups and specifically adsorbed ions, e.g., Ca^{2+} and PO_4^{3-}. Zeta potentials of these solids are, therefore, closely related to the solution pH. Figures 2 and 3 give measured zeta potentials of the polystyrene latex and haematite particles in 10^{-3} M $NaNO_3$, respectively [12].

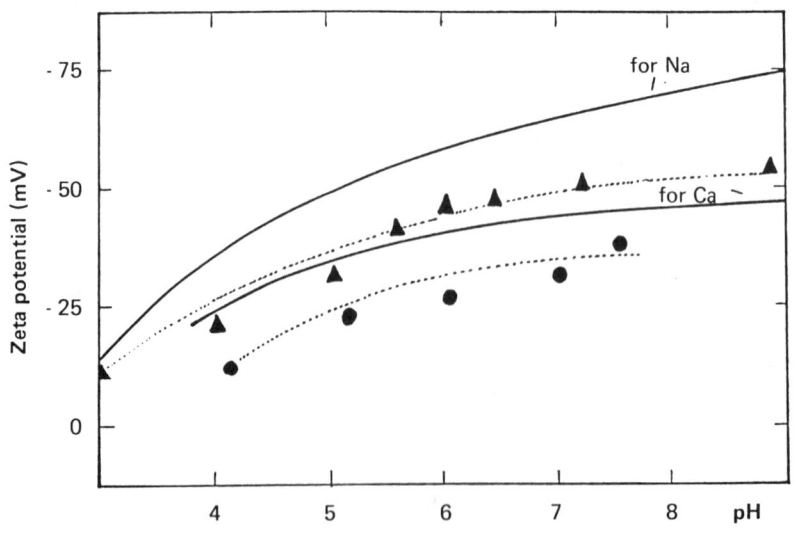

Figure 2. Measurement and prediction of zeta potentials of the latex particles ($a_p = 0.108 \mu m$) in the presence of 10^{-3} M $NaNO_3$ (▲) and $0.33 * 10^{-3}$ M $Ca(NO_3)_2$ (●) [12]. Values of parameters for zeta potential prediction are given in Table 2

Table 1. Parameter values used in the solution chemistry part of the MINEQL program

Reaction	Log (equilibrium constant)
$Al^{3+} + H_2O = Al(OH)^{2+} + H^+$	-4.99
$Al^{3+} + 2H_2O = Al(OH)_2^+ + 2H^+$	-10.13
$Al^{3+} + 4H_2O = Al(OH)_4^- + 4H^+$	-21.57
Aluminium hydroxide solubity at surface:	
$(Al^{3+})(OH^-) = K'_{SO}$	-33.10

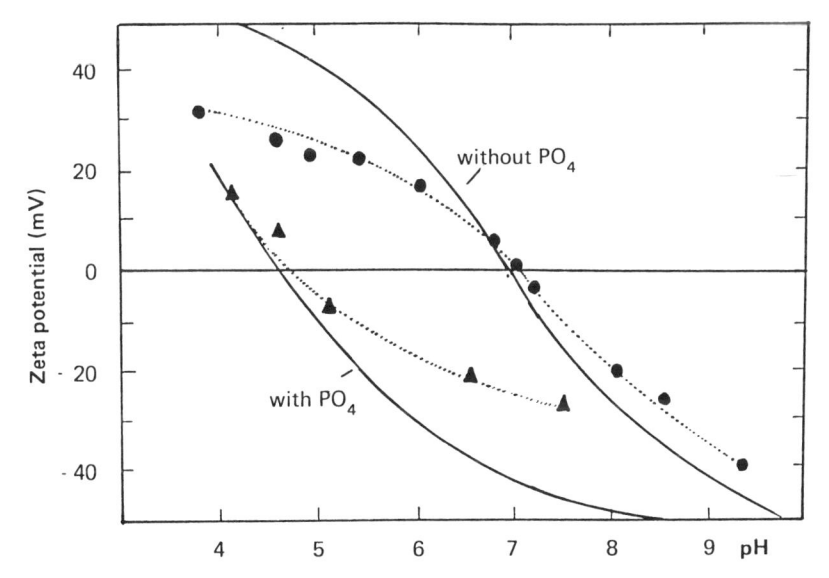

Figure 3. Measurement and prediction of zeta potentials of the haematite particles (a_p = 0.15 μm) in the presence of 10^{-3} M $NaNO_3$ (▲) and 10^{-3} M $NaNO_3$ + 10^{-4} NaH_2PO_4 (●) [12]. Values of the parameters for the zeta potential prediction are:

pK_{a1}^{int} = 4.6, pK_{a2}^{int} = 9.2, pK_{Na}^{int} = 8.4, $pK_{NO_3}^{int}$ = 5.4, C1 = 0.9 F/m², C2 = 0.2 F/m², N_S = 3.5 OH/nm².
Constant capacitance model for system with PO_4 (parameter values for const. cap. model from Sigg, 1979 [64])

The computer program MINEQL, written by Westall et al. [24] was designed to solve the composition of complex chemical systems which may involve solution species as well as precipitating and dissolving solids. Davis et al. [25] expanded the MINEQL program to include the effects of the electric double layer on ionization, electrolyte binding and complexation reactions for surface function groups. Prediction of zeta potentials of a solid with an ionizable surface can be carried out by solving a set of equations numerically with the MINEQL program. These equations include (1) equilibrium equations for the surface reactions in which the apparent equilibrium constant for a surface reaction (e.g., ionizable reaction) includes the intrinsic equilibrium constant (K^{int}) and an electrostatic term

$(\exp(-zF\psi_i/RT))$ where z is the difference in charge between the complexed and uncomplexed surface sites, and ψ_i is the potential at the locus of the surface complex; and (2) electrostatic equations for the Stern compact layer and the diffuse double layer. Because of the difference in the geometric description of the surface reaction process, particularly its effect on the expression for coulombic interactions, there are several models for calculating the surface chemical equilibrium, e.g., the constant capacity model and triple layer model [25, 26]. The triple layer model is more suitable for the prediction of zeta potentials of oxides because it yields results that are consistent with observed electrostatic characteristics of oxide surfaces, viz., the high surface charge and intermediate zeta potentials usually measured by potentiometric titration and electrophoresis [22]. In this model, the Gouy-Chapman-Stern-Grahame (GCSG) model is applied to describe the structure of the double layer (see Figure 4). In this scheme, H^+ and OH^- are considered to react in an inner compact layer while specifically adsorbed ions, including the indifferent electrolyte, are in a layer adjacent to the surface. Two different capacitances are assumed for the inner and outer regions of the Stern layer.

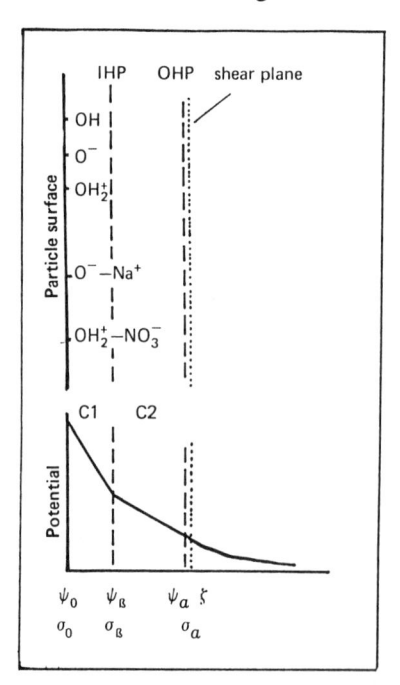

Figure 4. A schematic diagram of the electric double layer structure of particles with aqueous ionizable surfaces in the triple layer model

A Surface Precipitation Model. As an effective coagulant, the Al(III) solution chemistry including hydrolysis, precipitation and coordination has been carefully studied both in thermodynamics and kinetics [33, 34]. It has also been well documented that hydrolyzed aluminium adsorbs on various hydrophobic colloidal surfaces [35, 36, 37]. A comparison of stability domain diagrams obtained using mineral particles with those obtained using natural humic substances [38, 39] suggests that adsorption of hydrolyzed aluminium is also an important mechanism in these systems.

Table 2. Parameter values used in the surface reaction part of the MINEQL Program for aluminium hydroxide-latex composite particles

Parameter	Symbol	Values	Reference
Intrinsic constants			
Base particle sites	pK_{a2}^{int}	5.1	Wang [12]
	pK_{Na}^{int}	4.3	Wang [12]
Al sites	pK_{b1}^{int}	5.7	James and Parks [22]
	pK_{b2}^{int}	12.3	James and Parks [22]
	pK_{Na}^{int}	9.0	James and Parks [22]
	$pK_{NO_3}^{int}$	6.9	James and Parks [??]
Surface site densities			
Base particle	N_S^S	1.38 sites/nm^2	Wang [12]
	N_S^a	8.00 sites/nm^2	James and Parks [22]
Complete coverage	r	10^{-5} mole Al/cm^2	estimated
Base particle specific surface area			
	S_S	26 m^2/g	Wang [12]
Integral capacitances			
	C_1	1.4 F/m^2	Davis et al. [25]
	C_2	0.2 F/m^2	Davis et al. [25]

Based on the concept of the nucleation and precipitation of hydrolyzed metal ions adsorbed onto the base particle surface [40 – 49]. Letterman and Iyer [50] developed a patch site distribution model in which a partial coating of the particle by adsorbed hydrolyzed aluminium leads to the formation of a composite surface. They then presented a modified MINEQL program in which the composite surface is modelled by using two double layers, one for the areas of uncoated primary particles and one for the areas covered by the aluminium hydroxide. The double layers have a common dependence on the characteristics of the bulk solution and are used to calculate the net electric double layer properties of the composite particle, e.g., zeta potentials. The basic points of the zeta potential calculation using the surface precipitation model are: (1) The concentration of adsorbed aluminium hydroxide (M_p^{Al}), i.e., the adsorption of hydrolyzed Al(III), is assumed to be equal to the concentration of Al(OH)$_3$ precipitate predicted by the MINEQL model in which the adsorbed Al(OH)$_3$ solubility product (K_{so}') is estimated by considering the effect of the electric double layer. Then the surface coverage (β) by the adsorbed Al(OH)$_3$ can be calculated by choosing a suitable value for the complete coverage (r). (2) The composite surface consists of a base particle site (e.g., SOH, SO$^-$, SO$^-$Na$^+$ etc.) and aluminium sites (e.g., AlOH, AlO$^-$, AlOH$_2^+$, AlO$^-$Na$^+$ etc.) and the surface charges and potentials of the base particles and adsorbed aluminium hydroxide are calculated separately by using the MINEQL model. Finally, (3) the effective diffuse layer charge of the composite particle (δ_d) is calculated using an area-weighted average of the diffuse layer charge densities for the two parts of the surface, i.e., $\delta_d = \delta_d^s(1-\beta) + \delta_d^{Al}\beta$.

Attachment Efficiency Factor. The attachment efficiency factor (α) ist defined by the ratio of the single collector efficiency (η) to the transport efficiency (η_0):

$$\alpha = \eta/\eta_0 \tag{2}$$

For favourable filtrations, the particle deposition on a single collector is only determined by the physical transport of suspended particles, so $\eta = \eta_0$. Thus, theoretical predictions of α values for the favourable filtration should be unity, i.e.,

$$\alpha = 1 \tag{3}$$

When the effect of surface and solution chemistry is unfavourable for particle deposition (i.e., for unfavourable filtration), the particle deposition is controlled by the particle attachment step and the effect of the colloid chemical interaction between particle and collector becomes important. Because the grain size is usually much larger than the particle size in deep bed filtration, the colloidal interaction can be considered as a sphere-plate interaction and Equations 3–7 are used to calculated potential energies of the total colloid interaction, the double layer interaction and the London attraction.

Theoretically, one could extend the general transport equation describing mass transfer within the porous media to include the term for the potential energy of colloidal interaction and to predict the effect of colloidal interaction by examining numerical solutions of the equation. However, there are so many parameters involved that no general solution can be obtained for a spherical collector [30]. For the submicron particle deposition and when the high energy barrier is sufficiently high in the colloidal energy profile, an approximate analysis can be made for the α prediction. In this case, the colloidal interaction occurs over a very thin layer near the collector surface and the colloidal interaction region is much thinner than the outer region, i.e., the laminar diffusion boundary layer. Thus, the effect of colloidal interaction can be considered as a boundary condition in solving the convective diffusion equation governing particle transport. In this socalled surface force boundary layer model [9, 10], a fictitious first order surface reaction governed by the energy barrier is assumed to occur at the collector surface to control attachment of the transported particles onto the surface.

For unfavourable deposition of submicron particles, an approximate expression of the single collector efficiency (η), based on the work of Spielman and Friedlander [9], can be presented in the following form:

$$\eta = \eta_D[q/(1+q)]\,S(q) \tag{4}$$

with

$$q^{-1} = 0.176\,\eta_D N_{Pe} N_R K \tag{5}$$

Here $S(q)$ is a modification term for a spherical collector and sample values for this parameter are given in Spielman and Friedlander [9]. The dimensionless parameter, K, is defined by

$$K = D/a_p k_r \tag{6a}$$

here the parameter, k_r, represents the rate constant of the fictitious surface reaction. These authors obtained an analytical expression for k_r by matching the particle flux through the colloidal interaction inner region with that of the outer region of the laminar diffusion boundary layer in solving the convective diffusion equation. Dahneke [31] generalized Spielman and Friedlander's transformation to take account of the effect of hydrodynamic retardation. Rajagopalan and Kim [32] gave a useful expression for calculating K:

$$K = B(N_L)\exp(0.96 V_m) \tag{6}$$

with

$$B(N_L) = N_L^{-0.22} + 0.05\, N_L^{2.7} \quad \text{at } N_L < 2 \tag{7}$$

In order to account for the effects of interception and sedimentation on particle transport, the analytical expression for calculating α can be obtained:

$$\alpha = [q/(1 + q)]\, S(q) \tag{8}$$

with

$$q^{-1} = 0.176\, \eta_0 N_{Pe} N_R K \tag{9}$$

V_m in Eq. 6 is the maximum dimensionless potential according to Eq. 1, to be calculated by the MINEQL program [25] or the modified MINEQL program [12, 50]. The term $S(q) = 1$ (Eq. 4) fits for plate collectors or spherical collectors with $a_s \gg a_p$ (see Figure 1). The transport efficiency (η_0) may be calculated by a formular given by Rajagopalan and Tein [29] formulated for non-Brownian particles using the modified Happel sphere-in-cell model [14]:

$$\eta_0 = 4 A_s^{1/3} N_{Pe}^{-2/3} + A_s N_{Lo}^{1/8} N_R^{15/8} + 0.0034\, A_s N_G^{1.2} N_R^{-0.4} \tag{10}$$

The meaning of the parameters is as follows:

A_s is a porosity parameter, depending on the porosity f of the filter, $A_s = 38$

N_{Pe} is the Peclet number, $N_{Pe} = d_s U/D$ (11)

N_{Lo} is a dimensionless London force, $N_{Lo} = A/(9\pi\mu a_p^2 U)$ (12)
with the Hamaker constant [16], $A = 2.92 \times 10^{-21}$ J.

N_G is the ratio of the Stokes settling velocity of the particles to the filtration velocity,

$$N_G = 2(\rho_p - \rho) g\, a_p^2 / (9\mu U) \tag{13}$$

The other parameters used in Equation 6 to 10 are listed in Table 3.

Table 3. Values of physical parameters used in calculation of the attachment efficiency factors

Parameter	Symbol	Value
Suspended particle		
Particle radius [μm]	a_P	0.108
Density [g/cm^3]	ρ_P	1.05
Diffusion coefficient [cm^2/s]	D	2.1×10^{-8}
Dimensionless radius	N_R	5.2×10^{-4}
Filter bed		
Bead diameter [mm]	d_S	0.387
Bead density [g/cm^3]	ρ_S	2.5
Filtration vilocity [cm/s]	U	0.1
Bed depth [cm]	L	15.0
Porosity	f	0.4
Temperature [°C]	T	25 ± 1
Hamaker constant of latex-water-glass		
	A	2.92×10^{-21} J
London group		
	N_L	0.12

Equations 8 and 9 show that for the unfavourable filtration of submicron particles, $0 < \alpha < 1$, which means that the single collector efficiency, and thus the filtration efficiency, decreases remarkably for unfavourable filtration, compared with favourable cases. They also indicate that the model-predicted a is very sensitive to the energy barrier (V_m) and then dramatically decreases and rapidly approaches zero when the energy barrier is increased. In the meantime, because the colloidal energy profile (V_T) for the particle-collector interaction is quantitatively related to the solution chemical factors such as pH and specifically adsorbed ions through the relationship between predicted zeta potentials of suspended particles and collectors and these chemical factors of the triple layer model for aqueous ionizable surfaces, the effects of chemical factors on the attachment efficiency factor can be predicted using Equations 6 – 13.

Experimental Materials and Methods

The experimental work was focussed on the measurement of the electrophoretic mobility of latex particles and the deposition of these particles in beds of glass beads. Identical chemical conditions were utilized for both types of experiments.

Materials. All inorganic chemicals were analytical reagent grade and used without further purification. Water was doubly distilled and deionized (i.e., DI water).

The spherical monodisperse latex particles used in this study were supplied by the DOW Chemical Company, U.S.A. They are polystyrene latexes having surfaces modified with carboxyl groups. Their average diameter is 0.216 μm and density is

1.05 g/cm². According to the manufactor's information, the surface of the latex particle is smooth and has carboxyl functional groups covalently bonded to the polystyrene. Measured values of the densities of carboxylic groups (N_S) using conductometric titration by the manufacturer are 1.13 to 7.20 COOH sites/nm² for carboxylate modified latex particles with diameters from 0.038 to 0.807 μm. The value of N_S for the 0.216 μm latexes used in this research is 1.38 COOH sites/nm² measured using potentiometric titration [12].

The surface carboxyl groups of the latex particles become protonated (or ionized), and thus are uncharged only at low pH (e.g., 2), so the particles are negatively charged in pure water at natural pH levels. The intrinsic chemical constant of ionization of the carboxyl groups can be calculated, based on the titration results of the suspension, by a procedure of double extrapolation [25]. The intrinsic ionization constant (K_a^{int}) was thus obtained to be $10^{-5.1}$, i.e., $pK_a^{int} = 5.1$ [12]. The intrinsic chemical constant for the surface binding reaction between the carboxyl groups and Na⁺ was found to be $10^{-4.3}$, i.e., $pK_{Na}^{int} = 4.3$ [12]. These results are comparable to the values for these constants obtained by other investigators [51, 52, 53].

The specific surface area of the latex particles was calculated to be 26 m²/g by geometrical consideration. Dilute suspensions of particles were prepared by pipetting microliter volumes of the concentrated (10% solids) stock into the desired volume of glass distilled, deionized water (DI water).

Spherical glass filter media with particles having a mean diameter of 0.387 mm were used for filtration experiments. They are supplied by Ferro Corporation, Cataphote Division, Jackson, Mississippi, and manufactured from crushed fragments of a soda- lime glass (sodium-calcium-magnesium silicate). The hydroxyl functional groups exist at the glass/water interfaces. These surface groups may be produced due to dissociation of the chemisorbed water molecules at the aqueous surface. Due to bonding with different constituent ions, the surface OH groups of different oxides illustrate quite different acid-base properties. FeOH is amphoteric whereas SiOH is monoacidic. At pHs greater than 3, the aqueous silica surface is usually negatively charged. Based on potentiometric titrations of silica sols, the intrinsic ionization constant of the silanol groups of various silicas (e.g., aersol silica [54], pyrogenic silica [55], Ludox silica [56], etc.)was found to be in the range of $10^{-6.5}$ to $10^{-7.2}$, i.e., $pK_{a2}^{int} = 6.5$ to 7.2.

The total ionizable site densities of various silicas have been measured using several methods (e.g., ignition at high temperature, acid base titration, infrared absorption etc.). For $\alpha-SiO_2$ [22], N_S = 4.2 to 5.9 OH/nm² and for aerosol silica [54], N_S = 5.5 to 5.8 OH/nm². A theoretical calculation [57], based on geometrical considerations leads to an estimate of 7.8 silicon atoms/nm² at or very near the surface of the heat stabilized amorphous silica. But since all the silicon atoms cannot be exactly at the boundary (some must be above and some below), only half of the silicon atoms would bear OH groups; therefore there could be only 3.9 OH/nm².

The beads were cleaned by soaking in an ultrasonic bath in distilled water, 1.0 M HNO_3 and distilled water successively. After a final extensive rinse with DI water, the beads were ready for the experiments.

The filter columns were manufactured from cast acrylic tubing with a nominal inside diameter of 23 mm. The columns have detachable, flanged, influent and effluent sections and a nylon wire mesh media support. The bed depth was 15 cm and the filtration velocity was 0.1 cm/s. Solutions were prepared by dispersing reagent grade chemicals in DI water, and filtered through a 0.2 micron millipore filter. Al(III) stock solutions ($Al(NO_3)_3$) were prepared in concentrations greater than 0.1 M to prevent hydrolysis and subsequent aging of the solution [33].

Experimental Methods. Particle electrophoresis* was utilized to measure the electrophoretic mobility (u_e) of the latex particles for various pHs and concentrations of Al(III). The solution ionic strength was kept constant, using $NaNO_3$. It was 10^{-3} M. The solution pH was conditioned by adding HNO_3 and NaOH (they were all 10^{-1} M), and measured using a pH meter (model PHS-3) with a glass electrode and a Ag/AgCl reference electrode. The final concentration of solids in these samples was 2 mg/l.

The zeta potentials (ζ_p) reported herein were determined using the measured electrophoretic mobility values and the following equation [58]:

$$u_e = (\varepsilon \zeta / 6\pi\mu) f(\tau) \tag{14}$$

Here the Henry's modification function ($f(\tau)$) was taken to be 1.252 for $\tau = 11.23$. Henry's equation (Eq. 14) only allows one to calculate the zeta potential approximately. The surface potentials of the latex particles were less than 100 mV. These effects may be important in zeta potential calculations only at the solution pHs far away from the pH_{iep} of the particles.

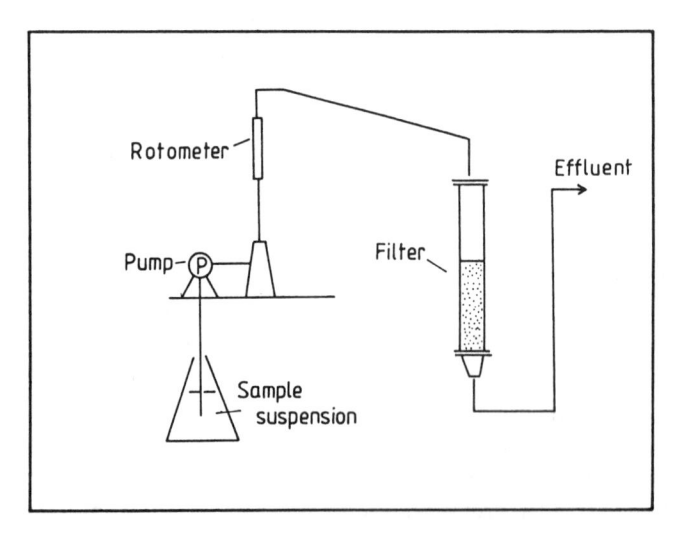

Figure 5. A schematic diagram of the apparatous used to perform the laboratory scale filtration experiments

---

* Zeta-Meter(III) with Pt electrodes and 1 cm × 1 cm cell, ZETA-METER Inc. U.S.A.

A schematic diagram of the apparatus used to perform the laboratory scale filtration experiments is shown in Figure 5. A dilute suspension of the latex particles (2 mg/l) of the desired chemical composition were prepared in a 2 liter pyrex bottle by adding stock solutions to DI water. This suspension was pumped at a rate of about 25 ml/min through a 250 ml vacuum flask (with head space) to reduce surging, a rotometer for monitoring flow rate.

To prevent aggregation of the suspended particles, the pH value of the suspension was kept away from the isoelectric points of the particles adsorbed by Al(III) ions. Before each filtration, the variation rate of the turbidity of the prepared suspension (dB/dt) was measured using a spectrophotometer*. If it does not change with time, i.e., $(dB/dt)_0 = 0$, the suspension is stable, and the filtration experiment can then be made.

The filter bed was prepared by adding the appropriate mass of dried glass beads to a clean filter column that was filled with DI water. The bed was backwashed with DI water and allowed to stand overnight. At the start of a filtration experiment, clear solution was pumped through the clean bed to equilibrate the media with the acid base condition and 10^{-3} M ionic strength to be used in the deposition phase of the experiment. The dispersion characteristics of the bed were studied using dye tests and the mean hydraulic residence time was estimated (it was about 8 min). Filtration lasted about 1 hr with effluent samples taken at 5 minute intervals. Particle concentrations in the influent and effluents were determined with turbidity measurements using the spectrophotometer. The pH of the prefilter suspension and the column effluents was measured using the PHS-3 acidity meter.

Results and Discussion

The line of reasoning of the model for the effects of adsorbed hydrolyzed aluminium on the latex filtration is as follows: firstly, the zeta potentials of the latex particles were predicted using the surface precipitation model for various pHs and concentrations of aluminium nitrate, and then, the colloidal energy of interactions between the particle and the glass bead was calculated. Secondly, the attachment efficiency factors under the same solution conditions were calculated. The experimental testing followed the same line. In all these cases, the initial particle concentration and the solution ionic strength were kept constant.

Zeta Potential Prediction. Figure 6 presents the measured zeta potentials of the latex particles. The measured values of CR2 of the particles were 4.3, 4.8, and 5.2, respectively, for the concentrations of aluminium nitrate of 10^{-4} M, 10^{-5} M, and 10^{-6} M and the corresponding measured CR3 values were 6.2, 7.2, and 8.1, respectively. The pH_{iep} of the aluminium hydroxide precipitate was measured to be 8.7.

The measured zeta potentials of the particles in the presence of aluminium nitrate are similar to those of the base particles at the pH values less than CR2. This

---

* UV-250 spectrophotometer with an integral sphere, Japan.

means that at pH < CR2, the Al(III) adsorption does not occur or it is so weak that its effects on the particle zeta potentials are also weak.

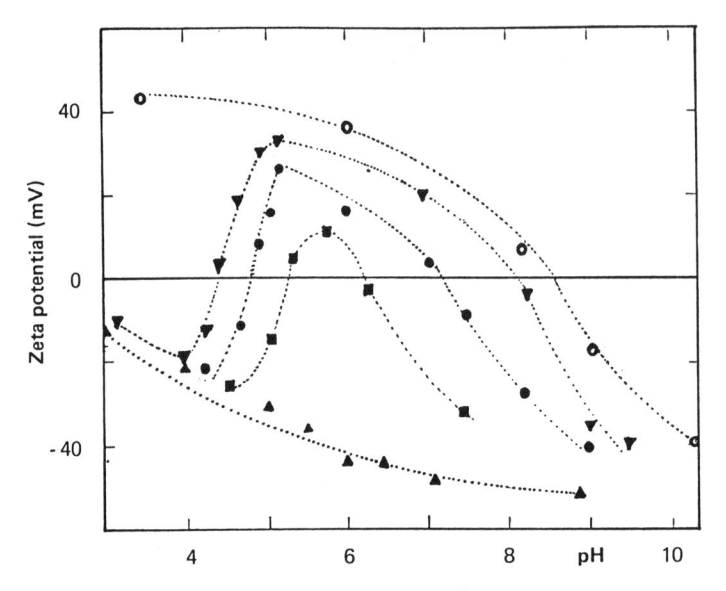

Figure 6. Measured zeta potentials of the latex particles when the concentrations of aluminium nitrate are 10^{-4} M (▼), 10^{-5} M (●), 10^{-6} M (■), and 0 M (▲), respectively. Datapoints (○) are for measured zeta potentials of aluminium hydroxide. The isoelectric points are named CR3, CR2 and pH_{iep}

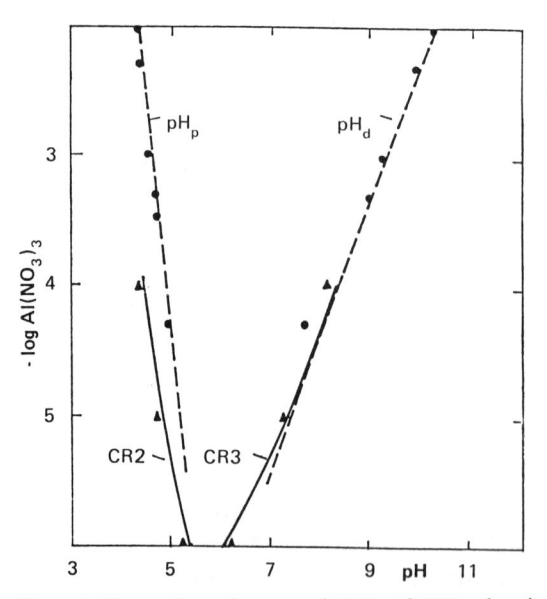

Figure 7. Comparison of measured CR2 and CR3 values isoelectric points for the latex particles in the presence of Al(III) with critical pH limits of precipitation of aluminium nitrate and of dissolution of aluminium hydroxide in aluminium nitrate solution (the critical pH limits were measured by Hayden and Rubin [33]). The solid lines are the prediction of these isoelectric points using the surface precipitation model with parameter values given in Table 2

It was considered that the critical pH range over which the abrupt adsorption of hydrolyzed aluminium species occurs is around the CR2. An alternative way of explaining the phenomenon is to regard adsorption as an interfacial precipitation of the aluminium hydroxide, occurring at the CR2. Comparison of the measured CR2 values and critical pH limits of precipitation of aluminium nitrate solutions, measured by Hayden and Rubin [33] (see Figure 7), shows that the formation of aluminium hydroxide precipitate on the latex surface occurs at a pH (i.e., CR2) below the bulk precipitation pH. James and Healy's explanation for this is that the solubility product of the interfacial precipitate (K'_{SO}) is less than that of the bulk precipitate due to the effect of an interfacial electrostatic field (i.e., the electric double layer). The value of K'_{SO} can be estimated, based on the CR2 for zero available surface or when the surface loading is infinitely large. In this research, an approximate value of K'_{SO} was obtained based on the measured value of CR2 in the presence of 10^{-4} M $Al(NO_3)_3$. For estimation of this value, it was initially assumed that interfacial precipitation occurs at pH = 4.3 and then the MINEQL computation for the hydrolysis- precipitation of hydrolyzable Al(III) was made. The hydrolyzed species and equilibrium equations involved in the calculation are given in Table 1. The estimated value of K'_{SO} is $10^{-33.1}$ while the K_{SO} for the bulk precipitation of amorphous aluminium hydroxide, measured by Hayden and Rubin [33], is $10^{-31.6}$.

The measured values of CR3 were found to occur at pH values below the pH_{iep} of aluminium hydroxide and to approach the pH_{iep} with increasing surface loading (see Figure 6). According to the "surface nucleation and precipitation" model, presented by James and Healy [42]. It may be said that the CR3 moves up as the thickness of the precipitated film increases and coincides with the behaviour of the bulk precipitate after about five layers have been established. However, it is not obvious why five layers of deposit should be required before the composite surface becomes electrokinetically indistinguishable from the pure hydroxide. Furthermore, James and Healy's model is only a schematic presentation and cannot predict these isoelectric points, especially the CR3, quantitatively. If the adsorption of hydrolizable metal ions can be viewed with the formation of the interfacial precipitate, it seems only natural that the surface coating pH is large enough, just as is the bulk precipitate. Figure 7 shows that the measured CR3 values are comparable to the critical pH limits of dissolution of the bulk precipitate of aluminium nitrate solutions. It can the be considered that the charge reversal at CR3 occurs due to a reduction in the surface coverage of aluminium hydroxide by dissolution of the interfacial precipitate layer. This provides a possibility for predicting the CR3 by calculating surface hydrolysis-precipitation equilibria (i.e., the modified MINEQL program [12, 50].

The solid line in Figure 7 predicts the CR2 and CR3 using the surface precipitation model. The values of parameters involved in the modified MINEQL calculation are given in Table 2. In this table, the complete coverage in $mole/m^2$ is an approximation. It can be obtained from measurements of adsorption isotherms. For adsorption of Al(III) on SiO_2, the complete coverage was measured as 9×10^{-6} $mole/m^2$ by Letterman and Iyer [50] and for adsorption of Al(III) on TiO_2, $r = 6.5 \times 10^{-6}$ $mole/m^2$ (Wiese and Healy [37]). According to the measured results of adsorption of Al(III) ions onto the PVC latex surface, given by Schull and Gutham

[60], the maximum coverage for the latex particles seems to be a little larger than that for SiO_2 and TiO_2. When the concentrations of aluminium nitrate are 1.1×10^{-6} and 5.0×10^{-5} M, the measured densities of adsorption of Al(III) on the PVC latex are 3.8×10^{-6} M/m^2 and 12.0×10^{-6} M/m^2, respectively. In this research, the value of maximum coverage was chosen to be 10×10^{-6} M/m^2. Using this value the predicted curves of CR2 and CR3 are closer to the measured values.

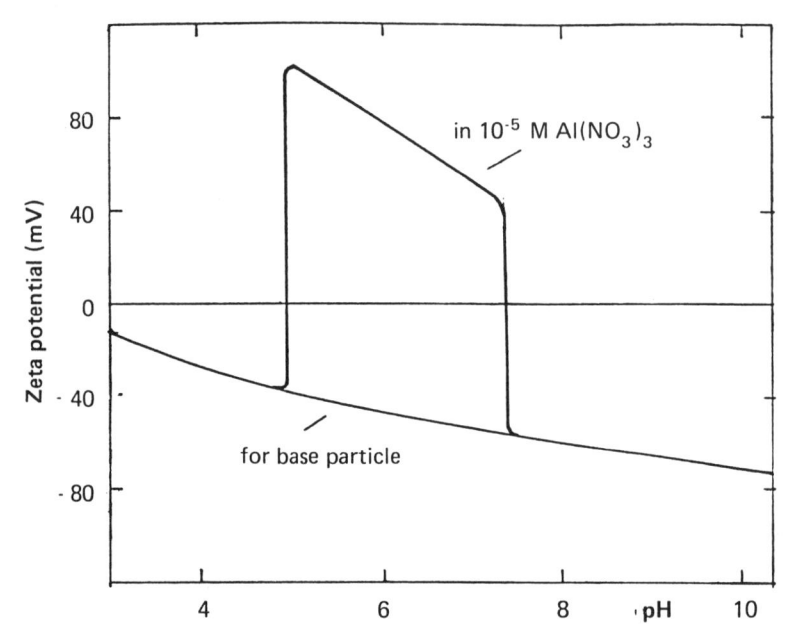

Figure 8. Predicted zeta potentials of the latex particles (2mg/l) in the presence of 10^{-5} M Al(NO$_3$)$_3$. The ionic strength is 10^{-3} M and parameter values given in Table 2

In the presence of 10^{-5} M Al(NO$_3$)$_3$, the predicted zeta potential curve of the latex particles is presented in Figure 8. The calculated zeta potentials using the surface precipitation model are much larger than the measured values. This is a common problem in the prediction of zeta potentials by the calculation of surface chemical equilibrium models (e.g., see Figures 2 and 3). However, the prediction using the surface precipitation model can at least give a qualitative description of the relationship between the zeta potentials and the adsorption of hydrolizable aluminium. In particular, the predicted CR2 and CR3 using this model are close to the measured values if the interfacial solubility product and the maximum coverage are chosen properly.

The zeta potentials of the glass beads at various pHs and in 10^{-3} M NaNO$_3$ have been predicted using the triple layer model in previous research [12]. The comparison between the prediction and the measured data available in the literature [61, 62] is given in Figure 9. Thus, the zeta potential predictions for both the particles and the collectors have been obtained at various pHs and the electric double layer force between them can be calculated for certain chemical conditions.

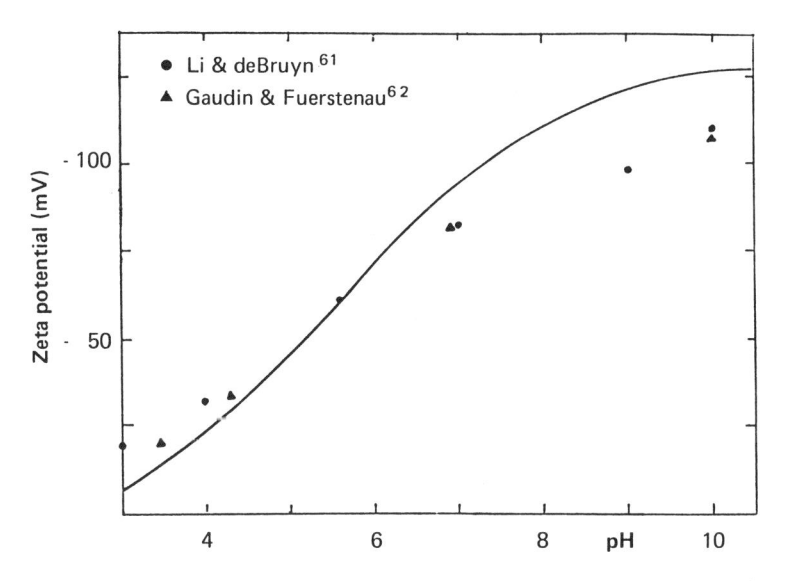

Figure 9. Predicted zeta potentials [12] of glass beads in the presence of 10^{-3} M $NaNO_3$ (—), and measured zeta potentials of vitrous silica [61, 62] in the ionic strength of 10^{-3} M

Figure 10 gives the two typical measured breakthrough curves for these cases. Curve 1 was obtained for filtration of the latex particles at pH = 5.1 and in the presence of 10^{-4} M $Al(NO_3)_3$. In this case, the particle zeta potential (ζ_p) was

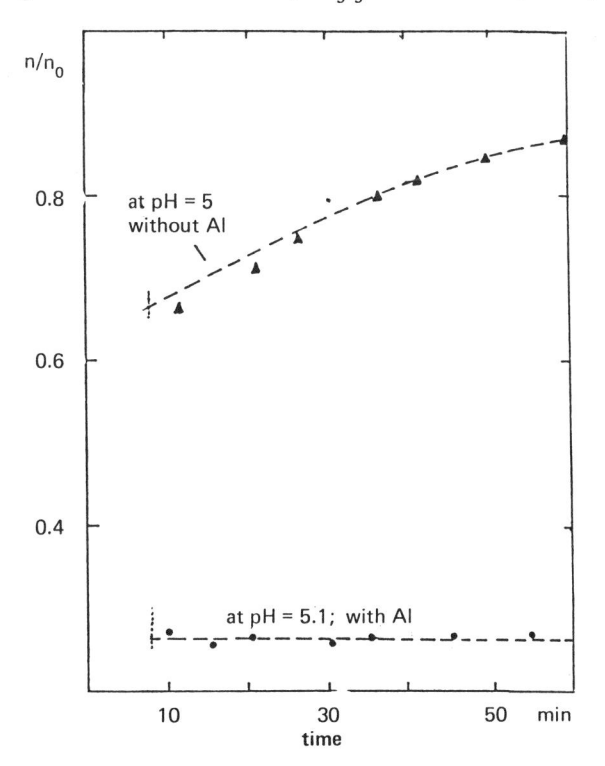

Figure 10. Measured breakthrough curves of the latex particles in filtration effluents for favourable chemical conditions

+36 mV and the zeta potential of the glass beads (ς_s) was about -50 mV (see Figure 9). The double layer force between the latex particles and the glass beads is attractive so this is a case of favourable filtration with $n/n_0 = 0.26$. After about 10 minutes, the filtration became stable and a plateau was reached. Curve 2 was obtained at pH = 5.0 and without aluminium nitrate, the particle zeta potential (ς_p) is -32 mV. The double layer force is repulsive and it is an unfavourable filtration case. The process is quite transient and no plateau can be found.

When a suspension flows through porous media in a packed bed, both deposition and longitudinal dispersion of suspended particles occur within the bed and affect the breakthrough curve. For favourable filtrations, deposition dominates the filtration process. For the initial deposition, Iwasaki's assumption is valid, the particle removal by the "clean" bed is independent of time and a plateau appears in the breakthrough curve. For unfavourable filtrations, particle deposition decreases and the effect of the dispersion on the effluent concentration becomes important, particularly in the initial period of filtration.

An experimental α can be determined from the following equation using n/n_0 values:

$$\alpha = -\ln(n/n_0)[\frac{2d_s}{3(1-f)\eta_0 L}] \qquad (15)$$

Here the transport efficiency (η_0) can be calculated using the transport model, Eq. 10.

Figure 11 presents the experimental α values of deposition of the latex particles on the glass beads.

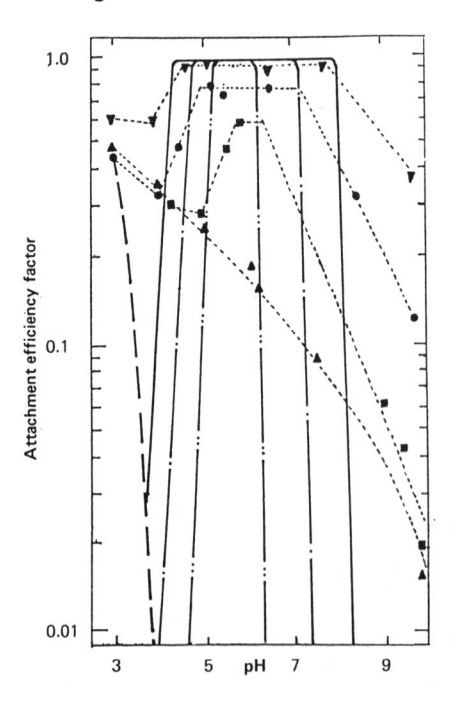

Figure 11. Comparison of measured α and predicted α (bold lines) for filtration of the latex particles in the presence of 10^{-4} M (▼), 10^{-5} M (●), 10^{-6} M (■), and 0 M (▲) Al(NO$_3$)$_3$, respectively. Physical parameters used in the model prediction are given in Table 3

Compared with Figure 9, one can find that the favourable region is just approximately above the pH range between CR2 and CR3. At pH < CR2, the experimental α values were near those without adding aluminium, particularly when the concentrations of aluminium nitrate were 10^{-5} M and 10^{-6} M. When the solution pHs were larger than CR3, the measured α values decreased again. In the case of 10^{-6} M $Al(NO_3)_3$, the measured α values were close to those without adding aluminium when the solution pHs were large enough (e.g., 9.0).

The model prediction for the attachment efficiency factor is also given in Figure 14 in which the predicted values for the concentrations of aluminium nitrate of 10^{-4} M, 10^{-5} M, 10^{-6} M and 0 M are represented by the solid line, the dot-dash line, the double dot-dash line and the dash line, respectively. In the pH region between CR2 and CR3, the model prediction shows that $\alpha = 1$ and it is independent of pH. At pH < CR2 and pH > CR3, the zeta potentials of the latex particles become negative and the filtration is unfavourable. In these cases, the energy barriers can be calculated, based on the zeta potentials of the particles and glass beads. Then, the attachment efficiency factors can be calculated with equations 6 through 13. For example, in the case of pH = 4.2 and $Al(NO_3)_3 = 10^{-5}$ M, the particle zeta potential (ζ_p) is -21 mV and the zeta potential of the glass beads is about -25 mV. The calculated energy barrier is 7.6 kT. The predicted α value is finally calculated to be 0.016. The experimental α was 0.31. For 10^{-4} M $Al(NO_3)_3$ and at pH = 4.0, the ζ_p and ζ_s are -18 mV and -21 mV, respectively. The calculated V_m is 4.2 kT. Thus, the predicted α is 0.034. The measured α is 0.49. For 10^{-6} M $Al(NO_3)_3$ and at pH = 5.0, the ζ_p and ζ_s are -16 mV and about -50 mV. The calculated $V_m = 8.0$ kT and the predicted α is 0.011 while the experimental α in this case is 0.18.

When the solution pHs are less than CR2, the predicted curves of α vs. pH are all close to those in the absence of aluminium nitrate because in these cases, the zeta potentials of the latex particles are similar to those when no aluminium nitrate is added. In the absence of aluminium nitrate and at pH = 3.0, the ζ_p and ζ_s are -13 mV and about -20 mV, respectively, and the calculated V_m is 3.3 kT. The predicted α is 0.53 while the experimental is 0.48. When the solution pHs increase, the predicted αs rapidly decrease.

Comparisons between predicted αs and measured αs show that the present model can qualitatively explain the effects of adsorption of hydrolyzed aluminium on filter performance.

In the favourable pH regions, the model prediction shows that the attachment efficiency factors are unity and independent of the concentrations of aluminium nitrate while the experimental αs were observed to decrease with decreasing aluminium concentration. For example, for 10^{-6} M $Al(NO_3)_3$, the measured αs were only about 0.6 in favourable filtrations. This anomaly may occur due to non-uniformity of the interfacial coating layer on the particle surface, especially when the concentration of aluminium nitrate is low (e.g., 10^{-6} M).

The second anomaly observed is that a catastrophic decline in the attachment efficiency is predicted to occur for certain chemical conditions while the experimental results show a more gradual reduction in α as the conditions become more unfavourable. This result is in agreement with other studies of the effects of chemistry on particle deposition [4, 6, 12]. Several hypotheses have been proposed to explain the observed failure of the "chemical model" to accurately predict deposition. These include the effects of surface roughness, heterogeneity of the surface of individual collectors and particles, heterogeneity among collectors and particles, and possible inadequacies of the equilibrium DLVO theory. A quantitative analysis of these effects is complex theoretically and difficult experimentally. The anomalies observed in this research may be due to heterogeneity of the surface charge properties of the latex particles caused by the adsorption of hydrolyzed aluminium species.

Conclusions and Applications

In this work, a fundamental approach for incorporating the effects of adsorbed hydrolyzable aluminium species on suspended particles in the theory for the deposition of submicron particles in filters has been described. The experimental results that illustrate the effects of adsorbed Al(III) species on particle deposition have been presented and compared with the theory. The following conclusions are presented based on the results of this work, and the possible applications of these results are also presented.

1) Favourable chemical conditions can be created by adding inorganic coagulants such as aluminium salts and by controlling the solution pH. This becomes more feasible and more controllable because the surface precipitation model can predict the charge reversal of suspended particles (i.e., CR2 and CR3) due to the adsorption of hydrolyzed aluminium. The modified MINEQL program for the surface chemical equilibrium calculation is available for this purpose.

2) In general, for favourable chemical conditions, transport models based on fundamental physical characteristics of the filtration system can predict particle removal by clean filter beds quantitatively. This conclusion has also been validated by previous research [12]. This provides a scientific basis for selection of the physical parameters such as filtration rate, depth of the filter bed and size distribution of filter media in design and operation of deep bed filtration. Particularly, it tells us the importance of controlling particle size distribution by pretreatment in improving filtrate quality.

3) The present model for effects of solution chemistry on particle deposition cannot predict quantitatively particle removal by filters for unfavourable filtrations. But both theoretical analysis and experimental results do indicate that fewer particles are removed under unfavourable chemical conditions, compared with the favourable cases. Particularly, the dramatic change in particle removal by filters occurs for a certain critical chemical condition which is indicated by the pH_{iep} of suspended particles. The pH_{iep} is the charge reversal point of suspended particles so it is also the critical point between repulsive and attractive electric double layer forces. Because specifically adsorbed ions can change the isoelectric point or points of suspended particles (e.g., the adsorbed cations (Ca^{2+}) or anions (e.g., PO_4^{3-}) can shift the pH_{iep}, and the adsorbed hydrolyzable metal ions (e.g., Al(III)) may cause several isoelectric points to occur), they will effectively

change the mode of filter performance and make the filtration more favourable or less unfavourable. This provides a means for improving the filtrate quality by conditioning the solution chemistry in filtration systems, especially by controlling the coagulant dosage and pH.

4) Because models for the attachment efficiency factor in unfavourable deposition of large particles (e.g., $a_p > 1\ \mu m$) have not yet been developed, a critical evaluation of the chemical effects on deposition of these kinds of particles is not possible. But it is is believed that the following common points exist in both theoretical analysis and experimental observation between depositions of Brownian and non-Brownian particles:

   a) For favourable chemical conditions, the transport models are valid and can predict filtration removal efficiencies of both Brownian and non-Brownian particles.

   b) The effects of pH and specific adsorption of other ions, especially hydrolyzable metal ions Al(III) on the zeta potentials of suspended particles follow the common pattern described by a combination of the surface chemical equilibria and electric double layer theories, regardless of particle size. The surface chemical equilibrium calculation can at least predict changes of the isoelectric points of particles due to the effects of solution chemistry as long as these chemical parameters such as surface intrinsic equilibrium constants, total ionizable site densities and ionic strength are available.

   c) Therefore, fundamental particle deposition theories that include chemical effects can predict quantitatively the onset of unfavourable conditions where reduced particle deposition is expected in filtrations of both Brownian and non-Brownian particles.

   Therefore, this research provides a scientific basis for evaluating the chemical effects and controlling solution chemistry for better filtrate quality for both contact or direct filtration and conventional filtration in which submicron particles are aggregated to larger size particles due to extensive flocculation and settling.

5) An improved understanding of the interaction between added chemicals such as coagulants and oxidants and natural dissolved species and natural particles, and of how chemistry affects surface interaction forces should improve the design and operation of filtration facilities that do not include settling tanks. This is especially important for the treatment of currently unfiltered, low turbidity, surface water supplies. The removal of the relatively low numbers of natural stable particles from these supplies by filtration will be controlled by, and possibly be very sensitive to, the chemical conditions utilized.

6) Heavy metal pollutants and aluminium coagulants have similar characteristics in solution and have similar surface chemistries. For example, they all experience extensive hydrolysis in natural waters and water treatment processes and chemical coordination with inorganic and organic ligands. They may exist in various complexed forms such as hydrolyzed species and inorganic and organic complexes. They all interact with particle surfaces (i.e., adsorption) and affect the particle surface charge and potential in a similar way. It is therefore believed

that somehow the heavy metal pollutants and aluminium coagulants affect the particle-particle interactions in natural waters and water treatment plants in the same way. An improved understanding of the effects of heavy metal ions on coagulation and filtration of natural particles will be important in studying coagulation in natural waters and removal of these pollutants by water treatment.

7) The results obtained in studying the effects of phosphate anions on zeta potentials of natural particles and on deposition of these particles in filter beds may provide some hints for exploring effects of natural organic substances (e.g., humic acid) and organic pollutants on interactions among these particles and between the particles and filter grains or soils. It has been found that adsorption of some dissolved organic species on natural particles and their effects on particle surface properties are similar to interactions with phosphate anions [23, 63]. In dealing with the effects of organic species on particle interactions, hydrophobic and steric interactions may become important.

Acknowlegement

The author gratefully acknowledges the caring support and advice of professor Charles R. O'Melia, Department of Geography and Environmental Engineering, The Johns Hopkins University, Baltimore, MD 21218, U.S.A.

References

[1] Yao, K.M. (1968), Ph.D. dissertation, University of North Carolina, Chapel Hill, NC
[2] Rajagopalan, R., Tien, C. (1979): Progress in Filtration and Separation, Vol. 1. In: Wakeman, R.J. (ed.), Elsevier
[3] Ghosh, A.M., Jordan, T.A., Porter, R.L. (1975), J. Environ. Eng. Div., Proc. ASCE, EE1, Vol. 71, pp. 71
[4] FitzPatric, J.A. (1972), Ph.D. dissertation, Harvard University, Cambridge, Mass.
[5] Kallay, N., Nelligan, J.D., Matijevic, E. (1983), J. Chem. Soc. Faraday Trans., Vol. 79, pp. 65
[6] Gregory, J., Wishart, A.J. (1980), Colloids and Surface, Vol. 1, pp. 313
[7] Derjaguin, B.V., Landau, L.D. (1941), Acta Physicochim., SSSR, Vol. 14, pp. 633
[8] Verwey, E.J.W., Overbeek, J.Th.G. (1948): Theory of Stability of Lyophobic Colloids. Nth Holland, Amsterdam
[9] Spielman, L.A., Friedlander, S.K. (1974), J. Colloid Interface Sci., Vol. 46, pp. 22
[10] Ruckenstein, E., Prieve, D.C. (1974), J. Chem. Soc. Faraday II, Vol. 69, pp. 1522
[11] Onorato, F. (1979), Ph.D. dissertation, Syracuse University, Syracuse, New York
[12] Wang, Z.S. (1986), Ph.D. dissertation, The Johns Hopkins University
[13] Payatakes, A.C., Tien, C., Turian, R.M. (1973), American Institute of Chemical Engineerings Journal, Vol. 19, pp. 58
[14] Happel, J. (1958), American Institute of Chemical Engineerings Journal, Vol. 4, pp. 197
[15] Spielman, L.A., Goren, S.L. (1970), Environ. Sci. Technol., Vol. 4, pp. 135
[16] Hamaker, H.C. (1937), Physica (Utrecht), Vol. 4, pp. 1058
[17] Hogg, R., Healy, T.W., Furstenau, D.W. (1966), Trans. Faraday Soc., Vol. 66, pp. 1638
[18] Usui, S. (1973), J. Colloid Interface Sci., Vol. 44, pp. 107

[19] Chan, D., Healy, T.W., White, L.R. (1976), J. Chem. Soc. Faraday Trans. I, Vol. 72, pp. 2844
[20] Dukhin, S.S., Lyklema, J. (1987), Langmuir, Vol. 3, pp. 94
[21] Spielman, L.A., FitzPatric, S.K. (1973), J. Colloid Interface Sci., Vol. 42, pp. 607
[22] James, R.O., Parks, G.A. (1982): Surface and Colloid Science, Vol. 12. In: Matijevic, E. (ed.), Plenum Press
[23] Stumm, W., Kummert, R., Sigg, L. (1980), Croat. Chem. Acta, Vol. 53, pp. 291
[24] Westall, J., Zachary, J., Morel, F.M.M. (1976), Technical Note No. 18, Raiph M. Parsons Laborary, M.I.T., Cambridge, Mass.
[25] Davis, J.A., James, R.O., Leckie, J.O. (1978), J. Colloid Interface Sci., Vol. 63, pp. 480
[26] Stumm, W., Huang, C.P., Jenkins, S.R. (1970), Croat. Chem. Acta, Vol. 42, pp. 233
[27] Iwasaki, T. (1935), J. AWWA, Vol. 29, pp. 1591
[28] Levich, V.G. (1962): Physicochemical Hydrodynamics. Englewood Cliffs, NJ, Prentice Hall
[29] Rajagopalan, R., Tien, C. (1976), AIChE J., Vol. 22, pp. 523
[30] Prieve, D.C. (1986), personal communication
[31] Dahneke, B. (1974), J. Colloid Interface Sci., Vol. 48, pp. 520
[32] Rajagopalan, R., Kim, J.S. (1981), J. Colloid Interface Sci., Vol. 83, pp. 428
[33] Hayden, P.L., Rubin, A.J. (1974), in: Rubin, A.J. (ed.): Aqueous-Environmental Chemistry of Metals. Ann Arbor Science Inc., Michigan
[34] Vermeulen, A.C., Geus, J.W., Stol, R.J., De Bruyn, P.L. (1975), J. Colloid Interface Sci., Vol. 51, pp. 449
[35] Brown, D.W., Hem, J.D. (1975), U.S. Geological Survey Paper, 1837-F, Washington, D.C.
[36] Iler, R.K. (1964), J. Amer. Ceram. Soc., Vol. 47, pp. 194
[37] Wiese, G.R., Healy, T.W. (1975), J. Colloid Interface Sci., Vol. 51, pp. 434
[38] Rubin, A.J., Hanna, G.P., Environ. Sci. and Technol., Vol. 2, pp. 358
[39] Managravite, F.J. et al. (1975), J. AWWA, Vol. 67, pp. 88
[40] Matijevic, E., Managravite, F.J., Cassel, E.A. (1971), J. Colloid Interface Sci., Vol. 35, pp. 560
[41] Stumm, W., Hohl, H., Dalang, F. (1976), Croat. Chem. Acta, Vol. 48, pp. 491
[42] James, R.O., Healy, T.W. (1972), J. Colloid Interface Sci., Vol. 40, pp. 53
[43] Stumm, W., Morgan, J.J. (1981): Aquatic Chemistry. Wiley- Interscience, New York
[44] Wiese, G.R., Healy, T.W. (1975), J. Colloid Interface Sci., Vol. 51, pp. 427
[45] Matijevic, E., Bleier, A. (1977), Croat. Chem. Acta, Vol. 50, pp. 93
[46] Black, A.P. (1967), in: Faust, S.D., Hunter, J.V. (eds.): Principles and Applications of Water Chemistry.
[47] Bleam, W.F., McBride, M.B. (1985), J. Colloid Interface Sci., Vol. 110, pp. 335
[48] White, D.W. (1980), Ph.D. dissertation, Clarkson University, Potsdam, New York
[49] James, R.O. (1981), in Anderson, M.S., Rubin, A.J. (eds.): Adsorption of Inorganics at Solid-Liquid Interface. Ann Arbor Science Publishers, Ann Arbor, Mich.
[50] Letterman, R.D., Iyer, D.R. (1985), in: Gregory, J. (ed.): Solid-Liquid Separation.
[51] Ottewill, R.H., Shaw, J.N. (1972), J. Electroanal. Chem., Vol. 37, pp. 133
[52] Stone-masui, J., Watillon, A. (1975), J. Colloid Interface Sci., Vol. 52, pp. 479
[53] Yates, D.E. (1975), Ph.D. dissertation, University of Melbourne, Australia
[54] Schindler, P.W., Kamber, H.R. (1968), Helv. Chim. Acta, Vol. 51, pp. 1781
[55] Abendroth, R.P. (1970), J. Colloid Interface Sci., Vol. 34, pp. 591
[56] Bolt, G.H. (1957), J. Phys. Chem., Vol. 61, pp. 1161
[57] Iler, R.K. (1979): The Chemistry of Silica.
[58] Henry, D.C. (1948), Trans. Faraday Soc., Vol. 44, pp. 1021
[59] Penners, N.H.G. (1985): The Preparation and Stability of Homodisperse Colloidal Haematite.
[60] Schull, K.E., Gutham, G.R. (1967), J. AWWA, Vol. 59, pp. 1456
[61] Li, H.C., De Bruyn, P.L. (1966), Surface Sci., Vol. 5, pp. 203
[62] Gaudin, A.M., Fuerstenau, P.A. (1955), Trans. A.I.M.E., Vol. 22, pp. 66
[63] Zhang, Y., Kallay, N., Matijevic, E. (1985), Langmuir, Vol. 1, pp. 201

[64] Sigg, L. (1979), Ph.D. dissertation, ETH Zürich

Z. Wang
Department of Environmental Engineering
The Tsinghua University
Beijing
China

Polyelectrolytes for the Treatment of Tap and Filter Back Washing Water

J. M. Reuter and A. Landscheidt

1 Introduction

Polymeric organic flocculants have been used in water treatment and in sludge dewatering processes now for more than 30 years. Meanwhile, these products have been introduced in potable water treatment, too. In this case they are used in sedimentation and settling processes as well as in direct filtration plants in combination with inorganic coagulation auxiliaries. The advantages of using polyelectrolytes are: increased plant capacity, higher water quality, stable operation of the plant technology and reduced costs.

The purpose of this report is to give a short description of the structure and kinetics of the flocculants used in potable water treatment, to describe the optimal application parameters and to describe some special effects concerning, e.g., the possibilities for reducing the dosage of primary coagulation aids or removing algae from the water. Polymeric flocculants offer in addition significant advantages in the purification of filter back washing water, in the dewatering of the slurries from this process and, because of this, make it possible to operate the water treatment plant in closed circuits.

2 Structure and Reaction Kinetics of Polymeric Flocculants

As shown in Figure 1, polymeric flocculants are available in a broad range of ionic types and different degrees of load density as well as in different mole masses. The actual range covers the whole spectrum from highly cationic charged products to medium and low charged types, nonionic types and low, medium and high charged anionic polymers. The range of mole mass of these products is between about 5 to 15 million and corresponds to polymerization degrees of about 50,000 to 200,000 and even more.

The type and density of electrical charge at the particle surface will determine the charge and charge density required in the polymer, whilst the mole mass of the polymer is the parameter responsible for floc size, floc density and floc resistance. On the other hand, several application conditions will make it possible to produce large or small flocs and to influence floc density and floc strength. The ionic charge densities of the polymers used in potable water treatment range from nonionic

Polymerzation degree resp. molemass

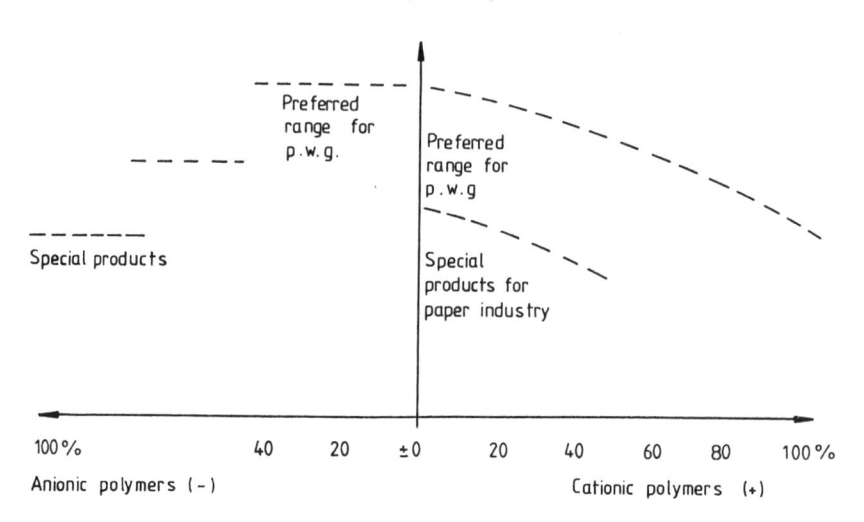

Figure 1. Available range of polymeric flocculants

polyacrylamide to copolymers with an anionic charge up to 40% and those with a cationic charge up to 30%.

Figure 2 shows the structure of a 30 to 35% anionic charged polymer. Under some special conditions cationic low molecular weight homopolymers are used, too. The selection of the optimal product depends on the characteristics of the raw water which has to be treated, on specific impurities, which shall be removed depending on the water treatment conditions, and on the local or constructional situation in the site.

$$\bar{n} \approx 150\,000 - 250\,000$$

$$\overline{MG} \approx 8 - 12 \text{ Mio.}$$

Figure 2. Model of an anionic polymer molecule with about 30 - 35 % of acrylate groups (COO$^-$Na$^+$)

3 Safety Regulations for Polymeric Flocculants to be Used in Potable Water Treatment

Polymeric flocculants or polyelectrolytes are considered to be technical auxiliaries in potable water treatment, because they do not remain in the treated water like an additive, but are removed from the water more or less completely. With this proviso, some governments have given official permission for the use of polymeric flocculants in potable water treatment: USA (Environmental Protection Agency, EPA) Great Britain (Department of the Environment), Federal Republic of Germany (Bundesgesundheitsamt, Bundesministerium für Gesundheit). Some other countries have adopted this position. As a general rule, acrylamide-containing polyelectrolytes may be used in potable water treatment if the monomeric acrylamide content is less than 0.05%, and if the dosage is limited to max. 1 ppm.

When production and application are strictly controlled, both conditions will be met without difficulties.

4 Preparation of Polymer Solutions

The potable water-grade polymers are normally produced as granular materials to powder. Before being added to the water, these materials have to be dissolved to make highly diluted aqueous solutions. The preferred modern technology is to prepare a stock solution of 0.3 to 0.5% by weight of the polymer in a batchwise-operating dissolving plant. This stock solution is diluted with an inline mixer in the pressure tube of the dosing pump on the way to the dosing point, in a ratio of 1:10. In this way, the concentration of the operational solution is decreased to 0.05 to 0.03%. This low concentration assures a very good solubility of the polymer molecules in the solution and a very good mixing of the polymer with the water, which is a basic condition for obtaining maximal polymer effectiveness.

In Figure 3 the cross section of a batchwise-operating dissolving and dosage plant is shown. Batchwise operation assures that the solutions are prepared with defined polymer concentrations and a defined state in solution, and the dissolving and stirring time will be exactly fixed.

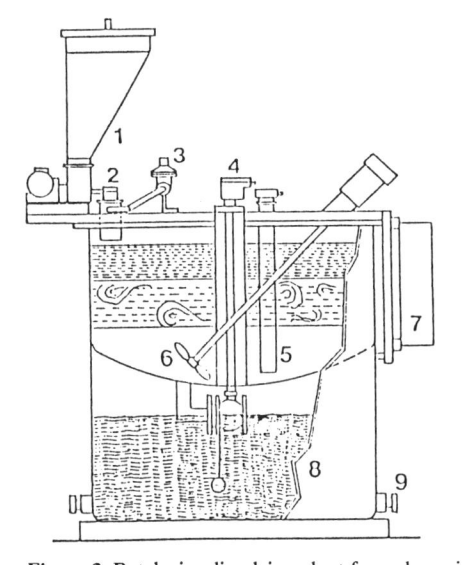

1 Storage bin with feeding system (i.e. screw feeder)
2 Wetting unit
3 Water regulator
4 Automatic valve between the two compartments
5 Electrode monitoring system for solution level
6 Stirrer
7 Switch board
8 Minimum level monitor (safety system for dosage pump)
9 Connecting pipe for dosage pump

Figure 3. Batchwise dissolving plant for polymeric flocculants

5 Technical Application of Polymeric Flocculants to Potable Water Treatment by Flocculation and Sedimentation

5.1 Treatment of the Water, Technology of Flocculation

The steps of flocculation using polyelectrolytes are:
- destabilization of colloidal and turbid materials by adding aluminium sulfate or iron chloride
- aggregation of destabilized colloids and fine particles to microflocs, precipitation of the metal ions as hydroxides
- flocculation of the microflocs to large aggregates using polyelectrolytes, growth of the macroflocs under optimized stirring conditions
- separation of the flocculated material by sedimentation in settling tanks with inclined plate settlers.

Typical technical data for the different process steps may be summarized as follows:
- Reaction time for the mixing of the primary coagulants with the water and the destabilization of the colloidal material should be 3 to 5 minutes at a peripherical speed of the stirrer of about 5 m/s with an energy input of 50 to 200 W/m^3 of water.
- Floc formation in a subsequent flocculation basin with a retention time of between 10 and 60 minutes, stirrer speed should be limited from 0.5 to 1.5 m/s at an energy input of 10 to 30 W/m^3.
- Sedimentation of the flocculated material with a retention time in the settling basin of 120 to 180 minutes. Due to the use of polyelectrolytes this time can often be reduced to 60 to 120 minutes, thus offering a remarkable increase in plant capacity.

Figure 4 shows an optimized system of destabilization and flocculation using polymeric flocculants. Here special "floc growth tanks" with slowly rotating horizontal mixers assure the formation of large, strong, dense flocs. Due to the slow rotation of the stirrers, which affects nearly the entire volume of the basin, there are many opportunities for contact between the flocs and the particles not yet flocculated. This allows the colloidal and fine particles to be brought together to form large size flocs.

Polymeric Flocculants. This technology uses anionic polymers as well as cationic polymers. The optimal product and the optimal application method should be evaluated using jar tests.

Anionic Polymers. Anionic polymers are sensitive to active aluminium and iron ions, which are not precipitated as hydroxides. The anionic acrylate groups will be blocked and precipitated by these metal ions as the water-insoluble aluminium salt or iron salt of polycarbonic acid. Since the precipitation of the aluminium ions depends on pH and on reaction time, both parameters should be carefully controlled before adding the anionic polymer solution. The specific effectivity

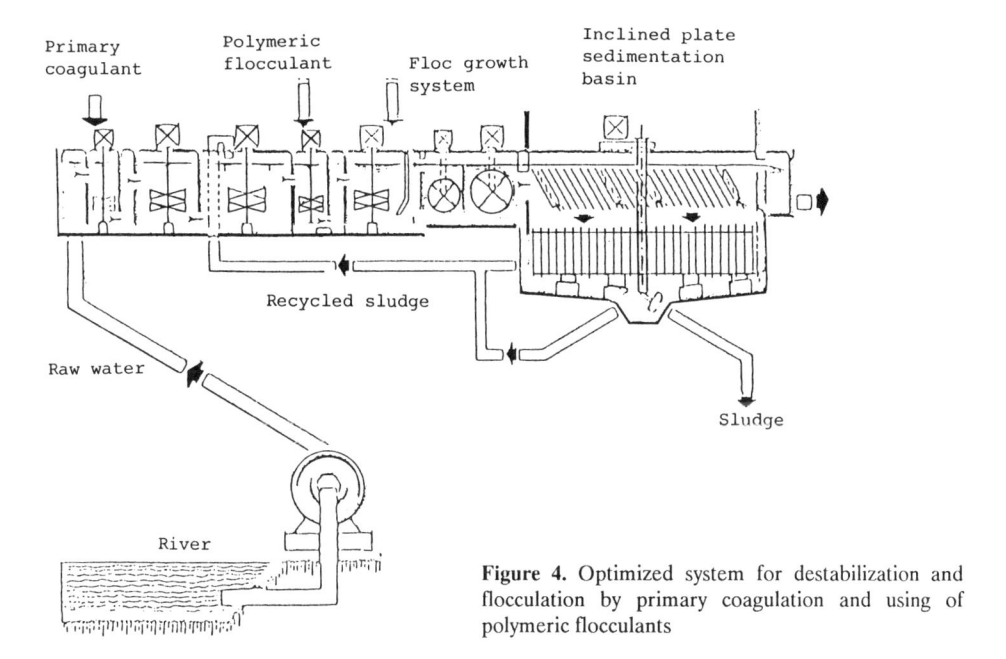

Figure 4. Optimized system for destabilization and flocculation by primary coagulation and using of polymeric flocculants

ranges of anionic polymers in relation to the anionic charge are plotted in Figure 5. The higher the concentration of non-hydrolyzed aluminium ions, the lower the anionic charge of the polymer must be, but, on the other hand, the specific effectivity of the polymer decreases. Under well controlled conditions the anionic polyelectrolytes will form large, voluminous flocs with high settling velocity. These flocs will build up a filter layer in the settling basin. The influent has to pass through this filter bed and this results in an additional purification and cleaning effect. Small

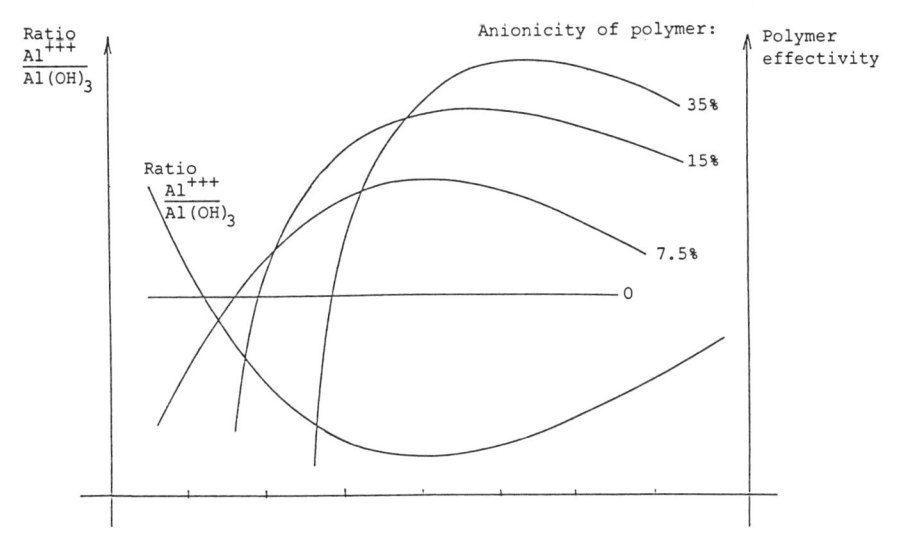

Figure 5. Polymer effectivity in relation to degree of hydrolysis of alumina ions for polymers with different anionicity

flocs, colloidal material and even macromolecules will be adsorbed and separated from the water in the filter layer.

Cationic Polymers. Cationic polymers are not sensitive to active inorganic cations in the water. Therefore the dosage range is independent of the degree of hydrolysis of the aluminium or iron ions. Cationic polymers will also react directly with negatively charged primary impurities in the water. Therefore it is often possible to reduce the amount of aluminium sulfate or iron chloride in primary coagulation when using cationic polyelectrolytes. In addition, when the algae content in the water is high, the cationic polymers will flocculate very specifically the algal material and assure an effective removal from the water. The flocs formed with cationic polymers will not be as voluminous and large as those formed with anionic types. The decision whether to use anionic or cationic polymers should be made based on jar tests, if not already predetermined by the local conditions in the plant.

As long as the use of polyelectrolytes is already provided for in the planning stages, technological design and construction of the plant, there is no problem in optimizing flocculation, as already demonstrated in Figure 4: dosage locations, reaction time, stirrer types and energy inputs are combined to produce an optimal interaction.

Very often in existing plants, however, which have been designed for operation without polyelectrolytes, many difficulties have to be overcome and compromises must be made if, at a later stage, the operation of the plant has to be upgraded through the use of polymeric flocculants. Furthermore, these are in general very large plants with very high capacities, and subsequent conversion of the installation or addition of a reaction tank or stirring unit is impossible. In that case, it is necessary to determine what chemicals are optimal and how they can be used most effectively by observing and modelling the present conditions in the plant using jar tests and then transferring the test conditions and results to the plant. In these cases, the realizable optimum in the plant is often less than the possible optimum which could have been achieved had the plant been designed from the beginning for the use of polymers.

6 Potable Water Treatment Using Direct Filtration

The use of polyelectrolytes in direct filtration requires strict control of floc size and floc behaviour. Here again the raw water is pre-treated with primary coagulants like aluminium and iron salts, but their dosage is much less than when treatment is by means of sedimentation. Dosage is in the range of 5 to 10 ppm. The flocs must be very small so that they will not be separated on the surface layer of the filter bed, thus covering it in a short time, but they must enter into the filter body. In addition, the small flocs must have a high tendency to adsorb to the filter material so that they are separated from the water when passing through the filter bed and, in addition, will increase the adsorption capacities of the filter.

Use of polymers in direct filtration requires only very small dosages at concentrations of 0.05 to 0.2 ppm. The polymer solution should be highly diluted at concentrations of 0.05 to 0.02% and should be mixed into the pre-treated water under conditions of high turbulence. The advantages of optimizing the use of polymers in direct filtration are:

- reduced turbidity in the filtered water. Normally the final turbidity of the water is reduced to less than half of the value obtained by filtration without polymers.
- increase in operation time of the filters of about 25 to 50% and thus a corresponding increase in plant capacity.
- decrease in the required backwashing for two reasons: first, as the operation time of the filter is increased, the frequency of backwashing is reduced. Second, as the filtered impurities are flocculated and enlarged, filter washing requires less time and less volume of water for each rewashing cycle.
- backwashing water is easy to purify and to clear.

There is also the possibility of using primary and polyelectrolyte flocculation in the preceding sedimentation settling process as well as in the subsequent filtration step.

7 Practical Experiences in Potable Water Treatment with Polyelectrolyte Flocculation

Figure 6 is the flowsheet of a potable water plant, which was planned and designed from the beginning for using polymeric flocculants in the sedimentation process. The process was optimized by extended laboratory and pilot plant test work. Here an anionic polymer, PRAESTOL 2540 TR, with about 40% acrylic groups is used. The additional dosage is about 0.25 to 0.3 ppm of the polymer.

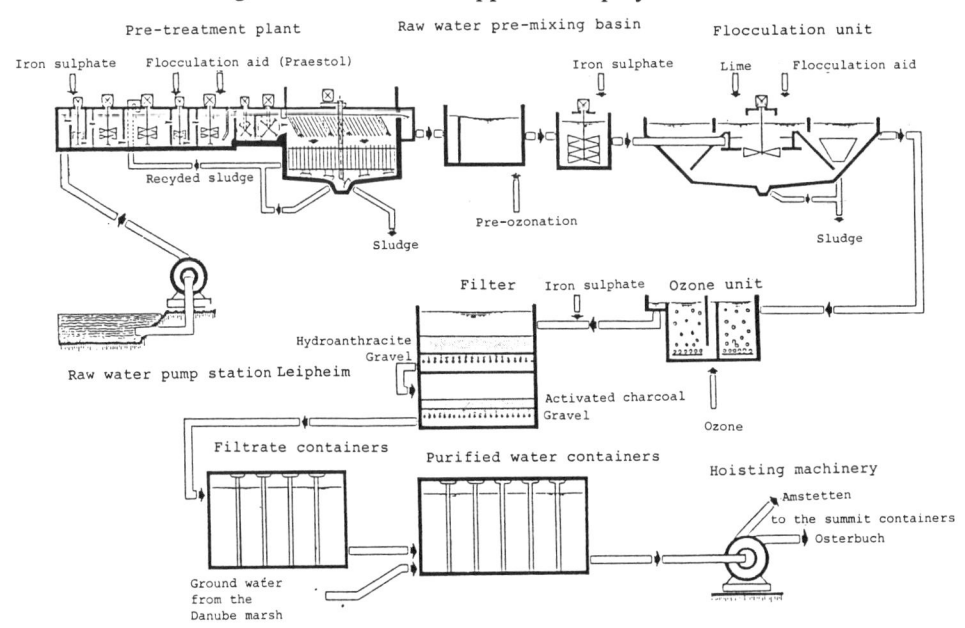

Figure 6. Flowsheet of a potable water treatment plant with optimized system for destabilization and flocculation

Table 1 presents the data for a large potable water plant in Africa, which utilizes flocculation followed by sedimentation and filtration. The raw water is taken from a river. The nominal capacity is 1.6 m/s. By using 0.4 ppm of the anionic polymer PRAESTOL 2540 TR, the dosage of aluminium sulfate could be reduced from 35 ppm to less than 30 ppm. Due to an increased clarification effect in the sedimentation process, the operation time of the subsequent open filter plant increased from about 6 hours to about 9 hours or about 50%, or to a capacity of 2.1 m^3/s from 1.5 m^3/s. When polymers are used, turbidity in the clarified water is reduced from 1.5 ITU to < 1 ITU. Comparative values for a plant in South America are summarized as follows: the raw water from a river is treated with 25 ppm of aluminium sulfate at a plant capacity of 2.5 m^3/s. The water is clarified in a sedimentation tank. The filter plant runs with a cycle time of about 14 hours and produces water with a final turbidity of about 1.2 ITU. After a series of tests, a 4-week test was started using the anionic polymer PRAESTOL 2530 TR at a dosage of 0.5 ppm. Aluminium sulfate doses could be reduced to 20 ppm and capacity was increased to 3 m^3/s. The operation cycles of the filter plant increased to about 24 hours and the final turbidity was reduced to < 0.9 ITU. Due to the very encouraging results, a stirring basin and an automatic dissolving plant for the polymer is now under construction and it is assumed that the polymer dosage, which is actually 0.5 ppm, will then be decreased to 0.3 to 0.35 ppm.

Table 1. Technical data and results for flocculation - sedimentation - filtration with polymeric flocculants

Site: Africa. Flocculation - sedimentation - filtration
Raw water from a river

Turbidity	200 ITU > 500 ppm	
Polymer	PRAESTOL 2540 TR (anionic)	0.4 ppm
Capacity [m^3/s]	1.6	2.1
Al$_2$(SO$_4$)$_3$ [ppm]	35 - 45	28 - 30
Operation period for filters [h]	5 - 6	8 - 10
Turbidity of clarified water [ITU]	1.5	< 1

Site: South America. Flocculation - sedimentation - filtratation
Raw water from a river

Turbidity	80 - 250 ITU	
Polymer	PRAESTOL 2530 TR [*] (anionic)	0.4 - 0.5 ppm
Capacity [m^3/s]	2.5	3
Al$_2$(SO$_4$)$_3$ [ppm]	24 - 28	20 - 22
Operation period for filters [h]	12 - 14	18 - 22
Turbidity of clarified water [ITU]	1.0 - 1.2	< 1

[*] Test with PRAESTOL 321 TR (cationic) now in preparation

At another site in South America, tests have been performed to increase the capacity of a large, existing filter plant by using polyelectrolytes. Due to the constructional design of the plant, the polymer has to be dosed directly after the addition of iron chloride after a reaction time of less than 5 seconds. In accordance with the results of earlier jar tests, the cationic polymer PRAESTOL 321 TR was selected for the plant tests. Figure 7 shows the data for the initial period of the first one-week test, which may be summarized as follows (see Table 2). Normal capacity of the plant is about 7 m^3/s. The iron chloride dosage is 12 to 15 ppm. Filter cycle time is 11 to 13 hours and turbidity of the purified water is 1.2 to 1.5 ITU. By adding 0.3 ppm of PRAESTOL 312 TR, the values changed as follows: dosage of iron chloride could be reduced to 8 to 9 ppm and operation time of the filters increased to 17 to 20 hours, accompanied by a decrease in turbidity to 0.9 ITU. Due to the increase in filter operation time, the total capacity of the plant was increased from 7 to 8.5 m^3/s or about 25%. The costs of the polymer are completely recovered by the decreased costs of the iron chloride. Additional advantages result from fewer backwashing cycles and by decreased turbidity of the clarified water.

At another site in South America, which operates with flocculation-sedimentation-filtration, tests have been carried out to determine whether the raw water, which comes from a lagoon, can be treated with satisfying results by direct filtration without operating the preceding sedimentation basins. Chemical treatment is precipitation with aluminium sulfate. Plant capacity is about 6 m^3/s. Facilities for using aluminium sulfate have been provided in a basin about 250 m before the water enters the plant, and the polymer PRAESTOL 2530 TR was dosed under conditions of high turbulence in a distributor tank, from which water was distributed to filters, by passing the settling plant. Test results confirmed that at this site the water can be clarified with optimal results by direct filtration, with the same capacity and with the same purification effectivity, without using the sedimentation basins, when using the anionic polyelectrolyte and under strict control of floc size and strength. Now the area occupied by the settling tanks will be used for the installation of additional filters, which is the most economical way to enlarge plant capacity by about 75%.

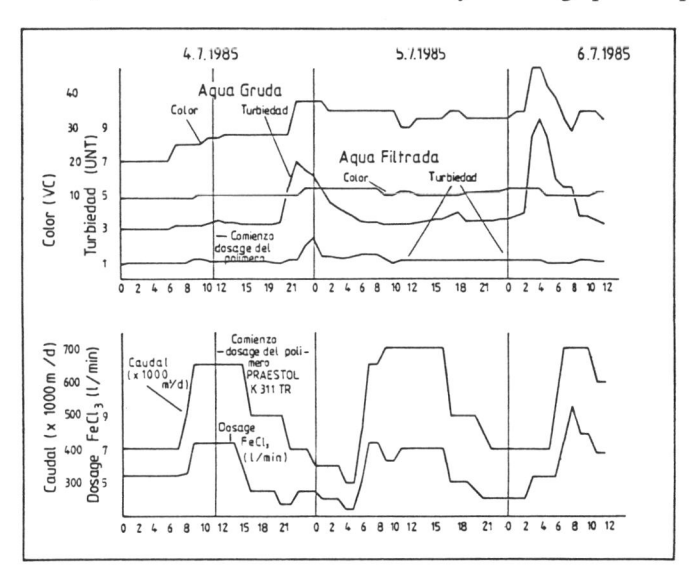

Figure 7. Data and results of a plant test using polymeric flocculation and direct filtration (PRAESTOL 321 TR)

Table 2. Technical data and results for polymeric flocculation in direct filtration

Site: South America. Direct filtration Raw water from a barrage		
Turbidity	8-20 ITU	
Polymer	PRAESTOL 321 TR (cationic)	0.3 ppm
Capacity [m^3/s]	6.5 - 7	8.2 - 8.5
FeCl$_3$ [*)] [ppm]	12 - 15	8 - 9
Operation period for filters [h]	11 - 13	17 - 20
Turbidity of clarified water [ITU]	1.2 - 1.5	< 0.9

Site: South America. Direct filtration (by-passing the sedimentation plant) Raw water from a barrage		
Turbidity	10 - 20 ITU	
Polymer	PRAESTOL 2530 TR (anionic)	0.4 ppm
Capacity [m^3/s]	5 - 5.5	5 - 5.5
Al$_2$(SO$_4$)$_3$ [**)] [ppm]	18 - 24	15 - 18
Operation period for filters [h]	9 - 12	9 - 12
Turbidity of clarified water [ITU]	0.9 - 1	0.8 - 1

[*)] Very short reaction time for hydrolysis
[**)] Reaction time for hydrolysis: 2-3 min.

8 Treatment of Filter Backwashing Water

In plants with combined flocculation-sedimentation and subsequent filtration technology, the filter wash water is treated together with the sludge from the flocculation-sedimentation stage. Sludge and wash water are fed to a static thickener. By adding polymeric flocculant, a solid-free overflow is obtained, which is often reused as filter wash water, or is combined with the raw water.

Those plants which operate with filtration as the only process step require an additional treatment plant for the filter wash water. Figure 8 shows the flowsheet for the purification plant. The wash water is homogeneous in solid content in a large storage basin, the volume of which must be calculated according to the operation cycles of the filters, the wash water volume for each wash water process and the total number of filters in operation. From this storage basin, the water is fed in equal volumes and solid load to a sedimentation tank. Here polymeric flocculants are dosed in order to assure high settling rates for the solids and clear overflow water. The settled and thickened sludges will be mechanically dewatered with centrifuges, sieve belt presses or chamber filter presses and will be disposed of. The water separated from the sludge is recirculated into the storage basin.

This technology results in a closed circuit: the only products leaving the plant are
– the filtered water, which is used for the public supply
– the dewatered sludge, which is deposited.

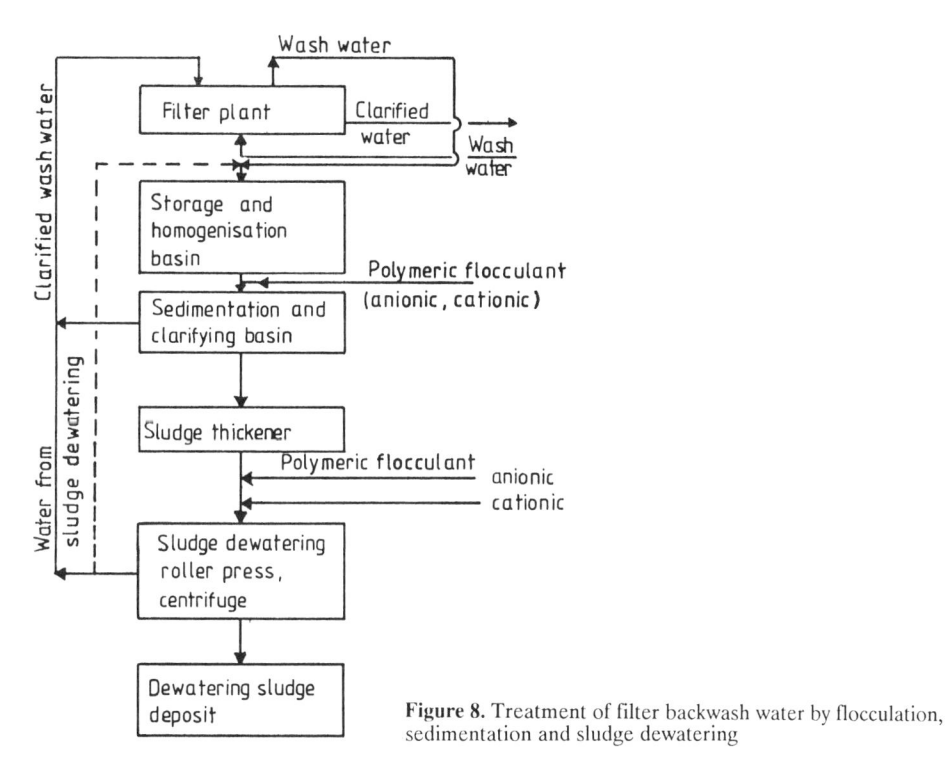

Figure 8. Treatment of filter backwash water by flocculation, sedimentation and sludge dewatering

9 Summary

Polyelectrolytes offer great advantages for the treatment of potable water. Sedimentation as well as direct filtration are improved in several respects. The cost of the polymers will mostly be recovered by reduced costs for primary coagulation chemicals, reduced backwashing cycles, increased capacity without adding mechanical equipment like sedimentation tanks or enlarging filter area, and, finally, by an improved water quality.

J.M. Reuter, A. Landscheidt
Chemische Fabrik Stockhausen GmbH.
P. O. Box 570
4150 Krefeld 1
Fed. Rep. of Germany.

New Coagulant Injection Process

C. Ventresque and G. Bablon

Introduction

Among the processes used in the treatment of potable water, one of the trickiest to handle is the addition of the coagulant. It is generally agreed that, in the case of iron and aluminium salts, mixing must be very energetic and completed within a very short time. This is an expensive process involving the use of a conventional stirring system (agitator and tank) and is not very efficient when dealing with large volumes.

At Neuilly-sur-Marne, a new coagulant injection system has been installed in a plant producing 800,000 m³ of water per day.

Description of the Treatment Train at the Neuilly-sur-Marne Plant

The Neuilly-sur-Marne waterworks supplies slightly under 2,000,000 inhabitants of the East suburbs of Paris (Figure 1). It belongs to the Syndicat des Eaux d'Ile de France, was designed and is managed by the Compagnie Générale des Eaux.

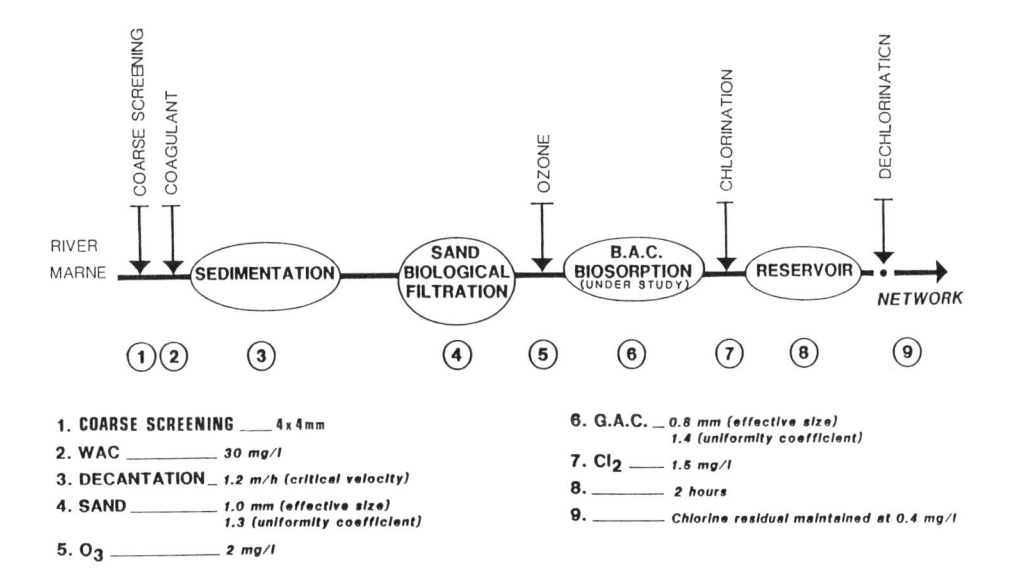

1. COARSE SCREENING ___ 4x4mm
2. WAC ___ 30 mg/l
3. DECANTATION ___ 1.2 m/h (critical velocity)
4. SAND ___ 1.0 mm (effective size)
         1.3 (uniformity coefficient)
5. O$_3$ ___ 2 mg/l
6. G.A.C. ___ 0.8 mm (effective size)
            1.4 (uniformity coefficient)
7. Cl$_2$ ___ 1.5 mg/l
8. ___ 2 hours
9. ___ Chlorine residual maintained at 0.4 mg/l

Figure 1. Description of the treatment at the Neuilly-sur-Marne plant

The raw water is drawn from the Marne river. After a coarse screening the introduction of the reagents is performed in a tank designed to avoid spurious interactions between reagents. Presently the coagulant used is aluminium polychloride, if necessary, ferrous sulfate, sodium hydroxide, carbon dioxide and powdered activated carbon which adsorbs micro-pollutants are added.

The water then flows into flocculation tanks where it is gently stirred and then proceeds into stories of settling tanks of the corridor type for at least two hours.

After settling, the water is sand filtered. Each filter is driven by a depression hood of the polhydra type. The regulation system ensures an even flow in each basin and a constant filtration velocity no matter how clogged the filter bed.

This filtration operation completes clarification and eliminates ammonia by biological nitrification. When clogged, filters are automatically backwashed with water and air scour.

An additional step, which is now under study, will allow a better removal of organic matter by means of ozonation followed by a biological activated carbon filtration.

The final treatment consists of chlorination at a rate of around 0.8 to 1.5 mg/l followed by a two hour contact time. The chlorine residual in the water is then adjusted to 0.4 mg/l with an injection of bisulfite.

With regard to coagulation, it was customary to feed the coagulant into a 140 m^3 tank, fitted with an agitator, for a stirring time of about 30 seconds (Figure 2).

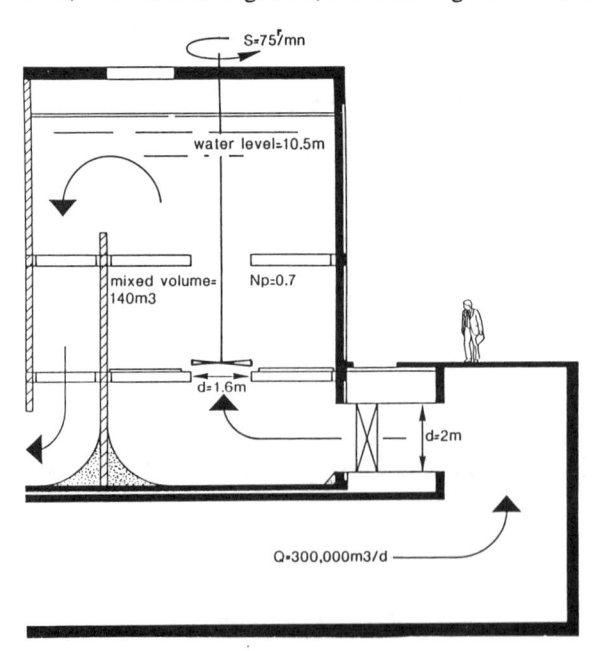

Figure 2. Injection of coagulant in pretreatment tank - hydraulic characteristics of agitator and tank

The average velocity gradient, calculated using the power number of the agitator, was about 400 s^{-1}. This coagulant mixing process turns out to be partly ineffectual, especially during the periods of algae growth reflected by an increase in the pH of the raw water. The effect of this is to accentuate the solubility of the aluminium (Figure 3). To avoid overstepping the mandatory limit of 200 µg/l, applicable in France at the present time, two solutions were examined.

- Lowering the pH by injecting acid or CO_2, but bearing in mind the high buffer capacity of the Marne water, this solution proved in practice to be very costly.
- Improving the system by which the coagulant is added in order to reduce, as far as possible, the proportion of particulates, or even dissolved aluminium at the outlet of the filters.

A study on the implementation of this second solution was initiated.

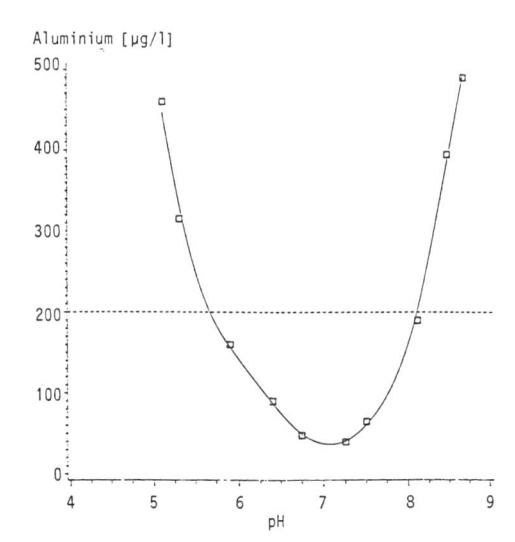

Figure 3. Solubility of aluminium in the Marne water

Equipment and Methods

Aluminium measurements were effected by means of a plasma torch, produced by the Instrumentation Laboratory 200. Ultrafiltrations were performed on 10 Amicon membranes and Nucleopore membranes of 1 µm to 10 µm. The variation factor, including filtration and dosing, is 7/100 at the outside.

The zeta potential of the water after the addition of the coagulant was measured on a Laser Zee Meter 400 made by Pen Kem Company. The measurements were taken in under 30 seconds to avoid the flocculation of destabilized particles in the measuring cell. All the zeta potential values given are the mean average of at least 5 measurements on the same sample. These measurements are subject to an error of 10% at the most. Turbidity measurements were carried out with a Hach 2100A device.

In order to highlight the parameters that enable the quality of the coagulation to be assessed in the case of water from the Marne, preliminary tests were performed on a continuously fed pilot reactor.

The characteristics of this reactor are comparable to those of the mixing system recommended by the water research center. Among other things, it enables the velocity gradient G to be varied. All the laboratory analyses were carried out on water sampled immediately after the reagent was added: measurement of the zeta potential, residual turbidity and aluminium after filtration on a 1 μm membrane. Figure 4 shows data on the water of the Marne. It shows that the dose of reagent to be used is in a sensitive zone that does not correspond to the disappearance of the zeta potential, but to its reduction to an extent sufficient to allow contact between the particles. The reduction of the surface potential reflects an improved agglomeration, shown here by a decrease in turbidity after filtration.

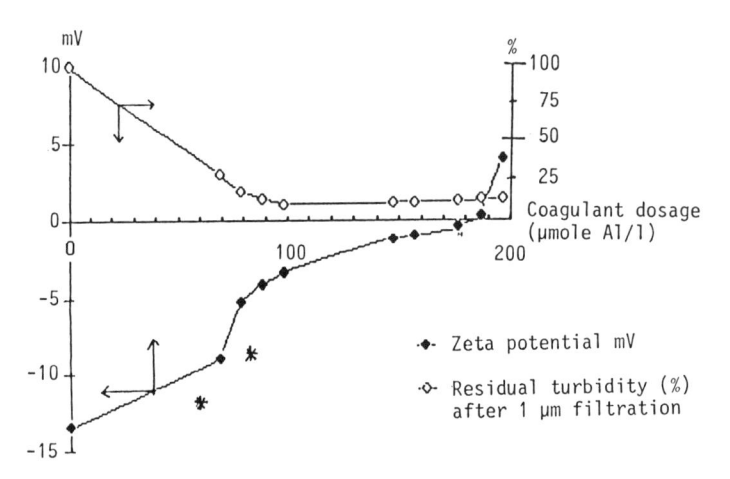

Figure 4. Development of zeta potential and of residual turbidity as a function of the coagulant dose for a velocity gradient of 500 s^{-1} (* indicates "works" point)

The influence of the velocity gradient G on the zeta potential of the coagulated water is very great in the case of Marne water. Figure 5 summarizes the curves determined in the laboratory: the increase in the velocity gradient makes it possible to lessen the potential in absolute terms, particulary insofar as the coagulant dose rate is low.

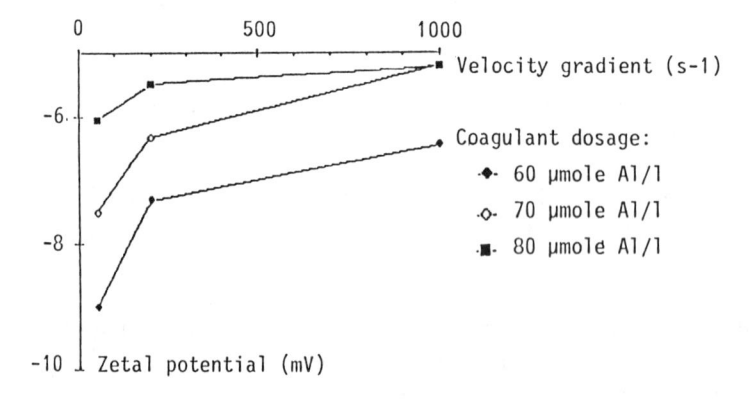

Figure 5. Zeta potential values of coagulated water in the laboratory as a function of the velocity gradient and the dose of coagulant

Stirring conditions do not only affect the quantity of agglomerated particles but also the dissolution in the effluent of the ions contained in the reagent. A part of the reagent is, of course, trapped in the ionic surface layer surrounding the suspended particles, but another part remains in a free ionic form and is then hydrolyzed. It would also appear (see Figure 6) that the "dissolved" fraction is also affected by the stirring conditions. The aluminium content after filtration decreases as the mixing energy increases.

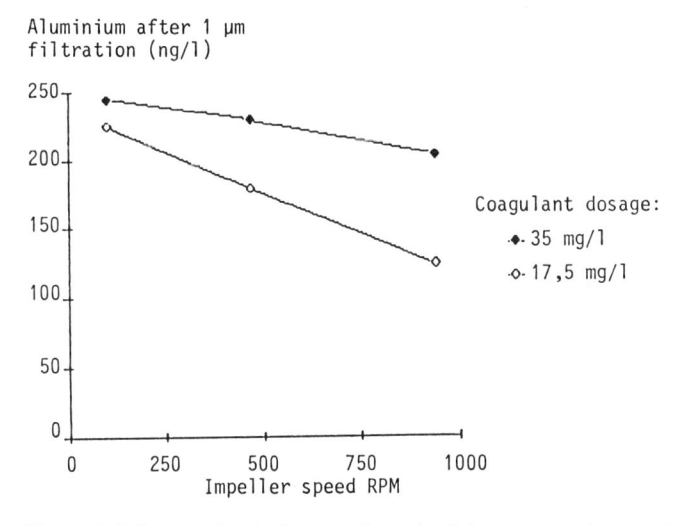

Figure 6. Influence of agitation speed on aluminium content in coagulated water after filtration on a 1 μm membrane

The zeta potential measurements marked with a star in Figure 4 were obtained at the outlet of the old agitation tank. The fact that the potential is higher in absolute terms than the values reached in laboratory testing can be explained by the inadequate stirring conditions in the full-scale reactor. A great variation in zeta potential values was also noted depending on the sampling location ($\pm$ 30%), which proves that the whole mass of water is not treated with equal efficiency when coagulant is added by means of a stirred system.

Description of the New Injection Process

Only the solution recommended by Chao and Stone (1971) could enable a degree of homogeneity to be achieved in a sufficiently short time to favour the phenomenon in which ions are fixed in the surface layer around the particles, instead of being used in the process of hydrolysis.

The coagulant feed installed in the plant includes three independent trains connected to a gravity circuit. The reagent flow rate is adjusted by low-pressure regulating valves. The flow diagram is presented in Figure 7.

The raw water flow rate can vary between 175,000 and 300,000 m^3/day. The coagulant injection chamber is located 15.4 meters above the axis of the raw water

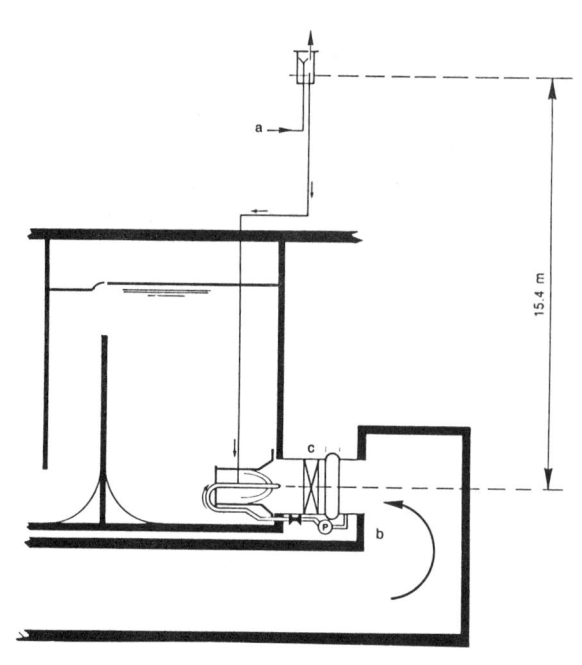

Figure 7. Coagulant feed layout:
a) coagulant input
b) raw water inlet chamber
p) reagent step-up pump
c) feed valve (dia. 2 metres)

main. The system recommended by Chao and Stone consists of an injection tube located on the raw water main and is equipped with injection ports. The flow of reagent is initiated by a recirculation pump fed with untreated water. The diameter of the main at the inlet to the tank is 2 m, and the length available for mixing, only 95 cm, is insufficient. We therefore made an extension with a diameter of 1.2 m. This size results from a compromise between the loss of available head and the effective mixing length. We calculated the diameter of the injection ports so as to observe a ratio $M = 7$ between the velocities. This ratio is defined as the injection rate to raw water flow velocity in the main.

As the raw water flow rate varies, the only solution for keeping M constant is to adapt a system regulating the discharge of the recirculation pump. We first wanted to ascertain whether or not it was possible to operate at a constant rate. Figure 8 shows the variations of M, when the coagulant dosing pump is working at 300 m³/h, as a function of the flow rate of the raw water to be treated. The mixing ratio can descend to about 5. But is this enough?

In order to settle this point, the mixing conditions using the type of injector fitted to our plant were compared for several values of M, with contact by stirring. The results are reported in Table 1.

Since the plant can be divided into distinct trains, the appearance of the floc and the interstitial water could be judged easily by sight and the two trains compared at the level of the flocculated water.

If we take as the lowest limit the coagulated water performance level obtained with the agitator, we find we must go down to a value of $M = 4$. The zeta potential becomes very variable at the outlet of the tank, indicating a poor level of homogeneity. The flocs are small and, above all, the interstitial water is cloudy. Because of these results, we decided to carry out the injection at a fixed dilution

Mixing ratio M

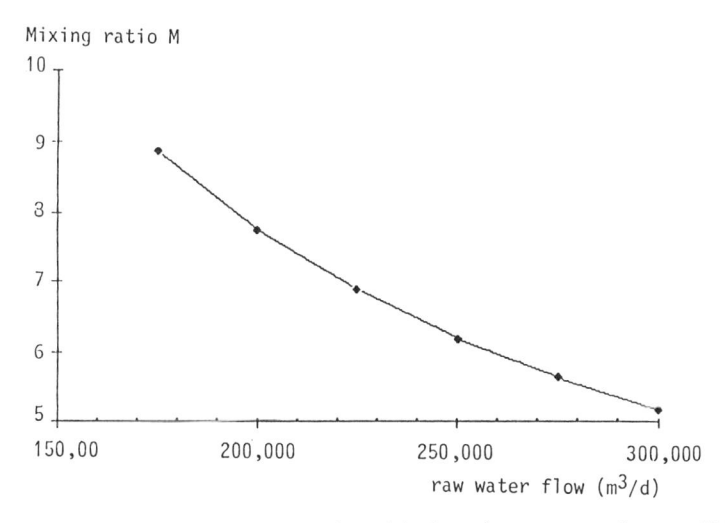

Figure 8. M ratio figures as a function of the incoming raw water flow rate if the discharge rate of the recirculation pumps is maintained at 300 m^3/h (M = reagent injection speed/raw water input velocity)

Table 1. Influence of M ratio on the flocs characteristics

	Injector	Agitator
Value of M	8	
Zeta potential	-3.9 mV	-7.6 mV
Appearance of floc	Large and dense	Small
Appearance of interstitial water	Clear	Cloudy
Value of M	6	
Zeta potential	-6.5 mV	-8.3 mV
Appearance of floc	Large and dense	Small
Appearance of interstitial water	Cloudy	Cloudy
Value of M	3.5	
Zeta potential	-6.2 to -7 mV	-8.1 mV
Appearance of floc	Small	Small
Appearance of interstitial water	Cloudy	Cloudy

flow rate which enabled us to cover a range between 175,000 and 300,000 m^3/d with M oscillating between 5 and 9. The problem of injecting the coagulant without having to change the present gravity feed system still remained, however. Details of this are shown in Figure 9. The coagulant is pumped up by means of a venturi placed directly upstream of the injection ports. The venturi tube must be designed to fit the recirculation pump, so as to ensure suction of the coagulant. The nominal rating of the dilution pump is 37 kW, which corresponds to an increase of 60% compared with the agitator.

Figure 9. Detail of coagulant reagent input

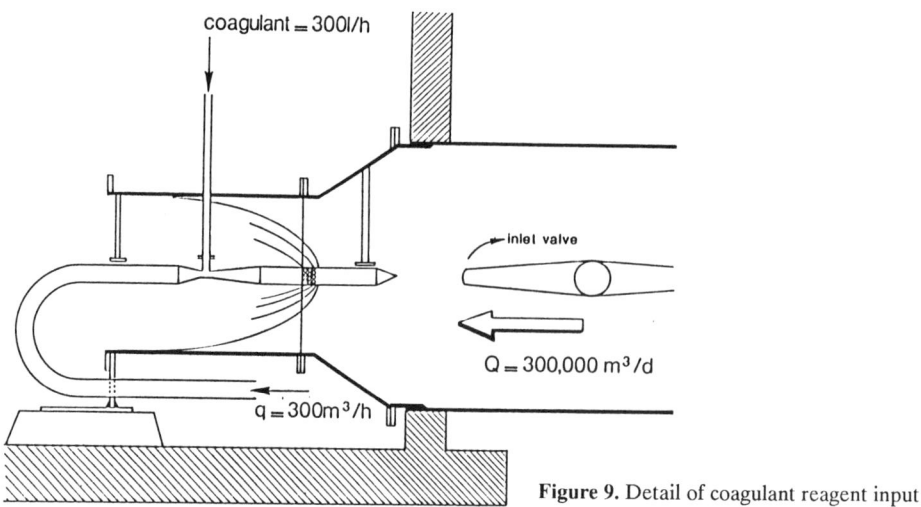

Comparison of the Two Types of Systems

According to the results reported in Table 1, the destabilization of the particles is greater with the injector, and the difference in the zeta potential works out to an average of about 2 mV throughout the year. As the plant can be divided into several parallel trains, we were able to make a very thorough comparison at the level of the sand filters. Two pilot sand filters having an area of 2 m² each were connected in gravity flow conditions to the two settled water channels in the plant. These pilots are fitted with sampling and pressure intakes, so that the profiles can be traced and their development monitored as a function of time. The filtering rates are adjusted to the maximum in the waterworks, i.e., 6 m/h with sudden switches to 7.4 m/h depending on electricity price scales. These changes in velocity cause the clogging front in the filters to progress in jumps rather than steadily.

Ultrafiltrations showed that the aluminium had two origins and this is confirmed at the level of the filters: the aluminium profiles in the filtering media run parallel to the rate of turbidity (Figure 10).

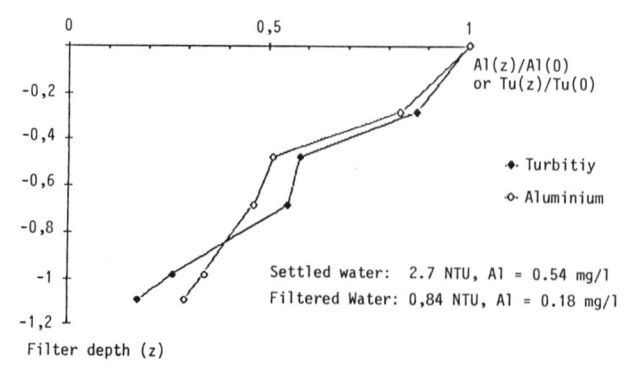

Figure 10. Removal of aluminium and turbidity in sand filters. Profiles fixed as the value in the influent as a function of depth

New Coagulant Injection Process 111

Figures 11 and 12 illustrate the turbidity and aluminium profiles in the two pilot filters. The water produced by the injector is of better quality, as is reflected by the concentrations of turbidity and aluminium. Consequently, the filtered water is also of superior quality.

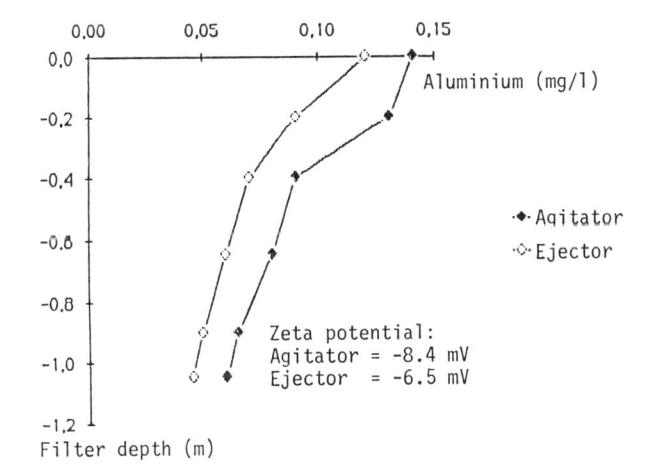

Figure 11. Total aluminium profiles in filters in terms of depth: comparison between the two mixing systems

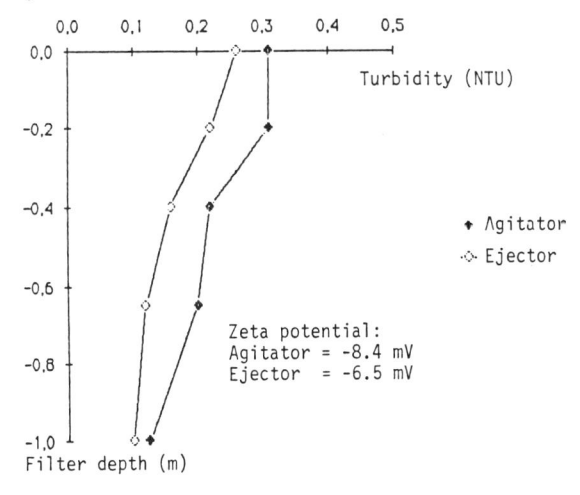

Figure 12. Turbidity profiles in filters in terms of depth: comparison between the two mixing systems

Conclusion

It has been proven that the conditions in which the coagulant is mixed with the water to be treated constitute an important factor in determining the efficiency of the separation treatments located downstream, and this is applicable down to the level of the filters.

If the volume of water to be treated daily is very large, mixing systems using agitators are not very effective.

The injection system can be adapted to large scale units by using a dilution of the coagulant, which allows treatment to be ensured in large diameter pipes. The mixing of undiluted coagulant with injection water must be done in less than 1/10th of a second before treatment, in order to avoid hydrolysis of the coagulant before it comes into contact with the water to be treated.

Moreover, the system presented here has a distinctive feature in that the reagent circulates by gravity flow without having to be put under pressure. The results obtained with such a system versus a conventional mixing system are appreciable at the level of filtration. The filters provide better removal of the suspended solids and aluminium, filtration runs are longer and so allow considerable saving on filter washing.

References

[1] Cornet, J.C. (1981): Détermination des Gradients Hydrauliques dans les Différentes Phases du Traîtement des Eaux. La Technique de l'Eau et de l'Assainissement *418*, 21 — 23
[2] Chao and Stone (1979): Initial Mixing by Jet Injection Blending. JAWWA *71*, 570 — 573

C. Ventresque, G. Bablon
Compagnie Générale des Eaux
52, rue d'Anjou
75008 Paris
France

Odour Control by Artificial Groundwater Recharge

R. Sävenhed, B.V. Lundgren, H. Borén and A. Grimvall

Abstract

The removal of odorous compounds during artificial groundwater recharge in sand and gravel ridges ("eskers") has been evaluated by gas chromatography with both instrumental and sensory detection ("chromatographic sniffing"). Seven Swedish water works were included in the study. It was shown that the off-flavour compounds in the raw water samples (geosmin, 2-methylisoborneol, 2-isopropyl-3-methoxypyrazine, 2,4,6-trichloroanisole, 1-octen-3-one, dimethyl trisulphide, and a number of unidentified muddy or musty odours) were all effectively removed during infiltration, thus proving that artificial groundwater recharge is not only a suitable method for water storage but can also be an effective method for removing odorous compounds.

Laboratory experiments with filtration of raw water through biologically active sand filters showed that a good oxygen supply is the key to both effective removal of naturally occurring off-flavour compounds and to prolonged running times for the filters.

Comparisons were made to alum coagulation/sand filtration and it was concluded that biological treatment methods often are superior to physical-chemical methods for effective odour control.

Introduction

Unpleasant tastes and odours ("off-flavours") in drinking water are most often caused by low-molecular weight organic compounds in very low concentrations (e.g. [17, 20, 28]). These trace organics can be present in the raw water, they can be formed during treatment, or they can be introduced into the water during distribution. Odorous compounds of biogenic origin, present in the raw water and not satisfactorily removed during treatment, are of great significance and have received world-wide attention [10, 16, 18, 21, 25, 26, 28]. Such odours are often described as earthy, muddy or musty; geosmin and 2-methylisoborneol are two well-known compounds from this group.

There are several reviews dealing with off-flavour abatement during water treatment [8, 9, 11, 14, 17]. A single treatment step is seldom sufficient for producing drinking water with acceptable sensory properties. Combinations of different techniques are often neeeded, and ozonation and water with acceptable sensory properties. Combinations of different techniques are often needed, and ozonation and activated carbon filtration are considered to be important links in such treatment schemes. Coagulation/sand filtration is generally rather inefficient for the removal of off-flavours. Artificial groundwater recharge, often used as a pretreatment step or as a method of water storage, has only received limited attention as a method for off-flavour control.

The present paper summarizes the results from investigations the goal of which was the removal of odorous compounds during passage of raw water through biologically active sand filters. The effects during artificial groundwater recharge in natural sand and gravel ridges (eskers) as well as during laboratory-scale sand filtration experiments were evaluated [13, 22]. The treatment evaluation was mainly based on gas chromatographic analysis with both instrumental and sensory detection ("chromatographic sniffing") of individual odorous compounds [21].

Combined Chemical and Sensory Methods for the Analysis of Odorous Organic Compounds in Water

The chemical and sensory methods which are available for studying off-flavours in water can be divided into three groups:

1) Sensory analysis of whole water samples.
2) Chemical analysis of selected compounds which are known or suspected to be of relevance for the water off-flavour.
3) Combinations of chemical and sensory methods for a systematic search for all important off-flavour compounds in the water samples.

Most of the present knowledge regarding off-flavours in drinking water is based on results obtained by using methods from the first two groups. For several reasons (e.g. a lack of knowledge regarding the chemical identity of the compounds of interest), methods from the second group have been restricted to only a few compounds, e.g. geosmin and 2-methylisoborneol [15, 25, 26]. The methods of the third group can be divided into two subgroups representing different strategies for the coupling of chemical data and sensory properties:

3a) Statistical analysis of relationships between chemical data and sensory properties of whole water samples [2, 24].
3b) Sensory analysis of individual compounds after enrichment and gas chromatographic (GC) separation, so-called chromatographic sniffing [20, 21].

Chromatographic sniffing implies that the GC effluent, by using some kind of "sniffing device", is directed to a funnel outside the chromatograph. The retention time, the perceived intensity and the perceived quality of the odorous compounds eluted from the GC column are reported by an odour observer. This technique has a fairly long tradition in various fields of aroma research [1, 6, 7], but its use for studying off-flavour compounds in drinking water has been rather limited. It has been used in the analysis of raw and drinking water in the Netherlands, first with packed columns [27] and later with glass capillary columns [19]. We have, in a systematic search for odorous compounds in drinking water, combined efficient enrichment techniques with gas chromatographic sniffing employing flexible fused-silica capillary columns [12, 21].

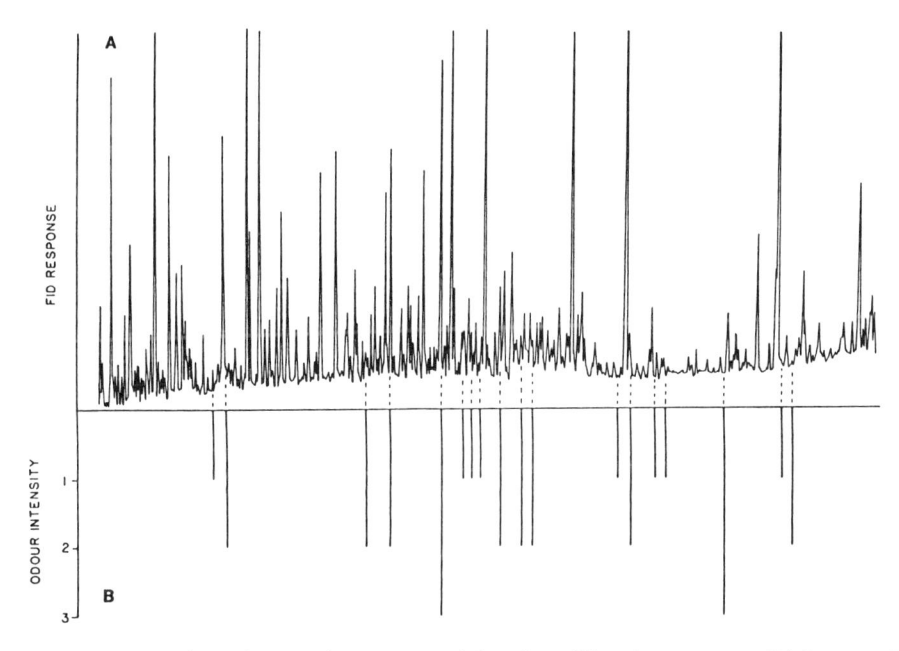

Figure 1. Comparison of a FID chromatogram (A) and a sniffing chromatogram (B) from a stripping extract of drinking water. Odour intensities: 1 = weak, 2 = medium, 3 = strong. A sample of 1 l from a water works utilizing alum coagulation/sand filtration was stripped at 60°C for 2 h

An obvious reason for utilizing chromatographic sniffing in the analysis of odorous compounds is the fact that the olfactory sense organ is a selective and very sensitive detector which differs significantly from the instrumental GC detectors. This is illustrated by Figure 1 where one ordinary FID (flame ionization detection) chromatogram and one "sniffing chromatogram" from the same stripping extract of drinking water are shown. The upper chromatogram is fairly complex and contains approx. 200 integrated peaks while the lower "chromatogram" contains only 18 "peaks" (odour perceptions). Furthermore, in the FID chromatogram, all perceived odours correspond to small peaks or have no corresponding peaks at all. Such observations are, in fact, not uncommon in aroma and off-flavour research (e.g. [3, 19, 23]).

Chromatographic sniffing is a very important tool for the chemical identification of odorous compounds [20]. Furthermore, by using this technique it is possible to determine the origin of such compounds, even if the chemical structures are unknown. In the present study, chromatographic sniffing has been used for studying the "fate" of odorous compounds during water treatment.

The chromatographic separation must, of course, be preceded by an enrichment step. Due to the low concentrations of the odorous compounds, the enrichment technique must have a high concentration factor and a low blank level. Furthermore, it is of great importance to evaluate the efficiency and relevance of the enrichment step by sensory analyses. Our studies have shown that the stripping technique using an "open system" [4, 5] at an elevated water temperature (60°C) normally gives a satisfactory enrichment of odorous substances in raw and drinking water samples. Sometimes a more efficient method such as liquid-liquid extraction using dichloromethane might be needed [12]. The results presented in this paper are mainly based on analyses using the stripping technique.

Artificial Groundwater Recharge in Sand and Gravel Ridges

In Sweden, approx. 50% of drinking water production is based on conventional surface water treatment, most often coagulation and sand filtration; about 25% of the production is based on artificial groundwater recharge (percolation of surface water through sand and gravel ridges, "eskers"); and approx. 25% of the production is based directly on groundwater. We have studied the off-flavour abatement effects during artifical groundwater recharge and, for means of comparison, during conventional surface water treatment. The following water samples have been analysed: (1) raw water (i.e. surface water) and infiltrated water from seven water works utilizing artificial groundwater recharge as the main treatment method; (2) raw water and treated water from one water works utilizing alum coagulation, rapid and slow sand filtration and chlorination; and (3) treated water obtained by laboratory-scale alum coagulation of raw water samples from three of the infiltration water works [22].

The percolation depths in the eskers were in the range of 5 – 20 m, and the residence times varied between one week and several months. In order to avoid clogging, the infiltration basins were drained regularly, with subsequent removal of the upper layer of the sand. A dry period of about one month was usually followed by a wet period of about five months. The surface water at some of the infiltration water works was pretreated by rapid sand filtration.

The raw waters for the eight water works (representing six different river basins in central and southern Sweden) were rivers and lakes with odours described as musty, muddy or earthy. The chemical oxygen demand (COD_{Mn}) ranged from 6 to 15 mg/l, and water production varied between 2 and 60 million m^3/yr.

A comparison of the two treatment techniques based on the COD values did not reveal any significant differences at all. At the water works using artificial groundwater recharge, the COD value was, on the average, reduced from 10 to 3 mg/l. At the water works using alum coagulation/sand filtration, the COD value decreased from 9 to 3 mg/l. It should be noted, however, that these results do not give any information at all about the removal of low-molecular weight organic compounds such as odorous substances. The organic content of natural water samples is totally dominated by humic acids and other high-molecular weight compounds.

The comparison of the treatment techniques showed that for the removal of off-flavour compounds, however, artificial groundwater recharge is superior to alum coagulation/sand filtration. This is illustrated by the chromatographic sniffing data summarized in Figure 2. Alum coagulation did not significantly affect the average odour intensities of the detected compounds (Fig. 2A). A small reduction was noticed during full-scale coagulation/sand filtration (Fig. 2B). Chemical analysis of some selected off-flavour compounds showed that this effect could be attributed to

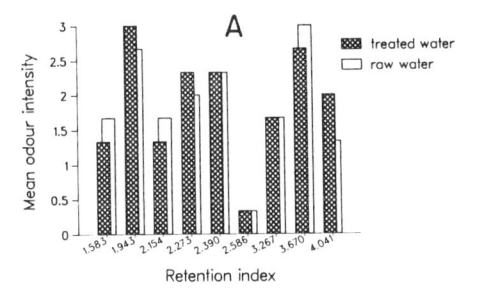

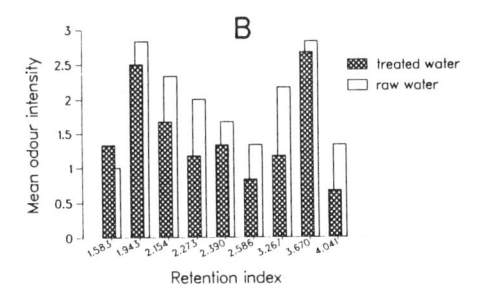

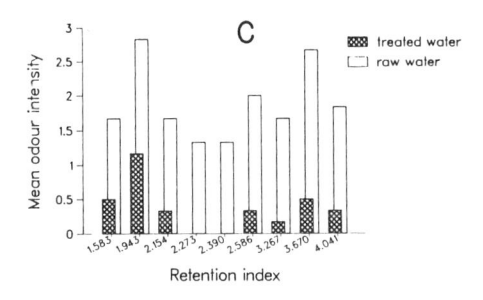

Figure 2. Effect of water treatment on the mean odour intensity of nine selected compounds. Odour intensity scale used in chromatographic sniffing: 1 = weak, 2 = medium, 3 = strong. Retention index based on C_6, C_8, C_{10}, C_{12} and C_{14} 1-chloralkanes.
Identified compounds: 1.583 = 1-octen-3-one, 2.154 = 2-isopropyl-3-methoxypyrazine, 2.586 = 2-methylisoborneol, 3.267 = 2,4,6-trichloroanisole, 3.670 = geosmin. A = Laboratory-scale alum coagulation (three different raw waters). B = Full-scale coagulation/sand filtration/chlorination (one water works, six pair of samples). C = Artificial groundwater recharge (six different water works)

the slow sand filtration step. Artificial groundwater recharge was found to be a very effective method, and very low odour intensities were observed after infiltration (Fig. 2C). The infiltrated waters were, in fact, usually almost odourless. The fundamental difference between conventional surface water treatment and artificial groundwater recharge in eskers is further illustrated by the chromatograms in Figure 3.

Chemical analysis showed that the concentrations of some selected odorous compounds (geosmin and 2-methylisoborneol) were not significantly altered by coagulation (this was found in both laboratory and full-scale treatment). Artificial groundwater recharge, on the other hand, had a dramatic effect on the concentrations of these substances. It was impossible instrumentally to detect geosmin and 2-methylisoborneol in any of the samples of infiltrated water, despite the fact that the average concentrations of these compounds in the raw water samples exceeded the detection limit by at least a factor of 50.

Laboratory-Scale Experiments with Biologically Active Sand Filters

The extensive removal of low molecular weight organic compounds during artificial groundwater recharge can mainly be explained by the fact that it is a treatment

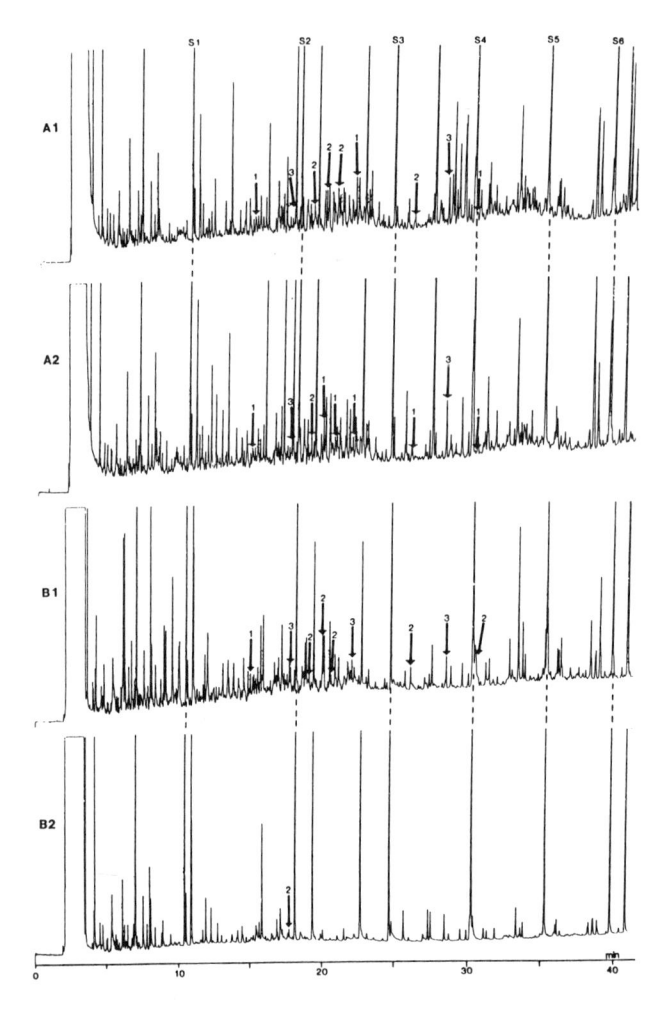

Figure 3. Chromatograms of water samples with odorous compounds marked by arrows. The figures at the arrows are the perceived odour intensities in chromatographic sniffing (see Figure 2). S1S6 are internal standards (1-chloralkanes) added to the water samples at a concentration of 50 ng/l. A1 = raw water and A2 = treated water from a water works utilizing alum coagulation/sand filtration/chlorination. B1 = raw water and B2 = infiltrated water from a water works utilizing artificial groundwater recherge

method based on biological processes. The activity of the sand filters, causing degradation of the organic material, would therefore be dependent on external factors such as temperature and oxygen status.

To study the influence of aeration on the removal of off-flavour compounds during artificial groundwater recharge, laboratory-scale filtration experiments were performed with sand columns running under full or reduced aeration. In one experiment, two identical columns (glass cylinders with a diameter of 10 cm packed with 50 cm of sand) were fed with surface water spiked with five well-known odorous compounds (geosmin, 2-methylisoborneol, 1-octen-3-one, 2,4,6-trichloro-anisole and dimethyl trisulphide) at a concentration of 50 ng/l for each compound. The threshold odour number (TON) of this water was approx. 500. During a period of six months the filters were run under full aeration. Thereafter, one of the filters was put under nitrogen atmosphere and fed with deaerated water. Within a few days a noticable difference in the performance of the two filters could be observed. The effluent water from the aerated filter still had a very weak odour (TON = 2), while the water from the disturbed filter gradually developed a muddy and earthy flavour. When tested after three months, the oxygen concentration in the effluent from this filter had decreased to about 15% of its original value, and the TON exceeded 200. Gas chromatographic analysis (FID) confirmed the results of the sensory evaluation. The removal of different model compounds for the two filters with different oxygen status is summarized in Table 1.

Table 1. Removal of selected off-flavour compounds during passage of spiked surface water through sand columns with different oxygen status. Initial concentration 50 ng/l for each compound. Chemical analysis by stripping enrichment and capillary gas chromatography

Compound	Removal full aeration %	Removal reduced aeration %
Geosmin	> 95	62
2-Methylisoborneol	> 95	55
2, 4, 6-Trichloroanisole	> 95	77
1-Octen-3-one	> 95	> 95
Dimethyl trisulphide	> 95	> 95

In another experiment without addition of model compounds to the influent raw water similar results were achieved. The muddy and earthy odour of the raw water (TON = 64) was almost completely removed in the fully aerated filter (TON = 2), but was easily discernible in the water from the filter with reduced aeration (TON = 16). Results from chromatographic sniffing indicated (1) that practically all off-flavour compounds in the raw water were either equally, or more effectively, eliminated under aerobic than under anaerobic conditions (see Fig. 4) and (2) that no new, important, off-flavour compounds were produced in any of the sand filters.

The high molecular weight organic compounds were not effectively removed by any of the sand filters. The two types of filtrated water as well as the untreated water had a TOC-value of approx. 9 mg/l. The maximum running time of the sand filters, however, proved to be strongly dependent on their oxygen status. In the filter that was run at reduced aeration clogging caused a 50% reduction of the water flow within six weeks. After increasing the oxygen supply again, organic substances from the filter were temporarily released, and clogging gradually decreased. In the fully aerated filter, no clogging or change in the removal of off-flavours could be observed in the course of 12 months.

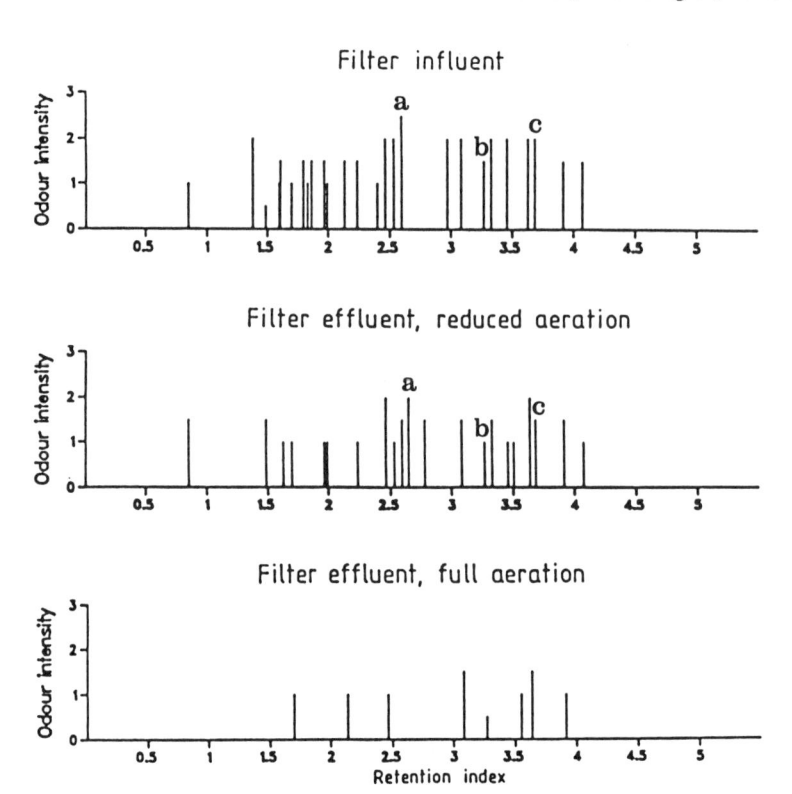

Figure 4. Sniffing chromatograms from surface water before and after treatment in biologically active sand filters. Water flow rate 25 mm/h. Odour intensities and retention index as in Figure 2. Enrichment by dichlormethane extraction, concentration factor 5,000. a = 2-methylisoborneol, b = 2,4,6-trichloroanisole, c = geosmin

Conclusion

During conventional surface water treatment (coagulation/sand filtration/chlorination) the content of high-molecular weight organic compounds (i.e. the colour or TOC-value) is normally reduced to an acceptable level. The treatment effect on low-molecular weight organic substances, including odorous compounds, is, however, not satisfactory. The two most interesting methods for an efficient removal of such compounds are activated carbon adsorption and filtration through biologically active sand filters, including artificial groundwater recharge. This paper has dealt with the latter method.

Our studies have clearly shown that artificial groundwater recharge in eskers is a very effective method for the removal of organic compounds with muddy and musty odours. The concentrations of these compounds were reduced to very low levels, often resulting in an almost odourless water after infiltration. Laboratory-scale experiments with filtration of raw water through sand filters (utilizing the same flow rate as during full-scale artificial recharge) have shown that a good oxygen supply is the key to both effective removal of odorous substances and to long running times for the filters. The excellent effects on the removal of many low-molecular weight organic compounds obtained with the well-aerated sand filters demonstrate the potential of treatment methods based on biological activity. Consequently, artificial groundwater recharge and slow sand filtration deserve increased attention in drinking water treatment.

Acknowledgement

The authors are grateful to Peter Balmér for valuable comments.

References

[1] Acree, T.E., Barnard, J., Cunningham, D.G. (1984): The Analysis of Odor-active Volatiles in Gas Chromatographic Effluents. In: Schreier, P. (ed.) Analysis of Volatiles, pp. 251 - 267. de Gruyter, Berlin

[2] Anselme, C., Duguet, J.P., Mallevialle, J., Suffet, I.H. (1987): Removal of Tastes and Odours by Ozonation. In: Proceedings from the 8th Ozone World Congress, pp. C102 - C127. Unionsverlag, Zürich

[3] Bemelmans, J.M.H., den Braber, H.J.A. (1983): Investigations of an Iodine-like Taste in Herring from the Baltic Sea. Water Sci. Technol. *15*(6/7), 105 - 113

[4] Borén, H., Grimvall, A., Palmborg, J., Sävenhed, R., Wigilius, B. (1985): Optimization of the Open Stripping System for the Analysis of Trace Organics in Water. J. Chromatogr. *348*, 67 - 78

[5] Borén, H., Grimvall, A., Sävenhed, R. (1982): Modified Stripping Technique for the Analysis of Trace Organics in Water. J. Chromatogr. *252*, 139 - 146

[6] Fuller, G.H., Steltenkamp, R., Tisserand, G.A. (1964): The Gas Chromatograph with Human Sensor: Perfumer Model. Ann. NY Acad. Sci. *116*, 711 - 724

[7] Gemert, L.J. van (1981): Coordination of Sensory and Instrumental Analysis. In: Maarse, H., Belz, R. (eds.), Isolation, Separation and Identification of Volatile Compounds in Aroma Research, pp. 240 - 258, Akademie, Berlin

[8] Heusden, G.P.H. van (1973): Progress in Removal of Taste and Odour. 9th Int. Water Supply Ass. Congr., pp. A1 - A5

[9] Hrubec, J., de Kruijf, H.A.M. (1983): Treatment Methods for the Removal of Off-flavours from Heavily Polluted River Water in the Netherlands - A Review. Water Sci. Technol. *15*(6/7), 301 - 310

[10] Jüttner, F. (1987): Physiology and Biochemistry of Biogenic Off-flavour Compounds. Second International Symposium on Off-flavours in the Aquatic Environment, Kagoshima, Japan, Oct 12 - 16, 1987

[11] Lin, S.D. (1977): Tastes and Odours in Water Supplies: A Review. Wat. Sewage Wks - 1977 Reference Issue, R141 - R163

[12] Lundgren, B.V., Borén, H., Grimvall, A., Sävenhed, R., Wigilius, B. (1987): Efficiency and Relevance of Different Concentration Methods for the Analysis of Off-flavours in Water. Second International Symposium on Off-flavours in the Aquatic Environment, Kagoshima, Japan, Oct 12 - 16, 1987

[13] Lundgren, B.V., Grimvall, A., Sävenhed, R. (1987): Formation and Removal of Off-flavour Compounds During Ozonation and Filtration Through Biologically Active Sand Filters. Second International Symposium on Off-flavours in the Aquatic Environment, Kagoshima, Japan, Oct 12 - 16, 1987

[14] McGuire, M.J., Gaston, J.M. (1987): Overview of Technology for Controlling Off-flavours in Drinking Water. Second International Symposium on Off-flavours in the Aquatic Environment, Kagoshima, Japan, Oct 12 - 16, 1987
[15] McGuire, M.J., Krasner, S.W., Hwang, C.J., Izaguirre, G. (1981): Closed-loop Stripping Analysis as a Tool for Solving Taste and Odor Problems. J. Am. Water Works Ass. 73, 530 - 537
[16] McGuire, M.J., Krasner, S.W., Hwang, C.J., Izaguirre, G. (1983): An Early Warning System for Detecting Earthy-musty Odors in Reservoirs. Water Sci. Technol. 15(6/7), 267 - 277
[17] Montiel, A.J. (1983): Municipal Drinking Water Treatment Procedures for Taste and Odour Abatement - A Review. Water Sci. Technol. 15(6/7), 279 - 289
[18] Persson P.-E. (1983): Off-flavours in Aquatic Ecosystems - An Introduction. Water Sci. Technol. 15(6/7), 1 - 11
[19] Piet, G.J., Zoeteman, B.C.J., Morra, C.F.H., Nettenbreijer, A.H., De Grunt, F.E. (1978): Organische Verunreinigungen und schlechter Geruch von Trinkwasser. In: Aurand, K. (ed.) Organische Verunreinigungen in der Umwelt, pp. 195 - 204. Erich Schmidt, Berlin
[20] Sävenhed, R. (1986): Chemical and Sensory Analysis of Off-flavour Compounds in Drinking Water. Ph.D. thesis, Linköping Studies in Arts and Science No. 3, University of Linköping, Sweden
[21] Sävenhed, R., Borén, H., Grimvall, A. (1985): Stripping Analysis and Chromatographic Sniffing for the Source Identification of Odorous Compounds in Drinking Water. J. Chromatogr. 328, 219 - 231
[22] Sävenhed, R., Borén, H., Grimvall, A., Lundgren, B.V., Balmér, P., Hedberg, T. (1987): Removal of Individual Off-flavour Compounds in Water During Artificial Groundwater Recharge and During Treatment by Alum Coagulation/Sand Filtration. Wat. Res. 21, 277 - 283
[23] Sevenants, M.R., Sanders, R.A. (1984): Anatomy of an Off-flavour Investigation: The "Medicinal" Cake Mix. Anal. Chem. 56, 293A - 298A
[24] Suffet, I.H., Maloney, S.W., Brock, G.L., Yohe, T.L. (1984): Instituting the Flavor Profile Method hand broad Spectrum Chemical Analysis to Understand Taste and Odor Problems. In: Proc. AWWA Water Quality Technology Conference, Appendix A. American Water Works Association, Denver, Colorado
[25] Vik, E.A., Storhaug, R., Naes, H., Utkilen, H.C. (1987): Pilot Scale Studies for Geosmin and 2-methylisoborneol Removal. Second International Conference on Off-flavours in the Aquatic Environment, Kagoshima, Japan, Oct 12 - 16, 1987
[26] Yagi, M., Kajino, M., Matsuo, U., Ashitani, K., Kita, T. Nakamura, T. (1983): Odor Problems in Lake Biwa. Water Sci. Technol. 15(6/7), 311 - 321
[27] Zoeteman, B.C.J., Piet, G.J. (1972/1973): On the Nature of Odours in Drinking Water Resources of the Netherlands. Sci. Total Environ. 1, 399 - 410
[28] Zoeteman, B.C.J., Piet, G.J., Morra, C.F.H. (1978): Sensorily Perceptible Organic Pollutants in Drinking Water. In: Hutzinger, O., Vanlelyveld, I.H., Zoeteman, B.C.J. (eds.) Aquatic Pollutants: Transformation and Biological Effects, pp. 359 - 368. Pergamon Press, Oxford

R. Sävenhed, B.V. Lundgren,
H. Borén, A. Grimvall
Department of Water
in Environment and Society
Linköping University
581 83 Linköping
Sweden

Section II

Industrial Discharges

Pretreatment of Industrial Wastewater: Legal and Planning Aspects - A Case Study

H.H. Hahn and K.-H Hartmann

Abstract

Pretreatment of industrial wastewater streams is not only necessitated by the requirements of sewer maintenance or by arguments of insufficient reduction of waste components in the course of biological treatment or recommended because of favorable economics. In many instances it has become mandatory for safe planning, design and operation of central municipal wastewater treatment. While there are many examples of favorable treatment through the combination of domestic and industrial wastestreams, the difficulties of changes in the spectrum of industries or in industrial production and the concomitant change in wastewater production may be even larger.

There are new administrative regulations in the FRG that permit the water authorities to demand in specific situations industrial pretreatment to a degree that the subsequent treatment of such industrial streams in the central municipal plant may no longer be a problem in terms of waste load or treatability. The purpose of this contribution is to describe these new regulations and to illustrate their effect by discussing one example of reorientation in combined industrial domestic wastewater treatment.

The first period of joint efforts between industry and municipality was characterized by combined treatment decided on the basis of the magnitude of the industrial stream and the treatability of these wastes. The second phase of this joint venture was overshadowed by a change in quantity and quality of the industrial wastestreams leading to serious difficulties in the operation of the wastewater treatment plant as well as sludge handling and disposal. The third phase, coinciding with a change in the pertinent regulations from the water authorities, witnesses a separation of both wastewater components in order to allow far-reaching pretreatment of the industrial streams. Extensive pilot plant work shows the technical and economic aspects and problems of joint central treatment versus decentralized pretreatment of industrial streams and subsequent handling in the central municipal plant. Arguments of safety in sewer operation as well as protection of treatment plant and receiving waters support the decision for pretreatment in a qualitative way.

Introduction

Pretreatment in this context is understood as the technical reduction of specific wastewater constituents as closely to their source as possible. Such pretreatment can include physical, physiochemical and chemical unit processes. Examples of such first steps of treatment are neutralization prior to discharge into the sewer and oil or grease removal units, both unit processes that are widely distributed. However, biological processes of pretreatment are rare or should even be avoided. It is characteristic for biological unit processes to reduce non-specifically (nearly all) wastewater constituents with high average efficiency and low adaptability from well-balanced wastewaters containing the necessary mixture of nutrients. Furthermore, such biological treatment is characterized by significant economy concerning scale. This means that only larger units are economically competitive. Thus, in most instances there will be no biological pretreatment.

Pretreatment, as defined above, is required by administrative regulations. Those responsible for the operation and maintenance of sewer systems permit the discharge into such systems only on the basis of discharge regulations. These regulations contain standards for an ever growing number of wastewater constituents. Table 1 shows a comparison of such discharge regulations in the Federal Republic of Germany for different dates. In particular the number of parameters regulated and the actual limit given for each parameter indicate a significant development. The immediate goal of such administrative ruling is the *protection of the sewer* lines themselves. An example illustrating the need for protective regulation is the possible discharge of highly explosive automobile gasoline with garage wastewaters; upon ignition such waste systems might destroy the whole sewer structure.

Second, pretreatment is necessary in many instances for the *protection of the wastewater treatment plant* against substances that may harm the treatment process. Wastewater systems with too acidic or too caustic a composition will not only damage the transport lines - and incidentally the treatment works - but also negatively affect biological processes in the plant. Another example for the need for regulating discharges into sewer systems is the deleterious effect of heavy metals on all types of biologically mediated reactions. Such negative consequences may be observed in trickling filters or activated sludge systems, in particular for higher concentration of heavy metals. But even relatively low concentrations of such toxic substances can cause the breakdown of the anaerobic sludge stabilization due to accumulation phenomena in bacterial systems leading to a rather high heavy metals content in treatment plant sludges.

Thirdly, and most important, discharge regulations must prevent the release of material into the sewer network that cannot be removed in the process of (central) wastewater treatment and thus could escape into *the environment*. These substances are banned completely from use in industrial processes or they are controlled by efficient pretreatment. To illustrate this aspect the whole group of halogenated hydrocarbons is to be mentioned. They may or may not disturb the biological treatment process, depending upon their concentration. They are removed to some degree in this biological stage by degradation incorporation and accumultation within the microbial system. But at the same time significant fractions of such

Table 1. Regulations for wastewater discharge into sewers (in part from ATV 1983 [1])

Parameter	Regulations from Baden-Württemberg	Proposal by Water Poll. Control Fed. 1970	Proposal by Water Poll. Control Fed. 1983
pH-Value	6 - 9.5	6.5 - 9.5	6.5 - 10
Halogen. hydrocarbons [mg/l]	5	--	5
Arsenic [mg/l]	--	--	1
Lead [mg/l]	2	--	2
Cadmium [mg/l]	1	--	0.5
Chrom [mg/l]	0.5	0.5	0.5
Copper [mg/l]	2	3	2
Nickel [mg/l]	3	5	3
Mercury [mg/l]	0.05	--	0.05
Selenium [mg/l]	--	--	1
Zinc [mg/l]	5	5	5
Tin [mg/l]	5	--	5
Cobalt [mg/l]	--	--	5
Silver [mg/l]	1	--	2
Ammonium (ammonia) [mg/l]	50	--	165
Cyanide [mg/l]	0.2	1	1
Fluoride [mg/l]	50	--	60
Sulphate [mg/l]	400	400	600
Sulfide [mg/l]	10	--	2

material leave the treatment plant and enter the receiving water. There again accumulation and subsequent persistance will lead to secondary problems.

There are aspects of a completely different nature that also make pretreatment an administratively and technically expedient measure. They result from the need to provide a central wastewater treatment facility for all wastewater producers within a community irrespective of the development of such wastewater sources. One of the most frequently observed problems in any central wastewater treatment facility is the change in amount and quality of wastewater arriving at the plant. In the past such changes usually meant an increase in wastewater load and complexity. Today one can no longer predict the general trend of such changes. Frequently the amount of discharged wastewater decreases, due to rising raw water costs. And equally frequently the composition of the wastewater becomes more and more heterogeneous and varies continuously in a direct response to changing production patterns in industry. Intensified pretreatment at the source of wastewater production will reduce amount and complexity of the sewage and thus allow *improved planning and operation of central treatment facilities* largely independent of industrial activities.

Finally it should be pointed out that the *preliminary treatment of industrial effluents* recommended or even prescribed *in developing countries* may be incorporated into a subsequent scheme of complete wastewater collection and central treatment by such a pretreatment concept. There are several advantages to this kind of development in modules. First, the development of a complete sewerage system will be a long process. Therefore, measures of immediate effect must be taken in particular with respect to more dangerous pollution, usually originating with industrial processes. Second, the requirements for pretreatment are significant incentives for the reduction of pollutant discharge. And thirdly, the complete incorporation of such preliminary treatment as pretreatment into a subsequent sewerage scheme avoids the negative effects of obsolete first stages. Figure 1 shows schematically the various possibilities of preliminary treatment of industrial wastewaters and the incorporation of such rudimentary treatment units into more complete schemes of sanitation. In addition regulations for dischargers, developed in industrial countries with the previously described objectives, could and should be used as standards for pretreatment. If this is the case, then there is the added advantage of protection for structures and processes in subsequently developed collection and treatment facilities.

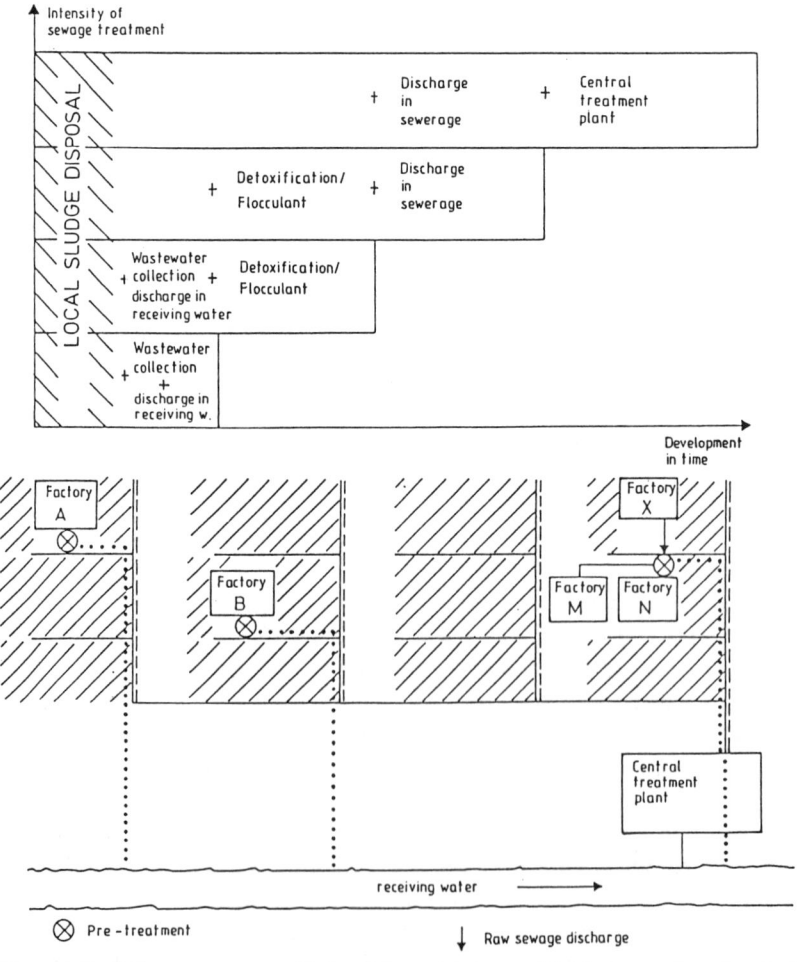

Figure 1. Preliminary treatment of industrial wastewaters and the incorporation of such elements into a global sewerage system

Present Situation in the Federal Republic of Germany

The Re-writing of Water Laws

The strongly federal character of the German union is reflected by the exclusive legislative power of the individual states concerning such issues as environment. There are legal frameworks issued by the Federal Government. These legal frames are to be filled in by specific state laws. Each change in the legal framework, either the introduction of new laws (for instance the wastewater charges law) or the updating of laws (for instance the general water management law), leads to a corresponding action by the state legislature.

The most recent up-dating of the general water management law (Wasserhaushaltsgesetz Novelle 1986 [2]) has introduced significant changes for the discharge of "dangerous substances" into receiving waters (paragraph 7a section 1) and into (public) sewer systems (paragraph 7a section 3). Table 2 compares the respective paragraphs for the two different dates.

Table 2. Excerpts from the Federal Water Management Law (1976/1986)

Wasserhaushaltsgesetz (Federal Water Management Law) (1976 version)	Wasserhaushaltsgesetz (1986 version)
...	...
Par. 7a Requirements for the discharge of wastewater	Par. 7a Requirements for the discharge of wastewater
(1) A permit for the discharge .... corresponding to the rules of *best practical* technology	(1) A permit for the discharge .... corresponding to the rules of *best practical* technology; if wastewater of certain origin contains substances wich might be poisonous, persistent, accumulating, cancerogenic, teratogenic or mutagenic (dangerous substances) then the requirements must correspond to the *best available* technology ....
(2) If existing installations do not meet this standards .... there are deadlines for the completion of the measures.	(2) If existing installations do not meet these standards .... there are deadlines for the completion of the measures.
	(3) The state legislature guarantees that prior to the discharge of wastewater containing dangerous substances into public sewers those meal sures described in section 1 sentence 3 are taken.
	...
Par.18b Design and operation of sewage systems	Par. 18b Design and operation of sewage systems
(1) Sewage systems are to be designed and operated .... according to 'best practical technology'	(1) Sewage systems are to be designed and operated .... according to 'best available technology' ....
(2) If existing systems .... paragraph 7a section 2 applies.	(2) If existing systems .... paragraph 7a section 2 applies.
	...

Significant changes result from an intensification of the treatment standards. The previous version of the water management law stipulated in paragraph 7a section 1 that prior to discharge into receiving waters all wastewaters must be "treated with the best practical technology". The up-dated version requires for "wastewaters with substances that are poisonous, persistent, accumulating, carcinogenic, teratogenic or mutagenic a treatment according to the best available technology" prior to their discharge.

This intensified pretreatment can be prescribed at the very point of wastewater generation, prior to any mixing with other wastewater streams. This part of the new legislation is even more stringent since industrial sewer systems historically have grown for the most part and therefore present a structure of combined sewers with little or no possibilities for individual treatment of the various wastewater streams. Thus, in many instances new sewers will have to be constructed and space will have to be provided for the pretreatment units.

For the discharge of wastewaters not directly into receiving waters but into sewer systems, state legislators are also asked to request a treatment according to "best available technology" if the sewage contains dangerous substances. This means in effect that pretreatment in the case of dangerous wastewater constitutents will be as extensive as if the waste was discharged directly into the aqueous environment. The regulation emphasizes the significant role that pretreatment is to play in the future. The intent of this up-dating is two-fold.

First, the protection of the aqueous environment is to be intensified. In the past many such "dangerous" substances have either passed with unaltered concentrations to the treatment plant designed according to best practical technology. Or they have affected the treatment process such that the overall removal efficiency was lower. In any case significant fractions of these substances entered the receiving water.

Second, this updating should serve as an incentive for the development of advanced technology in the area of avoiding the generation of wastes, in the field of treatment at the source and in the area of wastewater treatment in general. It has been observed in many instances that tightened regulations lead to noticeable advances in controlling technologies. And it has been observed likewise, that the request in the past only "best practical technology" has not brought enough relief for the environment.

Illustration of Consequences for Practice

The new development will cause significant changes in the way industrial wastewaters will be discharged in particular in the case of indirect discharge into public sewers. This requires, however, a definition of the term "dangerous" for each major wastewater constituent or for each major industrial wastewater producer. With the up-dating of the general water management law there were working-groups established whose task is to identify such dangerous substances.

Even though these deliberations are not finalized in all instances the possible consequences of these regulations are to be illustrated in the following paragraphs. The first example is to show a situation where there will be no noticeable changes even though the situation in the wastewater treatment plant is a precarious one. The second example is chosen such that positive and negative consequences of these new regulations can be demonstrated. And the third example describes a situation that was unsatisfactory under the old regulation and is now significantly improved.

Food processing industries in the Federal Republic of Germany frequently work on seasonal schedules. This leads to seasonal variations in the composition and total amount of wastewater produced. The effect is frequently a seasonal overloading of the municipal treatment plant. The design of treatment plants in the past has not taken into account such peak loadings, mostly for economic reasons. This is easily understood if one considers that peak loading occurs during less than three months of the year — for instance in the wine-producing industry (Hantge 1982 [3]). Here, a satisfactorily designed treatment plant has noticeable overcapacity for more than three quarters of the time, with all the operational problems resulting from it. These wastewaters, however, will not be classified "dangerous" for obvious reasons. Thus, there will be no pretreatment required of the kind discussed earlier. These situations will persist if no other solutions are found.

Manufacturing industries with related or even identical products have been concentrated in the past very much within specific areas. This has led to quasi-monolithic wastewater production and consequently to large problems for the central municipal treatment plants in terms of amount of wastewater as well as unbalanced composition. Leather related industries, as they will be discussed in the following paragraph, are illustrating examples. Their wastewater contains amongst other substances large amounts of chromium (in its three valent and therefore non-toxic form) and vegetable-based tanning agents (Joergensen 1979 [4]). Most likely the chromium constituents along with the sulfide will be classified "dangerous", leading to the need for advanced pretreatment. Presently discussed standards of emission are in the order of 2 mg/l for chromium. This reduction will have beneficial effects upon the central wastewater and sludge treatment facilities. Reduction can be accomplished by precipitation processes, followed by highly efficient phase separation processes, as pilot plant work and some technical experience show. It is not known, however, to what degree the other constituents of the industrial wastewater stream will be affected by this pretreatment. In particular the possible reduction in the concentration of tanning agents (which will most likely not be termed "dangerous") is unknown. Thus the future loading of the central treatment facility, in particular of the biological elements, is not known. This may cause technical and in particular economic problems in the design and operation of these plants.

Processes in the metal finishing industry and related branches produce heavy metal containing wastewaters (Hartinger 1976 [5]). These have been problematic for the central wastewater treatment plant due to an inhibition of the biological reactions in the treatment process (at high heavy metal concentrations) or in the phase of sludge digestion (already at intermediate concentrations due to accumulation in the treatment process). Most of the difficulties from heavy metal containing waste streams arise already at relatively low concentrations from the need to conceive other sludge utilization patterns: the agricultural use is no longer possible due to the noticeable accumulation in the organic fraction of the sewage sludge. All heavy metals containing wastewaters will be classified "dangerous" and pretreatment according to "best available technology" required. This will lead to a significant reduction in the concentration and load of heavy metals in the municipal sewage and consequently in the sewage sludge. Without any doubt this will lead to an improvement in the quality of sludge in view of agricultural use.

Wastewater Situation of B. - An Illustrating Case Study

The following discussion should illustrate the practical consequences of the changed water management legislation in the Federal Republic of Germany. The situation described in this case study is certainly extreme in terms of wastewater composition, changes in amount and composition and treatment concept. However, these kind of problems are by no means unique to the city of B. They are found in many other localities with a differing degree of seriousness. Thus, the case study is not intended as a criticism of a very unfortunate situation but rather as a very clear example of a host of existing problems.

Wastewater Origin and Wastewater Composition at B.

The situation in the city of B. resembles very much the above described structure of rather monolithic industrial activities, predominantly the leather industry. This means that large amounts of industrial wastewater of unbalanced composition (unbalanced from a point of view of biological treatment) are brought to the municipal treatment plant (see Figure 2). In fact it would be more appropriate to speak of an industrial wastewater treatment plant if one considers the rations of the two wastewater streams.

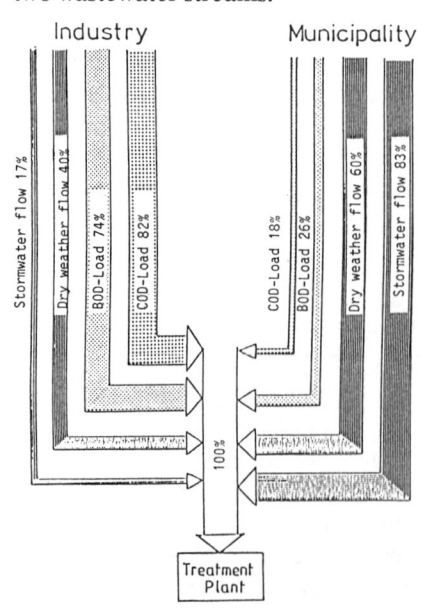

Figure 2. Origin and composition of wastewater of the city of B.

The drainage system is constructed as a combined sewer system. There are contributions to the stormwater runoff from the municipal as well as the industrial sector. Stormwater is discharged in part, in agreement with the German regulations, at stormwater outlets. The central treatment facilities of B. therefore have to cope with fluctuating hydraulic loading from stormwater events. The operation at the treatment works also suffers from any change in the industrial production process. The alteration of production causes changes in wastewater streams and loading. And since there is no pretreatment as of yet all these changes are directly conveyed to the central treatment facilities.

Present Wastewater Treatment at B.

The composition of the combined wastewater stream at B. is complex, due to the large contribution from one dominating sector of industry. The development of a suitable strategy for wastewater treatment was a complex task and led to the selection of a more advanced and less proven technology against the widely used concept of biological wastewater treatment. Figure 3 shows a schematic of the physico-chemical treatment concept which was finally selected. After conventional primary treatment there is a catalytic oxidation stage using granular activated carbon as a catalyst. The oxidation is supported by compressed air. The third stage is comprised of a precipitation and coagulation unit with a sedimentation basin. Arguments for this decision included treatment efficiency and cost.

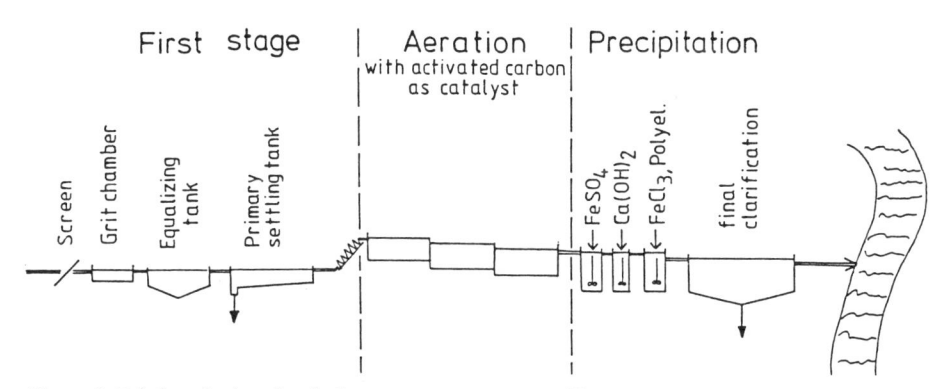

Figure 3. Existing physico-chemical wastewater treatment at B.

After a short period of successful operation efficiency data and cost data deteriorated. The difficulties in meeting official discharge standards were particularly disturbing and have led to many investigations, legal arguments and finally the decision to search for a completely different concept of wastewater treatment (see below). The reasons for this change in treatment plant performance may be manifold. Despite intensive analyses they have not been identified with any degree of reliability. A relatively sound assumption is that the wastewater characteristics have changed to such a degree that the formerly successful catalytic oxidation no longer occurs to any noticeable degree. Changes in wastewater characteristics are generally quite frequent. The fact that biological processes usually may be adapted without difficulties and in due time to changed situations has favored this treatment concept justly.

Development of a New Wastewater Treatment Concept

Based on the experience from previous pilot plant investigations and from the operation of the large scale plant, various alternative concepts for the treatment of these wastes were developed (see Figure 4). Beyond the purely physico-chemical approach mainly combined biological and chemical treatment alternatives were proposed. These treatment possibilities distinguish themselves mainly in the location of where the chemicals are added: in one alternative the chemical stage is ahead of the biological unit and in the other alternative the chemical stage is conceived as post-precipitation and coagulation.

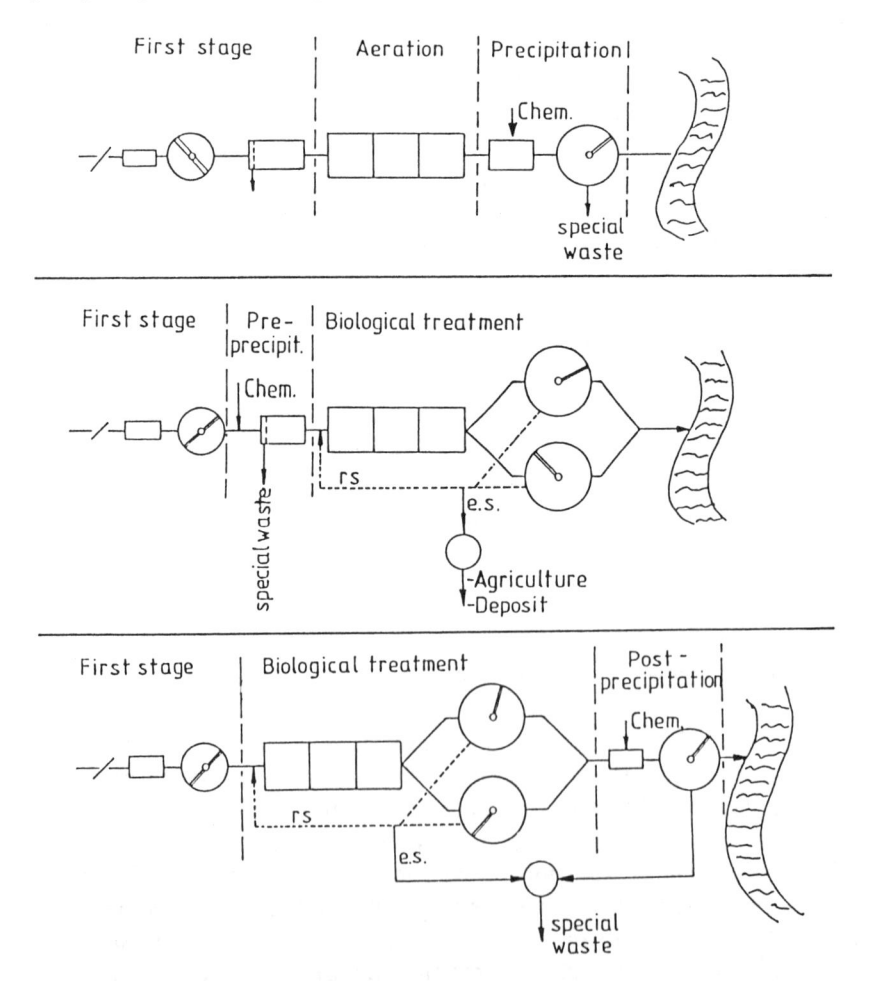

Figure 4. Alternatives for wastewater treatment at B.

Cost and efficiency datas for these latter treatment concepts derived on the basis of extensive pilot plant work are shown in Table 3. It can be seen that, with few exceptions, the efficiency data are comparable for the last two alternatives: nitrification profits from the early removal of organic load by means of pre-precipitation and also the chromium removal may be slightly higher in the case of pre-precipitation. On the other hand post-precipitation will most likely remove color better than the biological unit which usually brings out the color in a pronounced way. From a cost point of view, there are reduced chemical costs for post-precipitation and slightly higher investment costs for the construction of the complete tertiary stage (if it does not exist already as it is the case in this instance).

Table 3. Cost and efficiency data for various treatment alternatives (figures in parenthesis show absolute effluent units)

Type of treatment	COD remov. %	BOD remov. %	NH_4 remov. %	Cr remov. %	Invest (Mio.)	M&O&R DM/cbm
Katox (phys. chem.)	85 (150)	85 (50)	0 (50)	>85 (2)	15	0.31
Pre-precipitation and biology	>90 (100)	>95 (20)	>90 (5)	>85 (2)	23	0.31
Biology and post-precipitation	>90 (100)	>95 (20)	>80 (10)	>85 (2)	>23	0.25

The decision in this instance favored the pre-precipitation and coagulation alternative (Figure 5) for the following reasons:

— The biological treatment stage is better protected from unforeseen fluctuations in the composition and quantity of the incoming wastewater stream.
— Nitrificaton — the prerequisite of nitrogen removal (which is presently the foremost goal in pollution control in the Federal Republic of Germany) — will be slightly better in this alternative.
— The usually large amount of biological sludge produced will be less burdened with chromium due to the early removal of this heavy metal; this may permit agricultural sludge utilization.

The last mentioned argument of pretreatment was formulated before the updating of the federal water management law. This general development was not directly expected. However, it had to be anticipated as a logical extension of many public declarations. Fortunately for the city of B. the decision to design and build a new treatment plant was slightly postponed thus allowing a correction of the decision. The consequences of the presently required decentral pretreatment at the origin of the dangerous wastewater streams containing chromium could still be incorporated.

For many other communities this will not be so easily done where sewage systems and treatment concepts are firmly established. The foreseeable consequences in such situations are indicated in Table 4. It can be seen that with the reduction in pollution load — as must be anticipated — specific units will be larger than needed and therefore represent points of inactive investment and higher

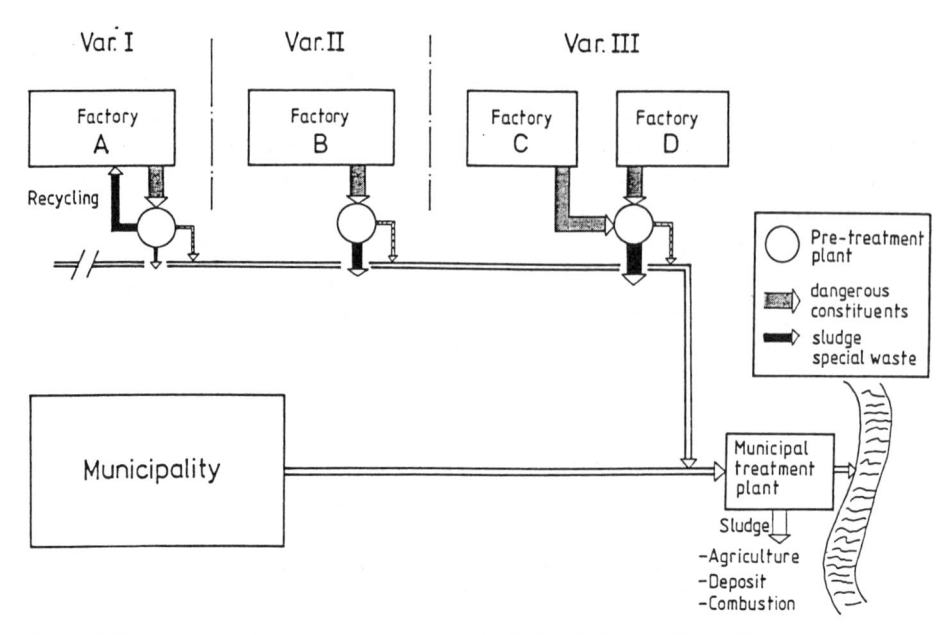

Figure 5. Future wastewater routing and treatment including sludge handling at B.

Table 4. Problems arising from introduction pre-treatment into established sewage systems

Unit process	Reactor dimensions controlled	Introducing pretreatment may cause	
		reduced hydraulic loading	reduced pollution load
Grit chamber	Hydraulic loading	Problems from intensified sedimentation	No technical difficulties
Primary sediment. chamber	Hydraulic loading	No technical problems *Inact. investment*	No technical difficulties
Biological elements (trickl./act. sludge)	Hydraulic loading; organic loading	No technical problems *Inact. investment* *High operat. cost*	Increasing stabil *Increased operat. cost* *Less sludge energy*
Secondary sediment. chamber	Hydraulic loading (solids loading)	No technical problems *Inact. investment*	(Increasing nitr. in biol. unit may cause) *Denitrif.* with insufficient sludge removal
Pre-/post-precipit. coagulation	Hydraulic and solids loading	No technical problems *Oper. adjustment*	No technical problems *Oper. adjustment*

operational costs. While pretreatment as a rule will only reduce the pollution load it must be expected that the hydraulic wastewater load might also be reduced by these measures. One reason for this will be a possibly intensified water recirculation due to improved water quality after intensification of pretreatment. Another reason might be the change in industrial production all the way to abandonment of incorrigible production schemes resulting from the extra financial burden of pretreatment.

Conclusion

The aforementioned change in the federal water management law represents a very ambitious legislation. Without any doubt this will lead to significantly increased expenditures in particular for certain industries. Thus, this legislation may not be directly transferred to other regions and countries.

The principles underlying the re-written water laws, however, are important ones and are based on various generally agreed upon technical and economic principles. These insights derive from the difficulties experienced by municipalities in providing central wastewater treatment facilities for all types of industries at all times without much influence on the (wastewater) production pattern.

Pretreatment should be developed to such an extent that dangerous substances cannot leave the production cycles. Substances may be dangerous for the structures of the sewage transport and treatment system. They may also be dangerous for the environment in general. This kind of treatment is best described as "best available technology" and will include all types of unit processes short of biological treatment. The latter one will not necessarily be the best alternative, due to the unbalanced composition of wastewaters at their point of origin. Furthermore the significant economy of the scale of these processes cannot be exploited when wastewater streams of small volume are to be treated.

Finally the specific problems of treatment of wastewater from industry in developing countries could and should be solved on an intermediate basis by establishing treatment concepts that later on can be integrated into a global sewage scheme as pretreatment units.

References

[1] ATV (1983) Arbeitsblatt 105, GFA St. Augustin
[2] Wasserhaushaltsgesetz Novellierung 1986, Bundestagsdrucksache Bonn
[3] Hantge, E. (1982): Abwasserprobleme bei der Weinerzeugung. Wasserkalender *64*, Schmidt, Berlin
[4] Joergensen, S.E. (1979): Industrial Waste Water Management. Elsevier Publishing Co., Amsterdam
[5] Hartinger, L. (1976): Taschenbuch der Abwasserbehandlung, Band 1. Hanser, München

H.H. Hahn, K.-H. Hartmann
Institut für Siedlungs-
wasserwirtschaft
Universität Fridericiana
7500 Karlsruhe
Fed. Rep. of Germany

Clean Technology in the Netherlands: The Role of the Government

A.B. van Luin and W. van Starkenburg

Abstract

Curbing the pollution of surface waters still remains the primary objective of water pollution control in the Netherlands. The Dutch Government wants to achieve this goal by stimulating, among other things, the development of clean technology. A number of projects involving clean technology are discussed in this paper. Moreover, the government's role in such projects is reviewed in light of current experience.

1 Introduction

The desire to protect human health and ecological and material interests from the adverse effects of pollution of the surface waters has led to measures to curb such hazards. Although significant results have been achieved over the last fifteen years, the pollution of surface waters in the Netherlands is still a cause for concern. Policies are therefore primarily aimed at ensuring a further reduction in water pollution [1]

Water pollution is caused by emissions which differ considerably in their basic composition and which originate from a great variety of sources. A general classification of pollution sources has been agreed upon, which differentiates between point and diffuse sources. In order to reduce water pollution further, it is essential that measures are taken to control both types of sources. In the case of point sources, this normally involves preventing the pollution from occuring at the source. This can be achieved by internal measures, such as the introduction of new or improved processes which reduce the level of environmental pollution, by keeping effluent streams apart to allow the individual waste flows to be treated separately, and by introducing ways of reusing raw materials.

Processes that have been developed to prevent pollution are commonly known as clean technologies. The responsibility for pursuing such initiatives should be borne by the companies causing the basic pollution. However, the Dutch government recognizes that it also has a role to play in stimulating these developments. Financial assistance is therefore provided to companies prepared to introduce clean technology. In this context, clean technology is defined as:

- the development of new production processes and the improvement of existing ones;

- the development of new wastewater treatment methods;
- the efficient management, handling and reusing of raw materials and energy.

Government subsidies are given for research projects for which the laboratory stage has been completed, but where practical implementation is restricted by the technical and/or economic risks involved. Projects that qualify for such assistance must have certain novel technological features, must be beneficial to the environment and have the potential to be applied on a relatively wide scale. It is hoped that government-subsidized research of this type should remove any such barriers. In addition, the Dutch government also supports demonstration projects involving clean technology.

The government provides a subsidy of between 30 — 60% of the research costs. The maximum amount available per project is Fl. 300,000. Projects which have already outgrown the research phase can be regarded as demonstration projects. In this case the government is prepared to subsidize the construction of the first plant of its type to be built in the Netherlands. Sums of between 20 — 35% of the investment required are available, with a maximum of Fl. 500,000 per project.

At present, some 60 research projects concerned with the development of clean technology are in progress or have been completed. A number of these projects are listed in Table 1.

Table 1. Projects involving research in clean technology

Anaerobic treatment of wastewater from slaughterhouses
Development of blue passivating baths (electroplating industry)
Wastewater treatment in the leather industry
Reduced levels of water pollution in the production of hide glue
Nitrogen removal by steam stripping
Biological nitrogen removal
Electrochemical removal of heavy metals
Replacement of cadmium by aluminium
Development of clean technology for phosphoric acid processes
Electrolytic treatment of toxic organic compounds
Measures to improve fish processing techniques

The first project listed above is concerned with the anaerobic treatment of wastewater. The existence of the Subsidy Scheme for Clean Technology Water is seen as being largely responsible for the development of anaerobic treatment techniques in the Netherlands. A number of the projects referred to in Table 1 will be outlined in more detail in Section 2. Following this, the general experience gained to date with clean technology projects will be discussed and reviewed.

The Dutch government attaches great importance to the promotion of technologies that prevent environmental pollution. These sentiments are further underlined by the findings of the World Commission on Environment and Development (Brundtland Commission), which are contained in its recent report entitled "Our Common Future" [2]. It was noted that experience in the industrialized nations has proved that anti-pollution technology has been cost-effective in terms of health, property, and avoidance of environmental damage, and that it has made many industries more profitable by making them more resource-efficient.

2 Clean Technology Projects

2.1 Electrochemical Treatment of Process Wastewater Containing Halogenated Organic Compounds

In the chemical industry large quantities of toxic or non-biodegradable wastewater are produced, which often contain halogenated organic compounds. If these compounds cannot be reused, treatment techniques are commonly employed which first involve concentrating the waste products present. Membrane separation, activated carbon adsorption or stripping combined with adsorption of the stripped compounds are examples of the approaches that can be used. However, these methods all have the serious drawback that a concentrated waste stream is produced, which requires further treatment.

One technique that does not suffer from this drawback is dehalogenation by electrochemical reduction. This method is particularly suitable for wastewater containing polar or ionic organochlorine compounds, which are generally difficult to decontaminate by adsorption or stripping. Tests have been carried out with eight representative chlorinated organic compounds chosen from the EC list of 129 priority pollutants in concentrations of 30 – 500 mg/l (Table 2), as well as on an industrial aqueous effluent stream to investigate the dehalogenation of various brominated organic compounds (10 – 100 mg/l). It was shown that electrochemical reduction was effective in removing all the chlorine and bromine atoms from the samples. In Figures 1 and 2 the results are given for pentachlorophenol.

Table 2. Selected chlorinated organic hydrocarbons from the EC list of 129 priority pollutants

2-Amino-4-chlorophenol	Pentachlorophenol
p-Chloronitrobenzene	2, 4, 5-Trichlorophenoxyacetic acid
Dichlorvos	Tetrachloroethene
Hexachloroethane	1, 2, 4-Trichlorobenzene

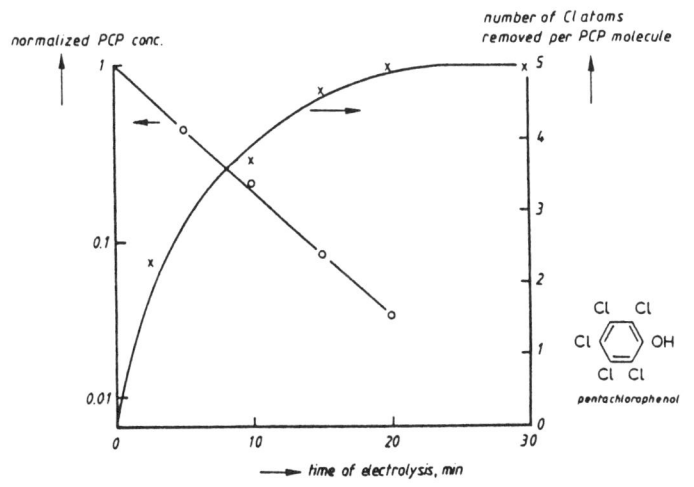

Figure 1. Normalized PCP concentration and yield of Cl-ions per PCP molecule during electrolysis

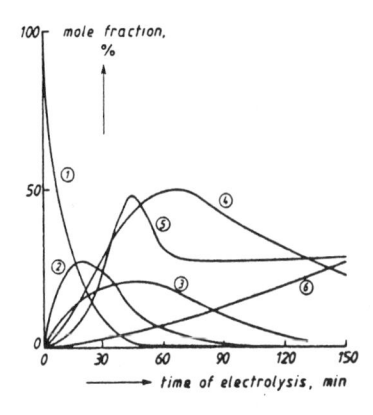

Figure 2. Mole fraction of phenols during electrolysis of 2 dm^3 of 50 mg PCP per dm^3 solution
1) PCP
2) Tetrachlorophenols
3) Trichlorophenols
4) Dichlorophenols
5) Monochlorophenols
6) Phenol

As a result, the toxicity of the wastewaters containing the selected compounds decreased by 95% or more of the initial EC-50 value. The biodegradability of the wastewaters was correspondingly improved making further biological treatment feasible.

Comparisons between the expected cost of the electrochemical treatment method and that of activated carbon adsorption have shown that they are of the same order of magnitude (i.e. Fl. 10 – 15 per m^3 of wastewater) [3].

2.2. Substitution of Cadmium in the Metal Plating Industry

The use of cadmium in the Netherlands has been drastically reduced over the last few years due to measures introduced to protect the environment. Cadmium is, however, still being applied in metal finishing processes because of its favourable corrosion properties. It is particularly useful for products and components that have to comply with rigorous quality standards.

A Dutch company HGA (Lelystad) has developed a method for electrolytically precipitating an aluminium layer on metallic surfaces, in collaboration with Siemens of Germany. An aluminium layer provides very good corrosion resistance and offers cathodic protection equivalent to that of cadmium. When exposed to the atmosphere the aluminium forms an oxide layer, which is resistant to a large number of chemicals.

With the exception of the pre-treatment and post-treatment stages, this process is completely different from normal galvanizing methods. Precipitation of the aluminium takes place at 100 °C in an organic electrolyte bath in the absence of air and water. It is important to note that this requirement effectively eliminates the possibility of air and water pollution occurring. Many metals such as iron, copper, stainless steel and brass can be treated using the process outlined above. Layers of comparable thickness to those formed with conventional galvanizing methods are readily obtained.

This new technique has already been successfully demonstrated in "rack machines" where components positioned on racks are galvanized. Two independent systems have been developed for the bulk processing of screws, bolts and nuts. One is based on agitating the components, while the other relies on the principle of a rotating disc. Plants utilizing these two techniques are being built primarily to produce aluminized components for the car industry. The price and quality obtained with aluminium is comparable to that of cadmium. It is expected that aluminized components will also tend to replace nickel-plated and galvanized products because of more stringent requirements for corrosion resistance [4].

2.3 The Development of Clean Technology Phosphate Fertilizer Production Processes

About 2 million tons of waste gypsum are produced in the Netherlands each year as a by-product of the manufacture of phosphoric acid, one of the raw materials used to make phosphate fertilizers. Until now this gypsum was discharged into surface waters. Since gypsum from phosphoric acid production is contaminated with heavy metals such as cadmium and mercury, government policy has been directed to stop such discharges.

However, to date it has not been proven possible to find a suitable processing or marketing route for phosphate gypsum.

The Dutch company DSM is carrying out research in collaboration with the Delft University of Technology to develop new processes for manufacturing phosphoric acid. Three possible process variants are being studied. All three result in cadmium concentrations of $12-18$ mg/kg P2O5 in the phosphoric acid and $0.4-0.6$ mg/kg in the calcium sulphate (on the basis of phosphate rock with $60-75$ mg Cd/kg P2O5). The present processes produce phosphoric acid with $55-65$ mg Cd/kg P2O5 and calcium sulphate with $1.8-12.9$ mg Cd/kg. Introducing the new types of production processes under consideration could lead to a reduction of about 13 tons per year (90%) in the emission of cadmium into the environment.

The costs involved in the various processing routes being studied amount to $25-65$ Fl/ton P2O5 more than those incurred with current manufacturing methods. Research in this area is continuing, aimed at the implementation of one of the processes on a commercial scale in 1994 [5].

2.4 Reduction in Chromium Discharge from the Leather Industry

The Dutch leather industry processes approximately 20,000 tons of skin material each year. About 370 tons of chromium (III) are used in the tanning stage, of which about 250 tons remain fixed on the skins. Until recently, about 120 tons of chromium were discharged in solution with the various aqueous effluent streams (6 kg chromium per ton skin).

An investigation has been carried out at six representative leather processing factories to study how discharges of soluble chromium (III) can be reduced. Studies of leather preparation techniques have shown that changes in the process and in the amount of chemicals used can appreciably reduce chromium emissions. In addition, a method has been developed that makes it possible to remove chromium from chromium-containing wastewater so that it can be reused for tanning purposes. Of particular importance is the fact that the waste streams from the actual tanning process and subsequent post-treatment stages are separated from the wastewater from the pre-treatment stages. This results in high, low and chromium-free effluent streams, respectively. Chromium is precipitated from the chromium-rich wastewater by adding magnesium oxide. The precipitated chromium hydroxide can then be used to regenerate chromium salts. This process is, however, not feasible for wastewater with low chromium concentrations. In this case the emphasis is placed on chromium removal only. Chromium-containing sludges in this case have to be removed.

The results of the investigation have already shown that the above-mentioned techniques for removing and reusing chromium from wastewater can be justified on economic grounds. The savings gained from the chromium needed for processing capacities in excess of 300 tons of skin material are sufficient to warrant the future implementation of such techniques. The removal and reuse of chromium will make it possible to reduce the total discharge of chromium(III) by the leather industry to less than 1 ton per year (50 g chromium per ton skin). As yet, schemes to reuse chromium on a commercial basis have not been applied in the Netherlands. Investigations are currently being carried out to determine to what extent treating skins with regenerated tanning fluids affects the quality of the leather. The introduction of other related measures has already reduced chromium discharges in the Netherlands to 2 tons yearly [6]. An important measure, necessary because of severe regulations, is the termination of tanning activities and the import of tanned skins ("wet blue").

2.5 Measures to Improve Fish Processing Techniques

A fish processing plant was confronted with the following problems:
— production had expanded dramatically, which had resulted in a corresponding increase in the volume of wastewater. The capacity of the existing physico-chemical treatment facilities had become too small;
— the costs associated with the discharge of wastewater had risen considerably.

Two possible solutions were identified to resolve the situation. On the one hand, existing end-of-pipe treatment capacity could be increased, while, on the other hand, pollution prevention seemed an interesting option to reduce the levels of pollution.

An analysis of these problems revealed that contact time of flushing water and the solid waste was a crucial factor in determining the extent of the pollution. This was of particular importance since the flushing water was used as a transport medium to remove the fish waste via a drain system.

Lengthy investigations were carried out into the options for disposing of the fish waste in a dry form. In a new factory this would have presented an ideal opportunity. However, in the case of this particular factory, this would have meant constructing a special conveyor belt system. This alternative had to be rejected in the final analysis because of veterinary reasons.

In the plan that was finally selected, the fish waste and water flows were separated at the point where the total effluent stream left the factory building. It was recognized that some extraction had already taken place at this point. However, research showed that the bulk of the pollution transfer to the water occurred within the first three minutes. The company therefore had to choose between either a faster removal of the waste material, which would necessarily have employed a much larger volume of less polluted water, or the introduction of an effective separation stage outside the factory building. It was finally decided to choose the latter option in combination with an aqua- guard separator. This equipment ensures that the waste material is effectively separated from the effluent stream by using a self-cleaning belt filter. In the following stage, the wastewater is treated with physico-chemical techniques involving coagulation and flocculation. Use of such an approach has enabled the company concerned not only to reduce the level of pollution drastically, but also to increase the amount of usable fish waste that is recovered in the process.

The investigation carried out at this company has indicated how fish processing plants could best be set up. Ideally, the waste material left after filleting should be removed as quickly as possible in a dry state. Air can be used as a suitable transport medium for this purpose. Other fish waste can be separated using the new techniques employed by the above-mentioned company. It was considered to be especially important in this context to ensure that the smallest possible amount of water is used in filleting the fish and in any resulting effluent streams [7].

3 Experience Gained with Research Projects Involving Clean Technology

After operating the Subsidy Scheme for Clean Technology Water for a number of years, certain observations can be made regarding its overall effectiveness:

1) It is clear that large companies have generally been able to take advantage of the scheme.
   Smaller and medium-sized companies are usually insufficiently aware of the possibilities that exist for obtaining governmental help in developing new technologies. Of the 60 projects mentioned previously, about 20% have been undertaken by smaller and medium-sized companies.
2) The procedures involved in applying for these types of subsidies are often considered to be obscure and time-consuming.
3) Information exchange needs to be improved. In the first place, this concerns the information transfer to individual companies about developments in governmental policy and future requirements. A further area where improvements could be made relates to the information generated by fundamental research carried out at universities and research institutes. Thirdly, the spin-off resulting from the knowledge developed within companies could be increased.

It has been observed that small and medium-sized companies seem to show relatively little initiative in developing and applying clean technology. One of the factors that is often cited as being responsible for this situation is the lack of knowledge about the technical potential that exists and the financial advantages of the application of such technologies. However, this is only a partial reflection of the

true situation. An evaluation of the results and spin-off generated from a dozen projects involving clean technologies in the Netherlands has revealed the importance of the following factors [8]:
— the time that is needed to develop and introduce new production processes;
— the sluggishness of the market in adapting these new developments;
— the uncertainty of market developments at home and abroad;
— unfamiliarity with possible solutions;
— the inaccessibility of information about such solutions.

Research in the USA has shown that there are several obstacles that prevent the maximum utilization of waste reduction concepts and approaches [9]. A summary of the different types of obstacles is given in Table 3.

Table 3. List of obstacles preventing the application of clean technology [9]

Poltical 60%
  1) Bureaucratic resistance (20%)
  2) Human conservatism (10%)
  3) Piecemeal legislation (10%)
  4) Media sensationalism (10%)
  5) Public ignorance and misinformation (10%)

Financial 30%
  1) Disposal Subsidies (10%)
  2) Scarce money (10%)
  3) Entrenched disposal industry (10%)

Technical 10%
  1) Lack of centralized reliable information (5%)
  2) Lack of assistance with the application of waste reduction approaches to individual firms'needs (5%)

If it is assumed that the situation in the Netherlands is not significantly different from that in the USA then it would appear that a complex array of factors are important rather than merely knowledge itself. Moreover, it is clear that the prospect of government subsidies alone is insufficient to encourage companies to develop new technologies.

Clean technology is first and foremost a state of mind [3, 9]. Changing the way people think about such things is an extremely slow process. It is important to remember in this context that even where a particular process is beneficial for the environment, it will not be used by industry if it is intrinsically uneconomical. The government has a fundamental role to play in trying to change attitudes. Concern about the deterioration of the environment both in a qualitative and a quantitative sense will strengthen the need for more stringent requirements. The market for clean technology will therefore increase, as long as it is implemented in stages where due account is taken of the market's sluggishness referred to earlier. The forces of supply and demand must be allowed to interact fully.

In addition, it is essential that the government projects a consistent policy in these matters. Industry and government should be able to agree on a common perspective. Financial incentives such as subsidies can play an important role in stimulating the market for clean technology. They should not be seen as providing the necessary conditions, but simply as a means of promoting such activities. The bureaucratic procedures associated with the granting of subsidies must be simplified in order not to frustrate the initiatives shown by industry.

In the process of changing the prevailing mentality and attitudes, the importance of information exchange should not be underestimated. This refers specifically to the information exchange between government and industry, between technologists engaged in fundamental research work and industry, and between manufacturing companies and potential users of environmental technology.

In order to inform industry at an early stage about the way government policy is developing with regard to environmental issues, a special commission has been established. The Commission for Environment and Industry is an advisory body of the central government, which also includes representatives from industry.

An active exchange of information is also needed between industry and the scientists/engineers engaged in fundamental research. Research within the framework of the Subsidy Scheme for Clean Technology Water is also carried out at universities and research institutes such as the Netherlands Organization for Applied Scientific Research (TNO). In order to create spin-offs from such projects it is essential that not only potential users will be found for the technology, but also that manufacturers express an interest in producing the requisite equipment. To a certain extent this introduces an additional barrier, which is not present when private companies develop their own environmental equipment. Experience has, nevertheless, shown that collaboration between research institutes and individual companies can produce extremely good results as long as contact is established at an early enough stage. More than ever before, manufacturers of environmental equipment must therefore be kept informed of the latest developments taking place at research institutes. It is of prime importance that companies that develop technologies for treating waste water should maintain an active dialogue with potential users. This will allow them to take into account the wishes of such customers at an early enough stage, and hence ensure the successful application of the technology concerned. The following aspects are seen as being particularly relevant to achieving this goal [8]:

- the timely preparation of a plan to commercialize the technology;
- the organization of the necessary financing for commercializing the project at an early enough stage;
- the publication of a clear and concise summary of results suitable for markets both at home and abroad;
- to organize a workshop to discuss the commercialization aspects of any significantly-sized project or group of projects.

A key element in any programme to stimulate information exchange is to make companies aware of the environmental and economic advantages of applying clean technology. This can be achieved by citing well-chosen examples, which contrast present-day and future environmental requirements. In this way it should be possible to solve the identifiable political, financial and technical bottlenecks that occur. It is crucial that, in future, sufficient attention be given to small and medium-sized companies. These sentiments are fully supported by the recent recommendations of the Brundtland Commission [2].

The Dutch government intends to promote clean technology so that environmental requirements may be accommodated, so that an equilibrium might develop between the forces of supply and demand. Secondly, the government will continue to participate in research and demonstration projects. Thirdly, it will attempt to create the conditions necessary to improve the level of information exchange. Indeed, the government has already organized a number of symposia at which various research results have been presented.

Although the government will continue to participate in a range of research and demonstration projects, it also intends to play a more active role in this area. Thus the government has recently initiated a multi-year research programme of its own. A number of items that need to be studied in the coming years have been included in the research plan. It is especially important that the transfer of information about clean technologies be intensified. Small companies, in particular, may be able to prevent a great deal of environmental pollution by introducing such technologies. Studies of specific branches of industry may also be of use in this respect, to identify those areas most suitable for pollution control and to indicate where changes in certain processes will be most beneficial.

Other main objectives of the research programme are:

— to improve treatment technologies;
— to curb the use of substances that have a detrimental effect on the environment such as heavy metal sand pesticides.

It is intended that the research programme will be conducted over the next five years and that it will be subject to annual review [10].

4 Summary and Conclusion

Curbing the pollution of surface waters still remains the primary objective of water pollution control in the Netherlands. The Dutch government wants to achieve this goal by stimulating, among other things, the development of clean technology. At present some 60 research projects concerning the development of clean technology are still in progress or have been completed. Based on past experience the government intends to play an active role in the stimulation of clean technology in future as well.

There is awareness of several political, financial and technical obstacles preventing the application of clean technology. The government must play a fundamental role in changing attitudes, among other things, by stimulating exchange of information, by providing greater clarity regarding future environmental requirements and by participating in research and demonstration projects.

A multi-year research programme has been developed and will be conducted. Special attention will be given to smaller and medium-sized companies.

References

[1] Ministry of Transport and Public Works (1986): Water Action Programme 1985-1989 (in Dutch)
[2] World Commission on Environment and Development (1987): Our Common Future. Oxford University Press
[3] Schmal, D. et al. (1987): Electrochemical Treatment of Process Waste Water containing Halogenated Organic Compounds Using Carbon Fibre Electrodes. Report of Phase 1, Netherlands Organization for Applied Scientific Research (TNO), Report No. R 87/69a
[4] HGA (Hegin Galvano Aluminium) (1987): Substitution of Cadmium by Aluminium in the Electroplating Industry. Lelystad
[5] Clean Technology Phosphoric Acid Processes. Results of a Feasibility Study, DSM/TU Delft, September 1987 (in Dutch)
[6] TNO/ISL (1984): Removal of Chromium from the Waste of the Leather Industry (in Dutch)
[7] Toet, W.A. (1986): Measures in the Fish Processing Industry. Ouwehand Katwijk, Sept. 1986 (in Dutch)
[8] Ministry of Housing, Physical Planning and Environment (Sept. 1987): Evaluation of Clean Technology Subsidy Scheme (in Dutch)
[9] Huisingh, D. and Bailey V. (eds.): Making Pollution Prevention Pay. Pergamon Press, New York
[10] Ministry of Housing, Physical Planning and Environment and the Ministry of Transport and Public Works (December 1987): Research Programme in the Field of Clean Technology (in Dutch)

A.B. van Luin, W. van Starkenburg
Institute for Inland Water Management
and Wastewater Treatment (DBW/RIZA)
P. O. Box 17
8200 AA Lelystad
The Netherlands

Synergistic Approach to Physical-Chemical Wastewater Pretreatment in the Food Industry

G. von Hagel

More than 90% of the food industry's operations discharge their waters into the municipal sewer system. This, in most cases, requires pretreatment. The reasons for pretreatment are very different from those in other branches of industry. Here, the biological stage may have to be protected from hazardous materials, and less biodegradeable substances may have to be eliminated at the point of origin.

In the food industry emphasis is placed on load reduction by removal of biodegradeable insoluble and soluble organic matter either to avoid or reduce municipal load surcharges and in particular process instabilities of downstream biological treatment plants.

There is a widespread uncertainty in the Federal Republic of Germany – and in other countries – as to when and where by reason of the need for installation of pretreatment facilities municipal surcharges will be applied. Until last year large cities did not impose any such charge, but this now becomes more and more widespread. Previously, only smaller towns and municipalities were trying to protect their sewage treatment plants by either enforcing pretreatment or imposing a surcharge.

The baseline above which municipal surcharges are considered is a COD concentration range between 300 and 600 ppm, approximately corresponding to the COD concentration of municipal sewage. Only above this level may municipal surcharges be expected and, if these are imposed, pretreatment for COD reduction commences.

Figure 1 shows the relationships between the necessity for pretreatment related to food industry wastewater and the applicable processes which may be used for pretreatment in different concentration ranges.

Above the baseline concentration of municipal sewage this range of pretreatment for removal of excessive organic loads may extend well into the tens of thousands of ppm of COD.

Basically, three methods of wastewater pretreatment are available and important to the food industry:

1. Anaerobic treatment above a COD concentration of roughly 2000 ppm
2. aerobic biological pretreatment, and
3. a combination of mechanical and physical-chemical processes.

Anaerobic treatability does not extend significantly below the 2000 ppm level. Aerobic treatment near the point of origin may be limited by space requirements, economical reasons, and other factors.

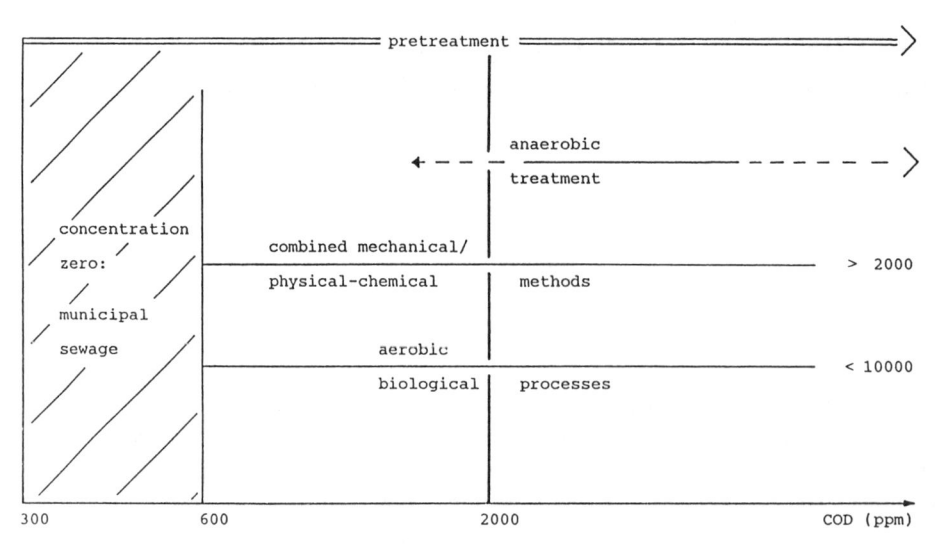

Figure 1. COD reduction in the food industry

Neumann [1] concludes, after comparing the three methods, that only two are applicable to the food industry.

1. Anaerobic treatment, preferably at high COD concentrations, and
2. a combination of mechanical and physical-chemical methods, excluding adsorption.

The application of flocculation/precipitation for food industry wastewater pretreatment is not particularly widespread today, but is gaining increasing importance.

Screening of food industry wastewaters with reliable fixed or movable fine screens has become the key step before chemical methods can be applied. The three basic principles of screening are shown in Figure 2.

Many wastewaters from the food industry cannot be readily used in chemical treatment without screening because of the wide range of particle sizes of the solids contained therein. This is where the synergistic effect in the treatment comes into force.

The mechanical part of the pretreatment sequence does not only reduce the amount of oxygen-consuming insoluble matter in the wastewater; it is the basic prerequisite for trouble-free economical application of subsequent flocculation/precipitation/phase separation steps. This is illustrated by Figure 3.

Screening and flocculation/precipitation have in common that they are easily controlled and can be put into operation at short notice. They are particularly well suited for application in production plants involved in seasonal harvesting of fruits and vegetables.

Figures 4 and 5 give an impression of modern screen installations where the totally enclosed type has become increasingly popular.

The synergistic effect of the combined application of fine screening and chemical treatment is applied now in an ever increasing number of food processing plants. Surprisingly the time for the development of this trend was relatively short.

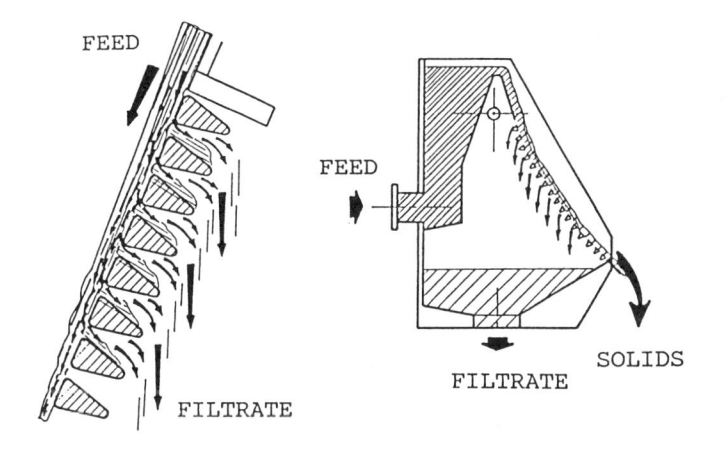

a) Fixed screen

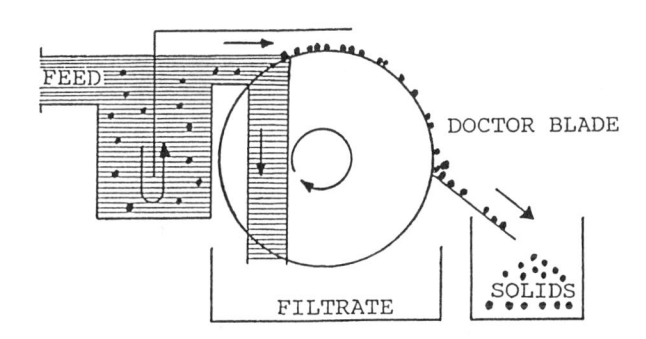

b) Rotating screen

c) Drum screen/thickener

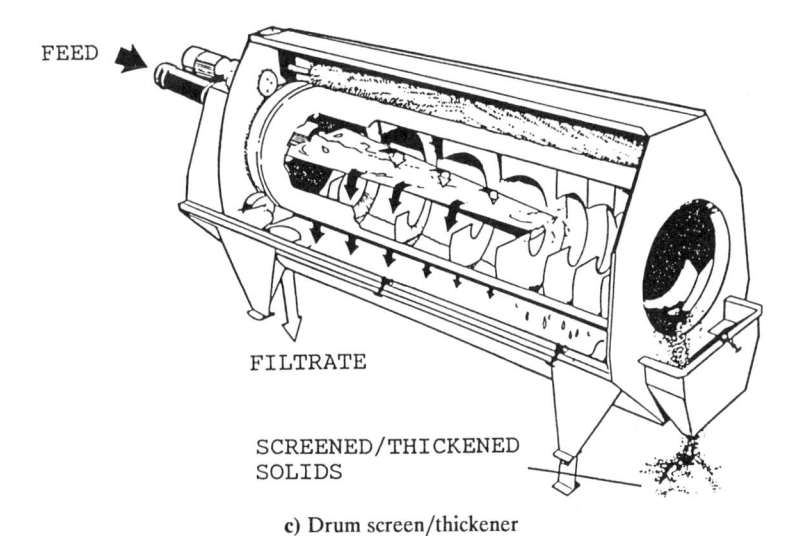

Figure 2. The three basic principles of screening

Figure 3. Screening before chemical treatment

Figure 4. Rotostrainer/rotopass screens in operation

Figure 5. Drum screen/thickener

Screens and sieves also reduce the amount of chemicals required. These devices can remove 35 to 50% of the settleable solids and 10 to 50% of the BOD_5/COD, due to the removal of suspended matter.

Chemical flocculation/precipitation removes 50 – 70% of the total organic load, expressed as COD. This high efficiency results from the removal of finely distributed and colloidal matter by adsorption on the hydroxide reaction product. It goes without saying that the applicability of chemical treatment must be ensured according to chemical flocculation/precipitation principles before designing a pretreatment system. At this point it must be made clear that treatability does not mean the application of any excessive amount of flocculant for pretreatment of wastewaters of any industry. Needless to say, the application of chemicals becomes uneconomical and senseless if certain limiting dosage values are exceeded. In this context sludge production is also a governing factor. The amount of thickened sludge generated by the process should not exceed 15% of the original wastewater volume. The comparative value for the chemical pretreatment of surface water is in the range of 1 – 2% of the raw water throughput.

Corresponding chemical dosages cannot be generalized and always depend on the origin of the wastewater. Pretreatment of such flows with flocculants should not require dosages above 200 to 400 ppm of iron(III) salts and 200 to 500 ppm of aluminium compounds.

The combination of fine screening and flocculation/precipitation makes the choice of the pretreatment process independent of the following factors:

1) Size and capacity of the food processing plant.
2) Time factors, such as seasonal harvesting of the raw materials
3) Status of the wastewater producer as direct or indirect industrial wastewater discharger.

For the sake of completeness, three other pretreatment methods applicable to the food industry are mentioned here:

1) Neutralization with acids, alkali or CO_2
2) Gravity separation on filter surfaces
3) Filtration with or without chemical dosing

To evaluate the present and potential future application of fine screening and chemical treatment in the food industry, Table 1 was compiled. It also shows which of the phase separation processes is predominant in this sector.

Using the official classification of the Federation of the German Food Industry (BVE), this industry is subdivided into 27 different branches (Table 1, column 1). This table also gives the number of food processing plants in operation in West

Table 1. Wastewater pretreatment in the food industry

Industry branch/ Production of	Plants in 1986	Fine screening	Phase separation	Neutralization	Classification of chemical treatment
Grinding mills	71	PT+	Starch	--	IV
Farinaceous products	17	--	--	--	IV
Food stuffs	65	P	--	Alk.	II
Starch and starch products	15	Optional	S	--	III, V
Potato products	41	P	S + F	--	III, V
Baker's wares	860	P	F	--	II
Durable baker's wares	77	P	F	--	II
Sugar industry	53	P	S	--	IV, V
Fruit & vegetable processing	215	P	--	Acid./alk.	V
Sweets	155	PT+	S + F	Alk.	I
Milk products and cheese	347	P	F	Acid.	I
Durable milk	49	P	F	Acid.	I
Oil presses, vegetable oil	18	P	F	Alk.	III, V
Margarine etc.	16	P	F	Alk.	II
Grease and fat	14	P	F	Acid.	I
Slaughter houses	151	P	F	--	I, II
Meat processing	280	P	F	--	I, V
Butcheries	394	P	F + S	Acid.	IV, V
Coffee and tea processing	47	Optional	F	Alk.	II
Breweries	481	P	F	Alk./acid.	III
Malting	48	PT+	--	--	--
Distilleries	18	Optional	--	Acid.	II; V
Spirits	112	--	--	Acid.	II
Wine and wine processing	32	P	--	Acid.	III
Mineral waters, soft drinks	273	P	--	Alk.	III
Others	109	P	F	--	III
Fodder, provender	245	P	F	Acid.	IV
Total	4016				

Germany in 1986 and provides an overview of the use of fine screens and sieves, the different types of phase separation, and indicates the necessity for additional neutralization.

Chemical treatment by flocculation/precipitation has the following classifications, as shown in column 6 of Table 1:

Category I – little application because of negative influence on end product
Category II – no application because it is uneconomical and/or less effective
Category III – application usual/widespread (not considering bulking sludge applications in aerobic biological treatment)
Category IV – no application of chemical methods although economical
Category V – long range prospects for additional COD removal, but as physical-chemical post-treatment after the biological stage.

Categories III and IV seem to be the most important. Of the total number of 4200 production plants in the German food industry, 1765 fall into categories III and IV where the application of chemicals for pretreatment is either usual, widespread, or nonexistent, although economically feasible. Proportions may be similar in other countries.

An estimated number of not less than 600 screen and sieve installations have been put into operation in 20 out of the 27 branches of the German food industry within a period of approx. 5 years. This gives a clear indication of the importance of the present development. However, only limited attention has been paid to the consequences of the synergistic effect resulting from the combination of mechanical methods and chemical treatment.

Flotation is the favourite phase separation process in at least 16 of 27 food industry branches with a potential in more than 3200 production facilities.

Category V, possible fields of advanced chemical post-treatment, comprises 8 industry branches with potential application in more than 1000 plants. This is the sector where the introduction of the BAT (Best Available Technology) requirements will probably be introduced first within the food industry.

There is a possibility of standardizing the equipment for chemical flocculation/precipitation as long as such systems are installed for pretreatment of wastewater flows with relatively low efficiency requirements. This also permits the application of less sophisticated flocculation systems, such as in combinations of tube-type flocculators and flotators.

References

[1] Neumann H. (1986) Personal Information, Behr's Seminar on Environmental Protection in the Food Industry, Wiesbaden

G. von Hagel
Ingenieurbüro f. Verfahrenstechnik
Bahnhofstr. 44-46
6200 Wiesbaden
Fed. Rep. of Germany

Separation of Heavy Metals from Effluents by Flotation

I. Zouboulis and K.A. Matis

Abstract

Different flotation techniques suitable for heavy metals separation from effluent, such as ion, precipitate and adsorbing colloid flotation, were described. Applications and potential of the process were discussed.

Introduction

In some cases, e.g. plating industries, about 10 – 90% of the plating metals are discharged in the form of effluent rather than being collected in the work places. The value of the metal lost is quite high (in U.K., 1971 prices, it was calculated as approximately 5 million pounds/year); but the failure to reuse the water may represent a much greater loss [1].

A range of processes is available to separate and possibly recover metals from dilute solutions — flotation is one of them — and with some modification could be successfully applied from a technical point of view, although from the economic standpoint this may be out of the question. There is a dearth of data on application of these methods to real situations and the information given on the overall economics of capital and operating costs for metal recovery are even more complicated. This is because in particular situations, factors such as the composition of the wastewater, size and location of the plant, the discharge limits and so on, would have to be taken into consideration [2].

The problems associated with the treatment of aqueous effluents are usually complex in nature. Certainly, processes for effluent treatment are chosen based on economics. Generally a metal bearing sludge or similar product is produced, which contains metals in a concentrated form, but which is rarely suitable for metals recovery and may present a pollution hazard on disposal [3].

From another point of view, the biological treatment nowadays represents more or less the dominant method in water and wastewater engineering. However, very often problems occur during operation, due to the presence of inorganic toxic compounds (usually in a dissolved form). An alternative process for the separation of these pollutants, i.e. a pretreatment method for the biological treatment, could be flotation.

Flotation is a separation process which has been in use for more than a century. Originally, it was applied in the field of mineral dressing (called froth flotation), where it has been established as one of the more significant methods for the concentration and beneficiation of ores and will continue to be so. A large variety of chemical species ranging from molecules and ions to macroorganisms can also be separated from one another and concentrated from solutions using several flotation techniques, or broadly termed adsorptive bubble separation techniques [4]. These are mainly based upon the differences in surface activity of the various substances that are present in a solution or suspension. In the seventies, the idea of utilizing rising gas bubbles for separation was applied successfully in effluent treatment.

Flotation is generally classified as dispersed-air and dissolved-air flotation, according to the way bubbles are generated. The former uses gas bubbles processed mechanically by such means as passing gas (usually air) porous media or mechanical shear of propellers. Figure 1 shows a classical arrangement, built in our laboratory. Electrolytic flotation also belongs in this category; here fine streams of bubbles are produced by electrolysis of the medium [5].

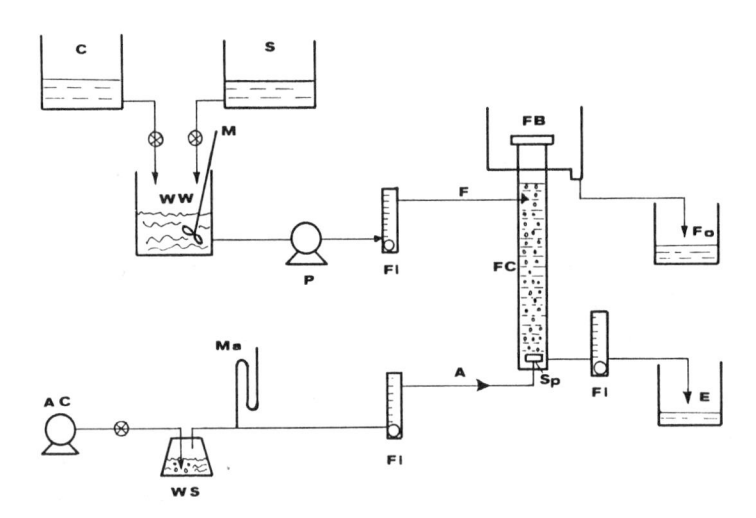

Figure 1. Apparatus of dispersed-air flotation (where c: conditioning (precipitation, coagulation, or other reagent addition), S: surfactant, WW: wastewater, M: mixing, AC: air compressor, WS: water saturator, Ma: manometer, Fl: flowmeter, P: pump, FC: flotation column, Sp: sparger, F: feed, A: air, Fo: foamate, E: effluent, FB: foam breaker)

Dissolved-air flotation uses bubbles of air generated under controlled conditions by release from a saturated solution of air in water [6]. Vacuum flotation and the so-called microflotation also belong here.

Ion Flotation

This technique involves the removal of surface-inactive ions from aqueous solutions by the addition of surfactants (or collectors) and the subsequent passage of gas in dispersion through the solution. As a result of this flotation procedure, a solid which contains the surfactant as a chemical constituent, appears on the surface of the solution. This permits separation and concentration in a small volume of collapsed foam.

If necessary, the foam layer can be broken up using various chemical, thermal or mechanical methods. Usually the surfactant is an ion of opposite charge to the surface-inactive ion (sometimes termed colligend) and thus, cations and anions are floated with anionic and cationic surfactants, respectively. The ion-surfactant product (sublate), when it ultimately reaches the surface of the solution, is always present as a solid. Prior to this, however, it may comprise groups of ions held to the bubble surface by the surface activity of the surfactant.

It is also possible for the surface-inactive ion (for instance, a weak undissociated acid) to require an activator as a ligand, to form a strong acid, and for the whole complex to float by the addition of a surfactant, which acts mainly by electrostatic interactions. This is another form of ion flotation.

This technique has been applied in the laboratory for removal of chromium [7, 8] and germanium [9]. The main parameters affecting the process, such as surfactant concentration, gas flow rate, bubble size, temperature, pH value of the solution, ionic strength and feed concentration, have been reviewed [10].

Precipitate Flotation

In ion flotation the concentrations are usually low and flotation occurs from a true solution. Raising the concentrations may lead to precipitation of the product to be removed, before gas is passed into the solution. Ion flotation carried out under these conditions is a form of precipitate flotation.

In addition, in ion flotation, varying the pH value of the solution may lead to a change in the nature of the process. The ion may be precipitated as a hydroxide, for instance, and then removed by precipitate flotation instead of ion flotation. An advantage of precipitate flotation is that it does not require a stoichiometric ratio of reagents as ion flotation does, rather much less. Thus it is suitable for large scale operations such as those required in wastewater treatment [11].

This technique is generally classified in three boundary categories: (a) Precipitate flotation of the first kind, where precipitation is succeeded by surfactants that do not form a chemical constituent of the product, but they exist on the precipitate surface. (b) Precipitate flotation of the second kind, where two hydrophilic ions precipitate each other giving a hydrophobic solid, without the use of a surfactant. (c) Precipitate flotation of the third kind, which is a form of ion flotation as described previously.

Precipitate flotation has been applied in the laboratory for the removal of zinc [11] and lead [12] as hydroxides, and copper and zinc as sulphides [13]. In the latter, a statistical approach has been followed.

An ion in an aqueous solution is known to be hydrolysed. The extent of hydrolysis, the formation of mono- or polymolecular species and their distribution is mainly a function of pH value and of the total ion concentration. Hence, the knowledge of metals speciation and aqueous chemistry is fundamental in flotation. A diagram, such as Figure 2 for zinc, can give an overview of the process [14].

Very often experiments and theory go together; this was the case, for example, for flotation of lead hydroxide by laurylamine (dodecylamine), presented in

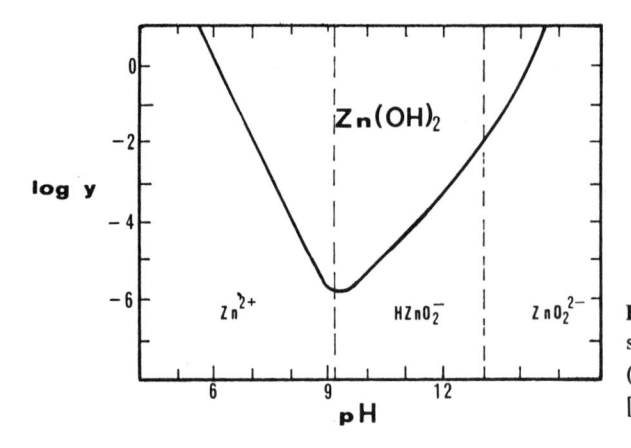

Figure 2. Effect of pH value on the solubility of $Zn(OH)_2$ at 298 K (where y is the sum of $[Zn^{2+}]$, $[HZnO_2^-]$ and $[ZnO_2^{2-}]$)

Figure 3. The theoretical curve shows the region of $Pb(OH)_2$ precipitation based on hydrolysis data.

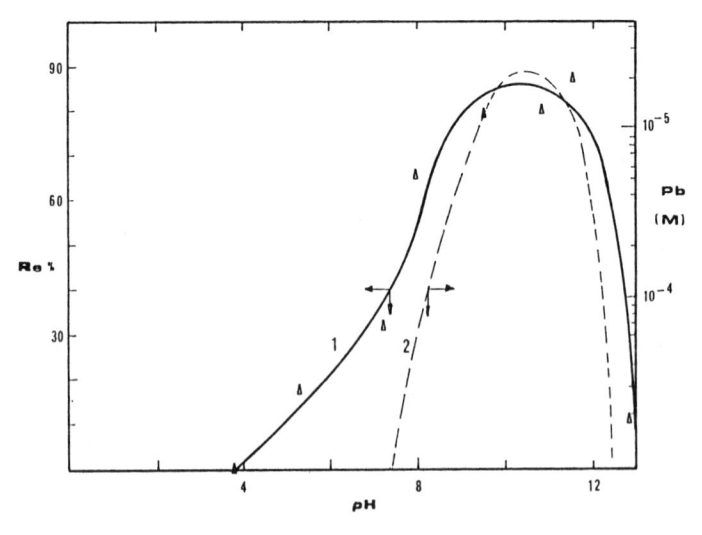

Figure 3. Effect of pH value of solution on precipitate flotation removal of lead (initial concentration of lead 10^{-4} M, laurylamine 2×10^{-4} M). The dotted line is a theoretical curve

Adsorbing Colloid Flotation

Ion flotation when first studied, was not restricted to simple ions, but also included charged colloids, which were often encountered in industrial wastewaters. This presented an opportunity to process certain types of uncharged pollutants that are normally removed by coagulation or adsorption (such as in the pretreatment step) for instance, with the addition of ferric or aluminium salts, which are hydrolysed.

Adsorbing colloid flotation is examined at this point in the paper only for historical reasons, since it was developed more recently. The technique generally involves the removal of a solute by adsorption on or coprecipitation with a carrier floc, which is then floated after the addition of a suitable surfactant. Adsorption, coprecipitation and occlusion are processes that are rather difficult to distinguish.

This technique has been applied in the laboratory successfully for the removal of arsenic [7, 15]. The forementioned applications are summarized in Table 1.

Table 1. Optimum flotation conditions for the removal of heavy metal studied

Arsenic (Adsorbing colloid flotation)		Chromium (Ion flotation)	
Concentration	: 10 ppm	Concentration	: 10 ppm
Adsorbant	: $Fe(OH)_3$, 60 ppm at pH:9	Flotation pH	: 7 - 8.5
Collector	: Sodium oleate, 15 ppm	Collector	: Dodecylamine, 8×10^{-4} M
Flotation pH	: 5	Removal	: over 80%
Removal	: over 90%		
		Germanium (Ion flotation)	
Lead (Precipitate flotation)		Concentration	: 10^{-4} M
Concentration	: 10^{-4} M	Flotation pH	: 7
Flotation pH	: 9	Complex	
Collector	: Dodecylamine, 2×10^{-4} M	formatation	: Pyrogallol, 3×10^{-4} M
Removal	: 90%	Collector	: Dodecylamine, 2×10^{-4} M
		Recovery	: 95%
Zinc (Precipitate flotation)			
Concentration	: 10 ppm		
Flotation pH	: 9.5		
Collector	: Dodecylamine, 2×10^{-4} M		
Removal	: over 98%		

In adsorbing colloid flotation (as also in precipitate flotation) it is not necessary for the surfactant to react with the solute and so, generally, it requires less reagents. It is noted that it is often possible to reclaim the surfactant relatively easily from the foamates, and hence it can be recycled back in the process [16].

The technique, however, suffers from the disadvantage of non-selectivity, as was the case for the system germanium-arsenic [9]. In contrast, as shown in Figure 4, ion flotation was effective for differential separation and germanium recovery. Germanium formed a tripyrogallol-germanic acid, which floated with the cations of laurylamine.

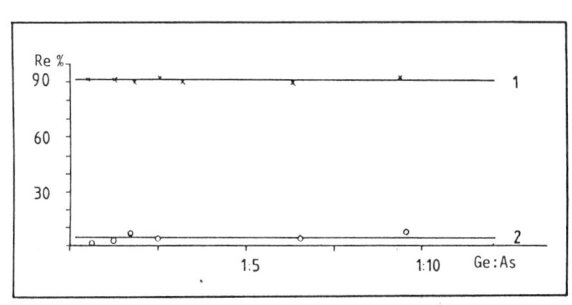

Figure 4. Ion flotation of Ge/As (initial [Ge] = 10^{-4} M, [pyrogallo] = 3.3×10^{-4} M, [laurylamine] = 2.2×10^{-4} M, pH value neutral). 1) Re% Ge; 2) Re% As

In this case, zeta-potential measurements were found to be very useful. Figure 5 shows the results with magnesium carbonate, in the presence of a common modified flotation agent, sodium hexametaphosphate. Magnesium carbonate fines have been used, for instance, in the adsorption of fatty acids, followed by flotation [17]. The applications of zeta-potential measurements were discussed previously [18].

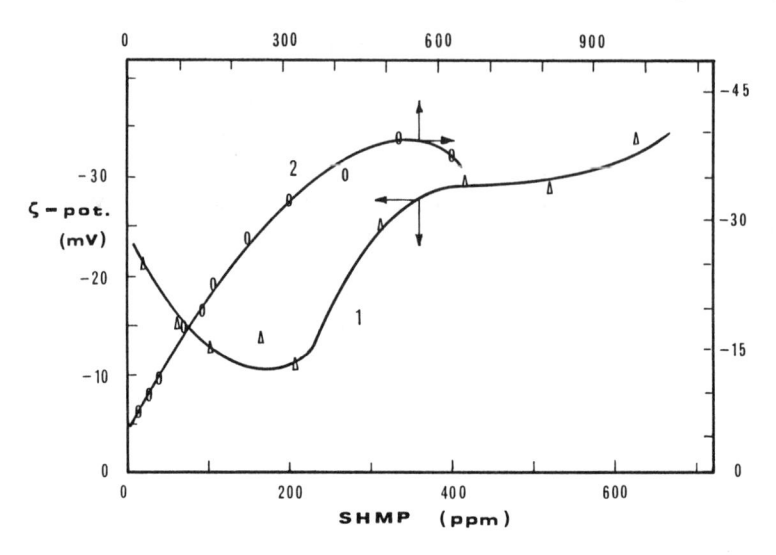

Figure 5. Zeta-potential measurments of $MgCO_3$ at varying sodium hexametaphosphate additions. 1) pH neutral; 2) pH 10

Discussion and Remarks

It is quite attractive to explore the possibilities of recovering relatively expensive materials from spent or leach solutions. Based on previous studies, for instance, a separation route was suggested [19], and is presented in Figure 6. Starting with an

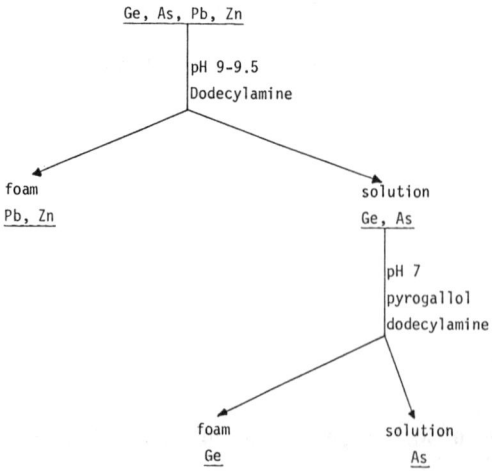

Figure 6. Suggested flowsheet for the separation of Ge from a mixture with As, Pb and Zn

aqueous germanium solution also containing arsenic, lead and zinc, elements that usually coexist with germanium during its production, with concentrations of 10^{-4} M (on the order of ppm), the following separation stages could be carried out: (1) pH adjustment to 9 – 9.5 and addition of an ethanolic solution of laurylamine, with a concentration proportional to the initial concentration of total lead and zinc. The solution is ready for precipitate flotation with about 3.3 cc/s of air as fine bubbles, with a duration of 300 s. The froth layer is expected to contain approximately 90% of the lead and zinc. (2) In the underflow, pyrogallol and laurylamine are added in stoichiometric concentrations to the initial concentration of germanium and the pH is changed to 7. Ion flotation is expected to concentrate around 90% of the germanium in the froth, leaving arsenic in solution.

Nowadays, hydrometallurgical processing is producing large volumes of dilute solutions and the application of flotation for further separation is considered to be promising. When flotation techniques can be applied not only as a clarification method for effluents, but also as a concentration and separation method for recovering metals or other useful materials from effluents, then it can be concluded that the materials recovered in such a way will cost less than, for example, the production of the same metal from conventional raw materials. Of course, it is not easy to separate the cost-effectiveness of metal recovery from the overall cost of effluent treatment.

To summarize, therefore, a review of the current laboratory research in this field was given.

References

[1] Pearson, D. (1971) Conf. Industrial Waste Problems. London
[2] Fletcher, A.W. (1975) in: Symp. Series *41*. I. Chem. E., U.K.
[3] Flett, D.S., Pearson, D. (1975) Chem. and Ind. *639*
[4] Lemlich, R. (ed.) (1972) Adsorptive Bubble Separation Techniques. Academic Press, New York
[5] Backhurst, J.R., Matis, K.A. (1981) J. Chem. Techn. Biotechnol. *31*, 431
[6] Matis, K.A., Gallios, G.P. (1985) in: Wills, B.A., Barley, R.W. (eds.): Mineral Processing at a Crossroads. Martinus Nijhoff, Dordrecht
[7] Matis, K.A., Stalidis, G.A., Zouboulis, A.I. (1984) 5th Europ. Conf. Envir. Pollution. EPRI, Amsterdam
[8] Matis, K.A., Zouboulis, A.I. (1985) Intl. Conf. Heavy Metals in the Environment, Vol. 1. CEP Consultants, Athens
[9] Matis, K.A., Papadoyannis, I.N., Zouboulis, A.I. (1987) Int. J. Min. Process. *21*, 83
[10] Zouboulis, A.I., Matis, K.A. (1987) Chemosphere *16*, (2/3) 623
[11] Zouboulis, A.I., Matis, K.A., Spathis, P.K. (1987) Techn. Chron. (Scient. J.) C-7(1):5
[12] Zouboulis, A.I., Matis, K.A. (1986) 11th Natl. Conf. Chem., Athens
[13] Stalidis, G.A., Matis, K.A., Lazaridis, N.K. (1987) Int. J. Min. Process. (to be published)
[14] Matis, K.A., Spathis, P.K., Zouboulis, A.I. (1986) 11th Natl. Conf. Chem., Athens
[15] Matis, K.A., Stalidis, G.A., Zouboulis, A.I. (1985) 10th Natl. Conf. Chem., Patra
[16] Beitelshees, C.P., King, C.J., Sephton H.H. (1979) in: Li, N.N. (ed.): Recent Developments in Separation Science, Vol. V. Florida
[17] Matis, K.A., Gallios, G.P., Zouboulis, A.I. (1987) 4th Intl. Symp. Envir.: Pollution and its Impact on Life in the Mediterranean Region. MESAEP, Kavala

[18] Gallios, G.P., Matis, K.A., Birda, E.S. (1987) Techn. Chron. (Scient. J.) C-7(3/4):21
[19] Zouboulis, A.I., Matis, K.A. (1987) 31st Intl. Conf. IUPAC, Sofia

A.I. Zouboulis, K.A. Matis
Lab. Gen. & Inorg. Chem. Technol. (114)
Aristotelian University
54006 Thessaloniki
Greece

Pre-treatment of Wastewater from the Automobile Industry

R. Klute

1 Introduction

The minimum requirements concerning the introduction of wastewater from the metal-working and metal-processing industries into lakes and rivers in the Federal Republic of Germany are defined in the 40th Wastewater Regulation, dated 5 Sept. 1984. In accordance with this, the requirements shown in Table 1 are given for wastewater produced during the manufacture of vehicles (land, air and water vehicles), machines, apparatus, containers, electrotechnical equipment and similar products in factories, in which different finishing processes and wastewater production areas are present next to each other [1].

These minimum requirements have been given in terms of concentrations up to now, and apply to the direct introduction of wastewater into lakes and rivers (direct polluter). In the amendment to the Water Resources Policy Act (WHG), which became effective on 1 Jan. 1987, the pollutant load itself and its minimization is addressed, in addition to limiting the concentrations of particular pollution parameters. Furthermore, the "generally recognized conventions of technology" are no longer the sole standard for determining the minimum requirements. Rather, the substances and groups of substances which are considered to be dangerous must meet the minimum requirements of the legislation which are "state of the art in technology", according to Para. 7a of the WHG. This also applies for the introduction of wastewater into a public wastewater treatment plant (indirect polluter), according to the amendment to the Water Resources Policy Act, and thus leads to consistency in the statute for direct and indirect introduction.

In an ordinance proposed by a regional study group for water (LAWA), substances and groups of substances were determined for which authorization must be obtained if a certain threshold is exceeded, before they can be introduced into a public wastewater treatment plant (see Table 2). In the meantime, some states in the FRG have adopted ordinances concerning the introduction of substances and groups of substances into wastewater treatment plants which are based on the LAWA's ordinance [2]. Most of the substances in Table 2 are either currently relevant to wastewater levy, such as cadmium and mercury, or will become important in future amendments to the Wastewater Levy Act, such as organic halogen compounds (AOX) and the heavy metals chromium, copper, nickel and lead [3].

Table 1. Minimum requirements for introducing wastewater from the vehicle manufactoring industry into lakes and rivers [1]

		Random sample	2 hr. mixed sample
Settleable substances	[ml/l]	0.3	-
Chemical oxygen demand (COD)	[mg/l]	-	400
Fish toxicity as dilution factor GF		-	8
Cadmium	[mg/l]	-	0.05
Mercury	[mg/l]	-	0.005
Aluminum	[mg/l]	-	3
Nitrogen from ammonium compounds	[mg/l]	-	30
Lead	[mg/l]	-	0.3
Chromium	[mg/l]	-	0.5
Chromium VI	[mg/l]	-	0.05
Cyanide, easily released	[mg/l]	-	0.05
Iron	[mg/l]	-	3
Fluoride	[mg/l]	-	5
Copper	[mg/l]	-	0.3
Nickel	[mg/l]	-	1
Nitrogen from nitrite	[mg/l]	-	5
Zinc	[mg/l]	-	2
Hydrocarbons	[mg/l]	-	5

Table 2. Ordinance concerning authorization to introduce substances and groups of substances into wastewater treatment plants

Substances or groups of substances	Threshold values required for authorization	
	[mg/l]	g/h
Total arsenic	0.05	1
Total lead	0.2	8
Total cadmium	0.02	0.4
Total chromium	0.2	8
Total copper	0.3	12
Total nickel	0.2	6
Total mercury	0.005	0.1
Adsorbable organic halogen compounds (AOX)	0.5	10
1,1,1-Trichloroethane, Trichloroethene, Tetrachloroethene and Trichloromethane	0.2	4
Free chlorine	0.2	4

The objective of amending the water legislation in the FRG is to achieve a significant reduction of pollutants which are introduced into the wastewater treatment plants or into lakes and rivers, in terms of tolerable concentrations as well as pollutant loads. These requirements can be met, on the one hand, by using more low-pollution technologies in industry. On the other hand, it is possible to eliminate more pollutants already at the source. In order to improve the level of purification possible at presently existing industrial wastewater treatment plants, a number of measures may be considered, for example:

– Designing the plants for a longer residence time in the case of hydraulic overload
– Improvement of solid/liquid separation
– Intercepting extraordinary pollutant loads in collection basins
– Dilution prior to introduction into the industrial purification plant
– Use of chemicals
– Separate pretreatment of wastewater streams prior to introduction into the industrial purification plant

This paper addresses the possibilities of improving the level of purification when treating wastewaters from the automobile industry. In particular, pretreatment of the wastewater streams using chemicals shall be investigated.

2 Types and Principal Treatment of Wastewaters from the Automobile Industry

There are mainly five different types of wastewaters produced by the automobile industry, which can be distinguished by their composition. These are:

1) Acidic/basic wastewaters
2) Wastewaters containing chromium
3) Wastewaters containing cyanide
4) Wastewaters containing lacquer
5) Emulsions of grease, oil and solvents in the water.

Usually wastewaters 1 – 3 continuously flow into the industrial purification plants in the form of rinse waters. In addition, there is the discontinuous production of wastewater when emptying used concentrates, cleaning rinsing baths, etc. Most of the wastewaters 4 and 5 are produced discontinuously, but the continuous and discontinuously produced wastewaters 1 – 3 may also be contaminated with these wastewaters.

Wastewater is typically produced in three situations in the course of usual production, namely, operation with change of bath, operation without change of bath, and change of bath and cleaning. This results in a significant variation of the concentration of pollutants in the wastewater.

Table 3 shows the variation of selected parameters for the acidic/basic wastewaters and the wastewaters containing chromium and cyanide. The data are based on a continuous, proportional removal of 4 hr. mixed samples at the inlet of the central detoxification plant of an automobile factory. The tests were conducted over one week and can be regarded as representative for the wastewater situation in

Table 3. Variation of selected pollution parameters at the inlet of the central detoxification plant of an automobile factory

Parameter	Acidic/basic wastewaters	Wastewaters containing chromium	Wastewaters containing cyanide
pH value	2.7 - 7.4	2.4 - 5.0	11.1 - 12.1
Settleable substances [ml/l]	<0.1 - 61	0.3 - 8	5.5 - 300
COD [mg/l]	240 - 2124	192 - 14,784	192 - 720
CN- [mg/l]	<0.1 - 1.4	<0.01 - 0.09	1.2 - 71
Total Cr [mg/l]	<0.5 - 5.6	10 - 83	0.3 - 2.3
Pb [mg/l]	<0.5 - 7.0	<0.5 - 20	<0.5 - 0.9
Zn [mg/l]	8.5 - 288	11 - 58	18 - 408
AOX [mg/l]	0.3 - 2.4	<0.1 - 2.6	0.3 - 2.1

the automobile industry. Since the analyses were carried out on 4 hr. mixed samples, it can be assumed that the instantaneous values vary significantly more.

In contrast to the three main types of wastewater mentioned above, the wastewaters 4 and 5 show an extreme high organic pollution due to suspended substances (grease, oil, lacquer components) and dissolved organic compounds. Table 4 shows the typical average values for the chemical oxygen demand (COD) and the concentration of halogenated hydrocarbons and aromatic organic compounds in some of these wastewaters.

Table 4. Typical concentrations of COD, halogenated hydrocarbons and aromatic compounds in wastewaters containing lacquer

Type of wastewater	COD (Mean value) mg/l	Halogenated hydrocarbons (Mean value) mg/l	Aromatics (Mean value) mg/l
Spray filler lacquer discharge	13.560	671	822
Spray lacquer, surface lacquer discharge	18.440	6	59
Water soluble lacquer discharge	13.700	0.015	0.460

Generally, the treatment of the wastewaters from the automobile industry proceeds as follows: the three main types of wastewater, namely, the acidic/basic wastewater, and the wastewaters containing chromium and cyanide, respectively, are collected separately and each is transported in its own conduit to the wastewater treatment plant. Here the wastewaters which contain toxic substances, such as cyanide or chromium (VI), are first detoxified. Cyanide is detoxified by oxidation of the cyanide ion under alkaline conditions, whereby sodium hypochloride and hydrogen peroxide are the most common oxidizing agents. The wastewaters containing chromium are detoxified by reducing chromium (VI) to chromium (III) under acidic conditions, whereby usually iron(II)-sulphate or sodium bisulphate (NaHSO$_3$) are used as reducing agents. Following detoxification, these wastewaters are combined with the non-toxic acidic/basic wastewaters and neutralized. When suitable neutralizing agents (slaked lime, soda lye, soda) are added, heavy metals precipitate out. The pH values required for precipitating most metals lie in the range 7 to 10.5 [4]. The flocs which are formed are subsequently separated by sedimentation. In order to improve the settling properties, a dose of an inorganic flocculation agent or an organic flocculation aid can be useful.

This wastewater treatment process is in accordance with the generally recognized conventions of technology and makes it possible to stay within the prescribed limits for introduction of wastewater into lakes and rivers shown in Table 1. Interruptions in operation may occur due to the discontinuous production of acidic, basic, cyanide-containing and chromium-containing concentrates. In order to avoid such interruptions, these concentrates are intercepted in sufficiently large storage basins and are added to the respective diluted wastewaters (rinse waters) continuously over a long period of time. Another cause of interruptions in operation at the central detoxification and neutralization plant is the production of wastewaters containing lacquer, as well as emulsions containing grease, oil and organic solvents. These interruptions, which are characterized by floc efflux and production of gas in the sedimentation basin, result in a reduced level of purification at the plant. In the next section, the possibility of pretreating these kinds of wastewaters prior to their introduction into the central wastewater treatment plant will be examined.

3 Pretreatment of Wastewaters Containing Lacquer by Chemical Precipitation/flocculation

Wastewaters containing lacquer are characterized by a heavy load of solids as well as chemically oxidizable substances having a COD on the order of 10,000 to 25,000 mg/l. The solids are in the size of colloids and have a negative surface charge. In addition, these wastewaters sometimes have a heavy load of halogenated hydrocarbons and aromatics (see Table 4).

Chemical precipitation/flocculation is regarded as a technically simple process, and efficient for the removal of turbidity and partially oxidizable substances as well. Using jar tests, a number of chemicals commonly used in chemical wastewater treatment were tested with regard to their effectiveness for pretreating wastewater. Specifically, the chemicals tested were the inorganic precipitation/flocculation agent Boliden AVR (aluminum/iron sulphate), Ferrichlor (iron chloride) and lime. The cationic polyelectrolytes Praestol 434K and 436K were used as flocculation aids.

The investigations were carried out using stirrers in 200 ml beakers. Chemicals were added at a stirring speed of 150 rpm. After 60 seconds, the speed of rotation was reduced to 50 rpm and stirring proceeded for 30 min. in order to form flocs which settle out well. After 30 min. settling time, samples were taken from the supernatant and analyzed for turbidity, measured with a Hach turbidity measuring instrument, and for COD, using the dichromate method. Turbidity and COD were the parameters chosen for evaluating the level of purification.

3.1 Wastewater from Fillers

Wastewater from fillers is produced in spray-painting areas in which the filler lacquer is applied to the body after the primer. The wastewaters are milky-brown. The pH value is around 8.5 to 9 and the COD value 10,000 to 20,000 mg/l. Figure 1 shows the results of the investigation for pretreating this type of wastewater.

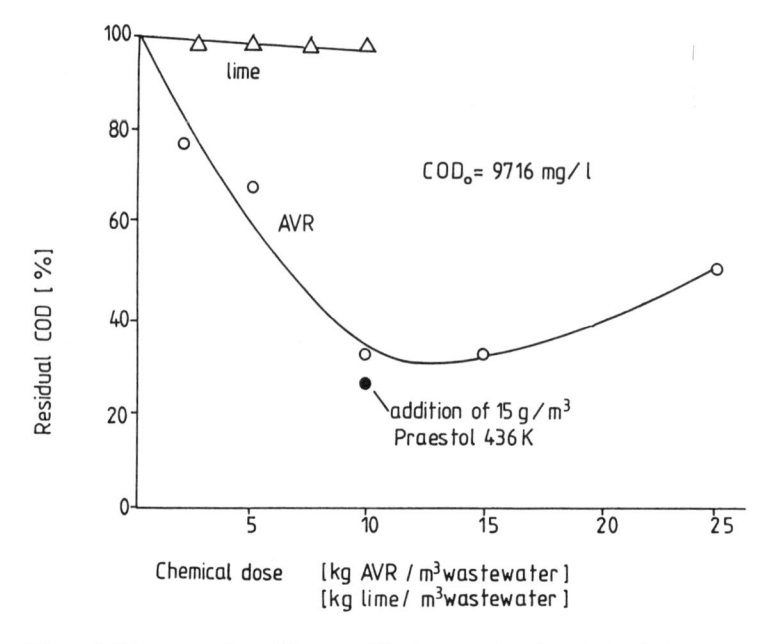

Figure 1. Wastewater from fillers - purification as a function of chemical dose

The use of lime produced no significant reduction in COD. In contrast, AVR reduced the COD by about 70%, and the concentration of chemicals required for this was around 10 kg AVR/m³ wastewater. At higher chemical doses, the level of purification in terms of COD decreased, and at the same time there was an increase in turbidity in the supernatant. By neutralizing the pH with hydrochloric acid and subsequently adding AVR, no further reduction in COD could be achieved, although the turbidity values of the supernatant were somewhat lower in this case. The addition of a flocculation aid (Praestol 436K) resulted in a further, albeit slight, reduction in COD, but a clearly lower turbidity value in the supernatant.

These results show that pretreatment of wastewater from fillers using precipitation/flocculation methods with aluminum/iron salts is effective, particularly with regard to a reduction in chemically oxidizable substances. In this case, however, the determination of the optimal dose of chemicals should be as exact as possible, since an overdose is expected to decrease elimination due to restabilization effects. With optimal process conditions, especially regarding type and concentration of flocculation aid, residual COD can be reduced to approximately 20 to 25%. The resulting supernatant is clear and only slightly colored. The volume of the precipitated sludge is approximately 2/5 of the original sample volume.

3.2 Wastewater from Metallic Surface Lacquer

The wastewater from metallic surface lacquer is produced in the spray-painting areas as a milky fluid having different colors. The pH value is around 9.5 and the COD 15,000 to 20,000 mg/l. The results of the investigations on wastewater from metallic surface lacquer are shown in Figure 2.

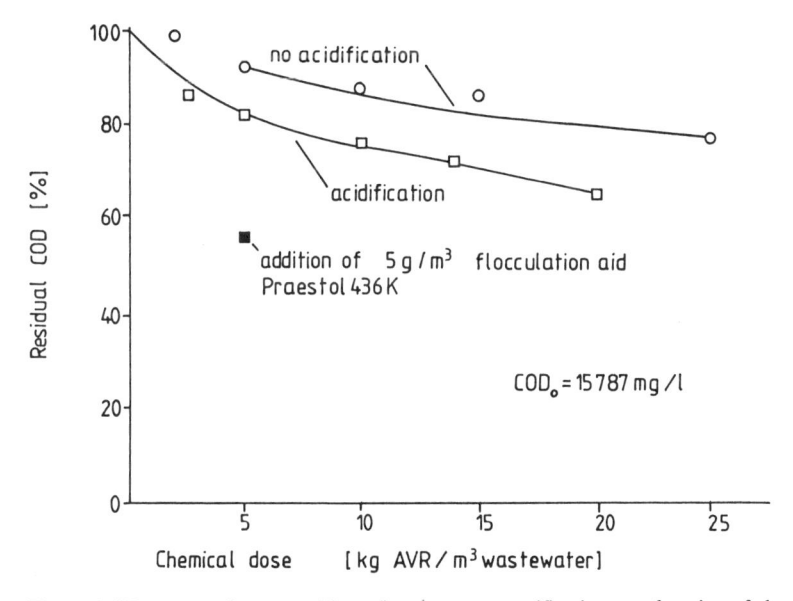

Figure 2. Wastewater from metallic surface lacquer - purification as a function of chemical dose

No COD reduction could be achieved for this wastewater using lime. The use of AVR or Ferrichlor resulted in an approximately 20% reduction in COD at higher concentrations of chemicals (25 kg AVR/m^3 wastewater) (see Figure 2). If the wastewater was first neutralized with hydrochloric acid, COD reduction was improved to about 35%. If the cationic polymer Praestol 436K was added subsequently, a further improvement of 45% removal of chemically oxidizable substances was achieved, with a significant reduction of residual turbidity in the supernatant to 20%. The volume of the precipitated sludge was around 50% of the original sample volume.

The above results indicate that pretreatment of wastewater from metallic lacquers by precipitation/flocculation with iron(III) or aluminum salts is effective. The first step in a possible pretreatment process could be neutralization with HCl, which is accompanied by a significant production of gas and foam. Subsequently, precipitation/flocculation agents as well as cationic polyelectrolytes (flocculation aid) could be added. The critical reaction time for precipitation/flocculation is 30 min., that for separation of sludge is 60 min. When the chemical dose is optimal, the resulting supernatant is clear and the sludge settles well. If the process is carried out correctly (with respect to pH, reaction times, stirring conditions, phase separation, for example), the level of purification can be improved so that residual COD is about 50%.

3.3 Wastewater from Water Soluble Lacquers

These wastewaters are produced when the gas tank is painted with water soluble lacquer. They are milky-white and contain small amounts of a fine black precipitate. The pH value of the wastewater is in the range from 6 to 7. The COD of the wastewater varied considerably during the investigation period; random samples gave values between 3139 and 23,210 mg/l. The precipitation/flocculation tests carried out on the slightly polluted wastewater are shown in Figure 3; those on moderately polluted wastewater in Figure 4.

Use of either the aluminum/iron salt AVR or lime resulted in a significant reduction of COD in the wastewater from water soluble lacquer. As shown in Figure 3, a concentration of 600 g AVR or lime per m^3 of wastewater produced residual amounts of COD of around 30%. When 5 g/m^3 of the flocculation aid Praestol 434K was also added, a further slight improvement in elimination could be achieved, due to the improved settling characteristics of the sludge. For another random sample which had a significantly higher COD value of 14,775 mg/l wastewater, AVR led to an even more obvious reduction in COD, which was around 80% without and 95% with the use of a cationic polyelectrolyte as a flocculation aid (see Figure 4). This

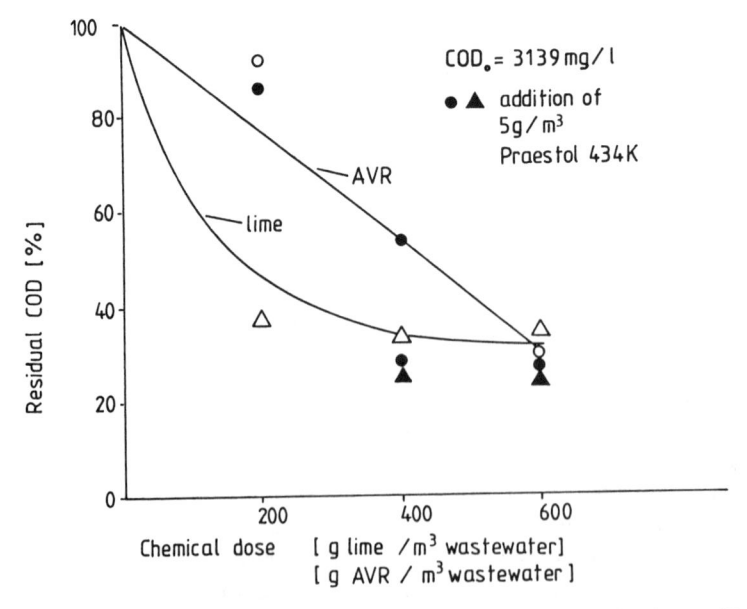

Figure 3. Wastewater from water soluble lacquer (moderately polluted) - purification as a function of chemical dose

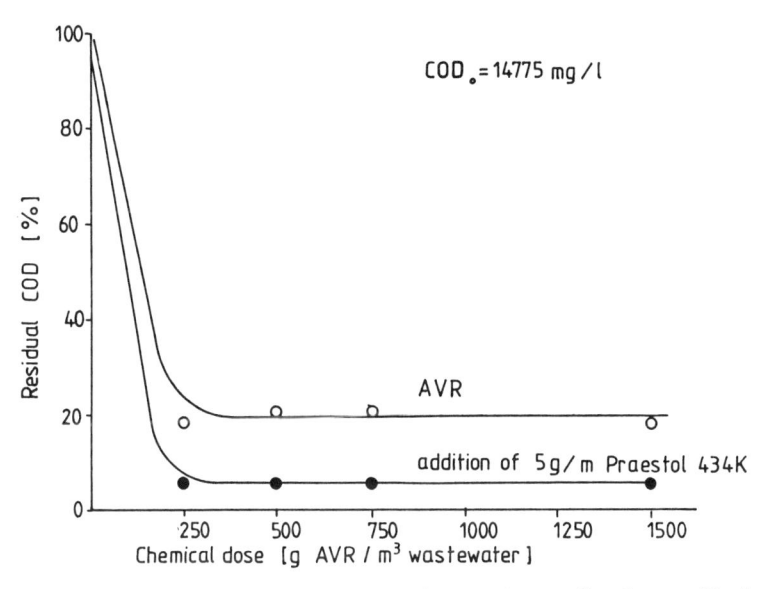

Figure 4. Wastewater from water soluble lacquer (very polluted) - purification as a function of chemical dose

level of purification could already be achieved with a chemical dose of 250 g AVR/m^3 wastewater. Interestingly enough, an overdose did not reduce the elimination efficiency.

These results suggest that there are various possibilities for pretreating the continuously produced wastewater from water soluble lacquers, whereby iron or aluminum salts or lime are effective as precipitation/flocculation chemicals. Addition of a flocculation aid improves the settling properties of the sludge and results in a further reduction in COD. The resulting supernatant is colorless and clear when the correct chemical dose is selected. The optimal reaction time for precipitation/flocculation is around 30 min.; that for sedimentation is about 30 to 60 min. assuming a flocculation aid is used.

4 Pretreatment of Wastewater from a Fleece Adhesive Plant by Chemical Precipitation/flocculation

This wastewater is produced discontinuously in the area where the body is assembled, and is milky-white in color. The pH value lies between 6 and 7, and the COD value varies for the random samples between 7300 and 31,000 mg/l. The investigations for pretreating this wastewater were carried out analogously to those for wastewaters containing lacquer. The results are shown in Figure 5.

This wastewater could be treated using AVR as well as lime. With lime, COD could be eliminated to residual amounts of about 40%; with AVR, the residual amounts were about 30%. A significantly higher elimination of COD could be achieved when both lime and AVR were used together, whereby 90% of the

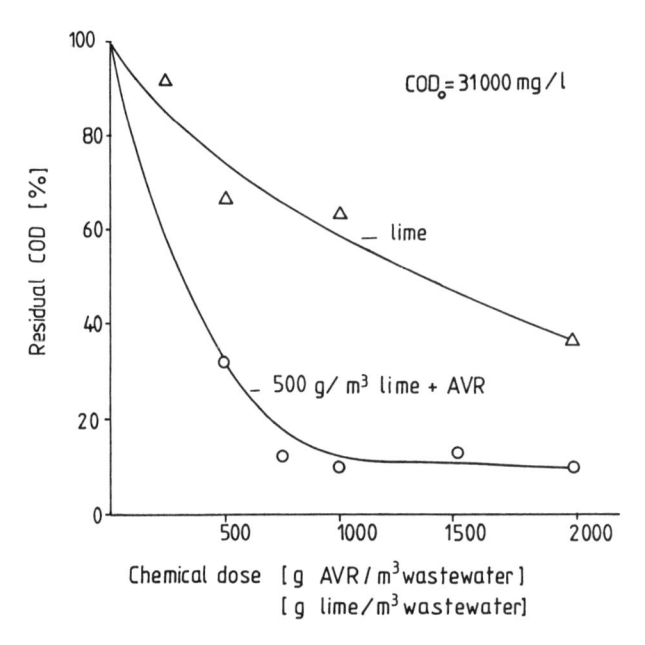

Figure 5. Wastewater from fleece adhesive plant - purification as a function of chemical dose

chemically oxidizable substances could be removed. Use of a flocculation aid did not achieve greater elimination.

These results seem to indicate that precipitation/flocculation results in a significant removal of substances, which are reflected in the COD value, from the wastewater of fleece adhesive; only about 10% remains as a residual pollutant load. The optimal concentration of chemicals depends on how polluted the wastewater was originally, but is on the order of 500 g $Ca(OH)_2$ and 1000 g AVR per m³ of wastewater in the present investigation. An optimal dose of chemicals results in a clear and colorless supernatant. The reaction times for precipitation/flocculation and sedimentation are 30 min. for each process.

5 Summary

There are mainly five different types of wastewaters produced by the automobile industry, which can be distinguished by their composition. Treatment of the wastewater is usually carried out in such a way that the acidic/basic wastewaters and the wastewaters containing cyanide and chromium, respectively, are collected separately and each is transported in its own conduit to the central detoxification and neutralization plant. In addition to these main types of wastewater, there is also the continuous as well as discontinuous production of wastewaters containing lacquer and emulsions of grease, oil and solvents in water. These wastewaters can cause permanent operational interruptions in the central detoxification and neutralization plant as well as a reduction in the level of purification.

Due to the amendment to the wastewater legislation in the FRG, stricter requirements for the introduction of wastewater from industry into lakes and rivers (direct introduction) and public wastewater treatment plants (indirect introduction) will be enforced. The interruption-free operation of the detoxification and neutralization plant is a prerequisite for the removal of dangerous substances and groups of substances in accordance with the "state of the art in technology". Thus pretreatment of wastewaters containing lacquer by chemical precipitation/flocculation was examined so that introduction of these wastewaters into the central wastewater treatment plant could proceed without difficulties.

Jar tests showed that the chemically oxidizable substances in these wastewaters, which gave a COD value on the order of 10,000 to 30,000 mg O_2/l, could be eliminated up to 90%, depending on the wastewater, by adding appropriate amounts of aluminum and iron salts resp. lime. The addition of an organic flocculation aid improved the settling properties of the flocs and further increased elimination to 95%. The procedures for pretreating wastewaters containing lacquer which were suggested here have since been applied in the automobile factory. Since then, operation of the central detoxification and neutralization plant proceeded without interruptions.

References

[1] Gesetz zur Ordnung des Wasserhaushalts (Wasserhaushaltsgesetz - WHG) in der Neufassung vom 23. September 1986. Bundesgesetzblatt I S. 1529
[2] Schmeken, W. (1988) Genehmigungspflicht für Indirekteinleiter bei Einleitung von wassergefährdenden Stoffen in Abwasseranlagen. Korrespondenz Abwasser 35, 134
[3] Schaal, H. (1987) Neue Anforderungen an die Abwasserbehandlung nach der Novellierung des Wasserhaushaltsgesetzes und des Abwasserabgabengesetzes aus der Sicht des Gesetzgebers. In: Bericht der Abwassertechnischen Vereinigung 37, 167
[4] Hartinger, L. (1976) Taschenbuch der Abwasserbehandlung für die metallverarbeitende Industrie. Hanser, München

R. Klute
Institut für Siedlungs-
wasserwirtschaft
Universität Fridericiana
7500 Karlsruhe
Fed. Rep. of Germany

Industrial Wastewater Pretreatment of a Dental-Pharmaceutical Company

F. W. Günthert and P.-M. Hajek

1 Introduction

The wastewater in the catchment of Lake Ammer, an important recreation area of Greater Munich, is collected in a separate system. The wastewater is treated in a mechanical-biological wastewater treatment plant by using the activated sludge process. The treatment plant with 60,000 population equivalents is situated at the mouth of Lake Ammer.

The ESPE GmbH company discharges its domestic and industrial wastewater into this municipal wastewater system. The organic substances, especially solvents, that are included among the different industrial wastewaters of this company, might cause damage in the wastewater system and a breakdown at the wastewater treatment plant.

For this reason, a pretreatment-purification plant for industrial wastewater, provided with balancing-, settling- and neutralization tanks, was already built in 1977. This plant does not meet the modern sanitary engineering standards and will be cited for excessive overloads if production is increased by the company. Because of this, the existing pretreatment plant was enlarged in 1986 and 1987.

2 Wastewater from the ESPE Company

The ESPE company employs 600 workers and produces dental- pharmaceutical and dental compounds. The company mainly produces tooth-impress, temporary and filling materials, denture compounds, local and topical anesthetics, medical instruments and technical attachments. More than 40 different products are produced using mainly the charge system. The water system of the ESPE company is shown in Figure 1.

Water is supplied by the municipal drinking water supply and private wells. The rainwater and the cooling water is directed immediately into the receiving water. The domestic wastewater is directed without any pretreatment into the wastewater system. The industrial wastewater from production, laboratory and technicum is directed to the existing and to the new wastewater pretreatment plant and then after a final inspection it will be transferred into the municipal wastewater system. The wastewater coming from the ESPE company is approximately 2%, on the average, of the total daily wastewater inflow to the municipal wastewater treatment plant and averages approximately 9% in terms of the COD load. The company may discharge an amount up to 5000 population equivalents into the municipal wastewater system.

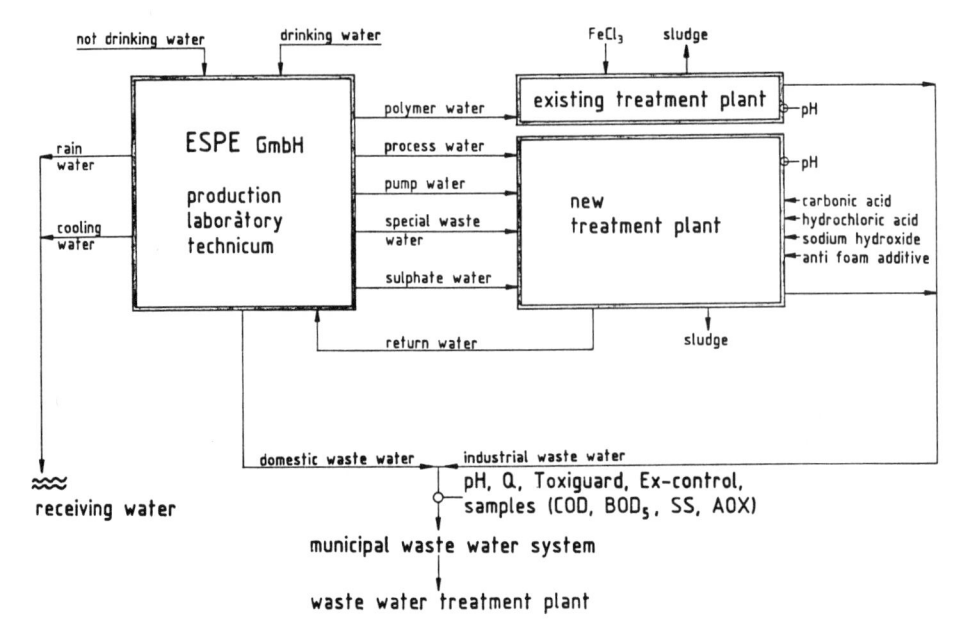

Figure 1. Wastewater system ESPE company

3 Pretreatment of the Different Kinds of Industrial Wastewater

Pretreatment of the industrial wastewater is carried out separately according to the following five categories:

— polymer-containing water
— process water
— pump water
— special wastewater
— sulphate-containing water

The pretreatment plant, existing and new, is shown in Figure 2.

3.1 Polymer-containing Water

Composition. During the production of tooth-impress, temporary and filling materials, washwater results, which contains low-molecular and water-soluble polymers and organic salts and small amounts of cyclohexane. The daily discharge amounts to 108 m^3/d and the daily load amounts to 175 kg COD.

Treatment. Up to now the water which contains polymer was treated with Fe(III)-chloride in a separate tank (V = 50 m^3). Therefore no adjustment could be made for

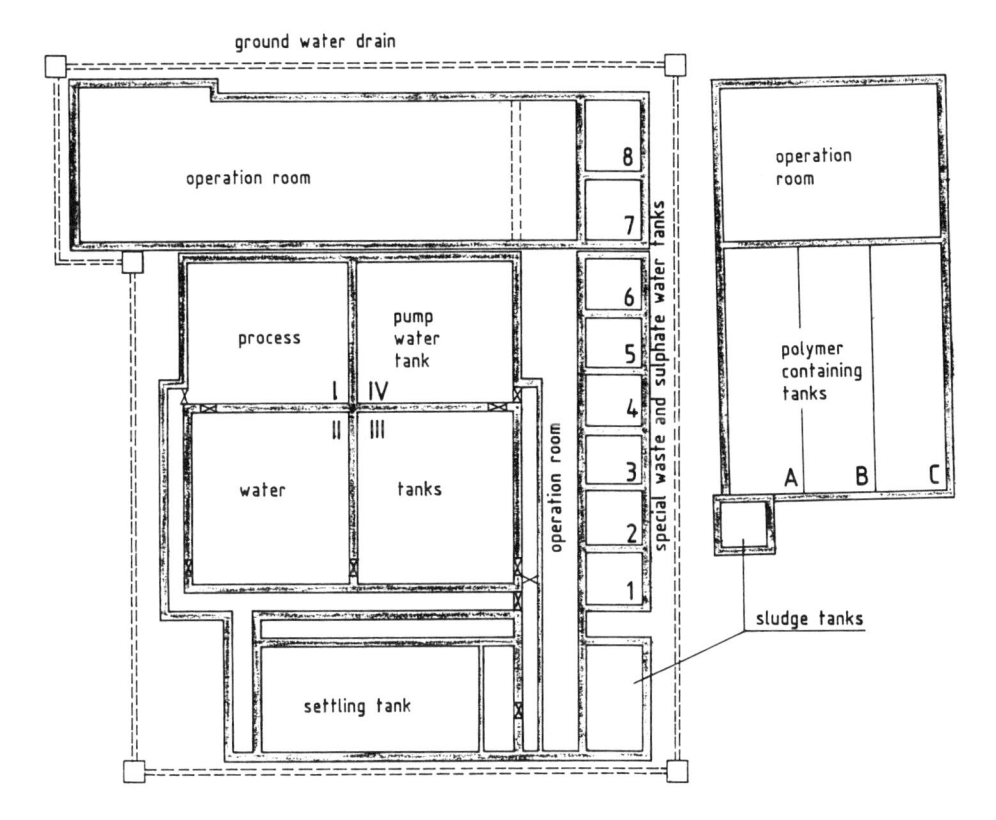

Figure 2. New and existing plant, ESPE company

quality and quantity, and holding back floating and settling sludge was possible only to a limited extent.

The existing treatment plant will be used for treating the polymer water (Fig. 3). The plant was rebuilt in 3 tanks which were series-connected with 50 m³ each. The polymer water will be precipitated with Fe(III)-chloride in the first tank in order to separate the floating and the settling sludge.

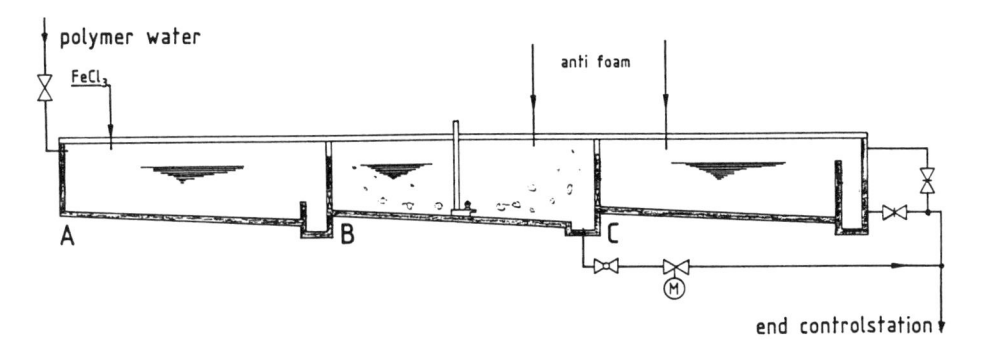

Figure 3. Treatment of polymer containing water

The dosage of Fe(III)-chloride will be regulated by the index of pH and by the inflow. Normally the first tank will always be full. Subsequently, the polymer water runs into the second tank and from there into the third tank. Both tanks are equipped with diving-aerators. When foam arises in the tank, anti-foam agents will be added. After this addition the inlet pipes will be rinsed with carbonic acid.

All three tanks will be filled in one production phase. After the wastewater treatment, tanks two and three will be emptied. The floating and the settling sludge will be disposed by a special waste disposal.

3.2 Process Water

Composition. The process water is produced in the laboratory, technicum and in production. It contains special organic chemicals, acids and lyes, and the composition varies. The daily discharge amounts to 70 m^3/d, the daily load from 60 – 120 COD.

Treatment. The existing tanks for the pretreatment of the process water could no longer guarantee a sufficient daily and weekly adjustment to the fluctuating wastewater quantity. Therefore, the process water will be passed through a new connection pipe (HDPE) which then leads to a rectangular settling tank that has a volume of 100 m^3 (Fig. 4a). The floated sludge will be retained by a scum board. At present there are no mechanical sludge scrapers. The floating sludge will flow through a scum channel into the sludge tank. The settled sludge will be removed from the sludge funnel into the sludge tank and then from there it will be transported together with the floated sludge to a special waste disposal facility. After the settling tank, the process water floats into one of the three buffer tanks which have a volume of 210 m^3 each (Fig. 4b). The three tanks can be filled separately or

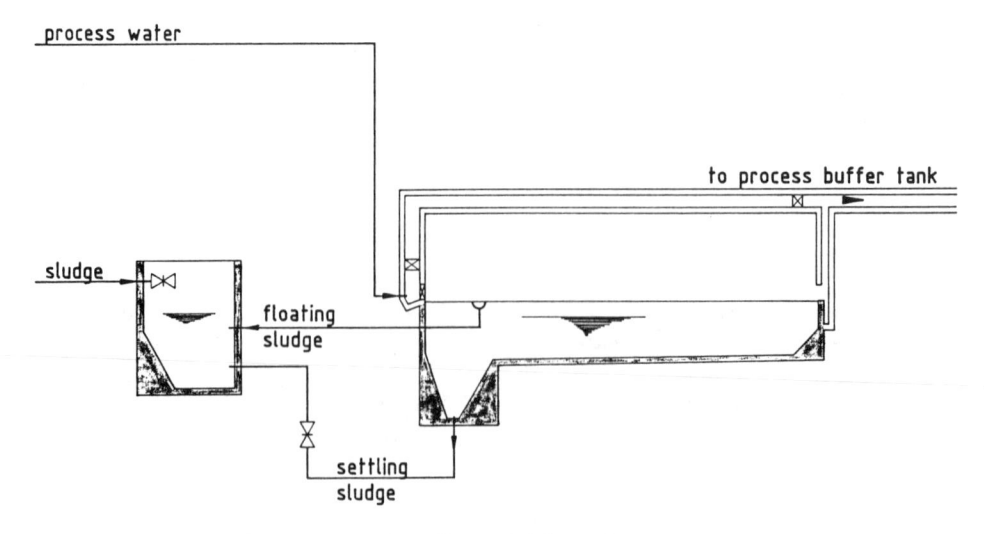

Figure 4a. Treatment of process water — settling and sludge tank

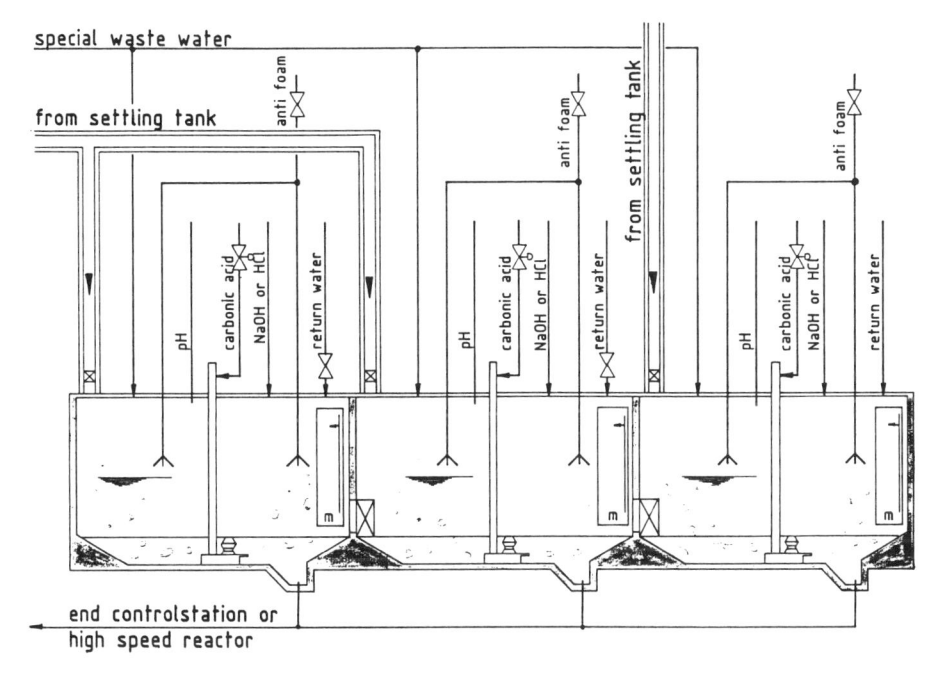

Figure 4b. Treatment of process water — buffer tank

they can be joined together. The tanks are equipped with an operation bridge, on which diving-aerators are installed. The diving-aerators have the function of keeping the process water fresh, mixing it and stripping solvents. The process water in the tank will also be neutralized.

An example of neutralization:

1) *tank I* is filled
2) index of pH in *tank I*: 3.5
3) switch on: neutralization
4) addition of sodium hydroxide
5) aeration on
6) sodium hydroxide will be pumped into *tank I*
7) when an index of pH 6.5 is obtained, switch off
8) automatic reset of aeration starts
9) rinsing and emptying of the inlet pipe

Neutralisation can also be achieved with a high speed reactor. The process water is pumped out of the tanks to the high speed reactor and then pumped back into the tanks. The pH-reading controls the addition of neutralizing agents.

The process water can only be discharged when the pH index is between 6.5 and 10.0. When the pH index is not between these values, the pumps automatically shut off and an alarm is sounded. When the pumps are emptying the tanks, the aerators are not in operation.

The pumps are stored in the operation room along with the pH meter, the neutralization plant and the unit which doses the anti-foam agent.

3.3 Pump Water

Composition. The pump water is produced by the ring pumps. The pump water can contain all liquids which accumulate in the course of production, even traces of solvents in the range of ppm, for example, tetrahydrofurane, acetone, toluene, cyclohexane and different alcohols. 150 m^3/d are produced daily and the daily load is 195 kg COD.

Treatment. Up till now the pump water was treated in the existing plant, but this is no longer sufficient. The pump water will now be directed into the buffer tank which has a volume of 210 m^3 (Fig. 5). This buffer tank and the process water tanks are both equipped with a diving-aerator and neutralization unit. The pump water can be treated in the same manner as the process water. Ordinarily it is discharged together with the process water. It also can be directed back into one of the process water tanks.

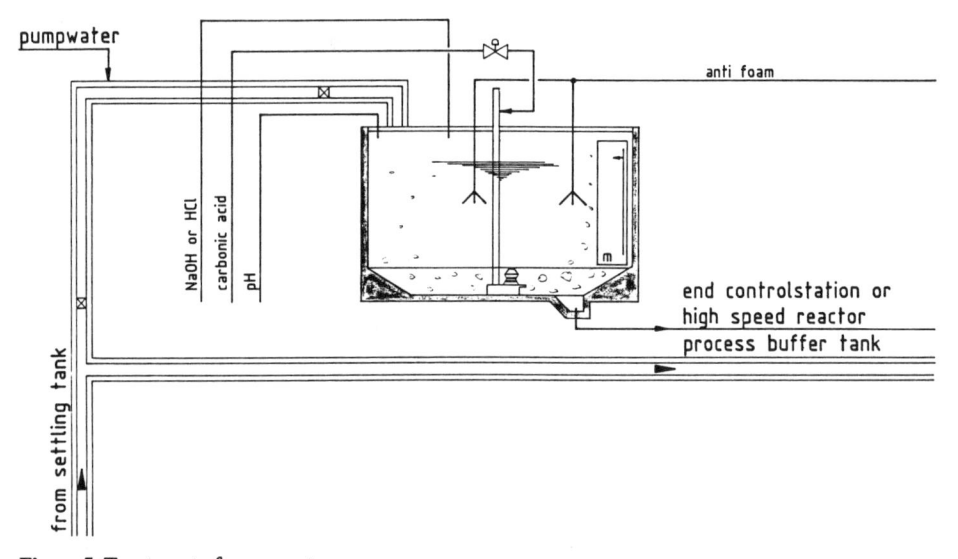

Figure 5. Treatment of pump water

3.4 Special Wastewater

Composition. In the technicum wastewater can be produced, for example in defective charges, which need to be treated specially before they can be discharged. This wastewater only occurs in exceptional cases, therefore no details about quantity and load can be given.

Treatment. A conscientious chemist must be able to recognize special wastewater in his laboratory. Therefore the most important thing is to prevent the production of wastewater. If special wastewater is produced, then the wastewater of the technicas I–IV can be directed to the four special wastewater tanks (Fig. 6). These buffer

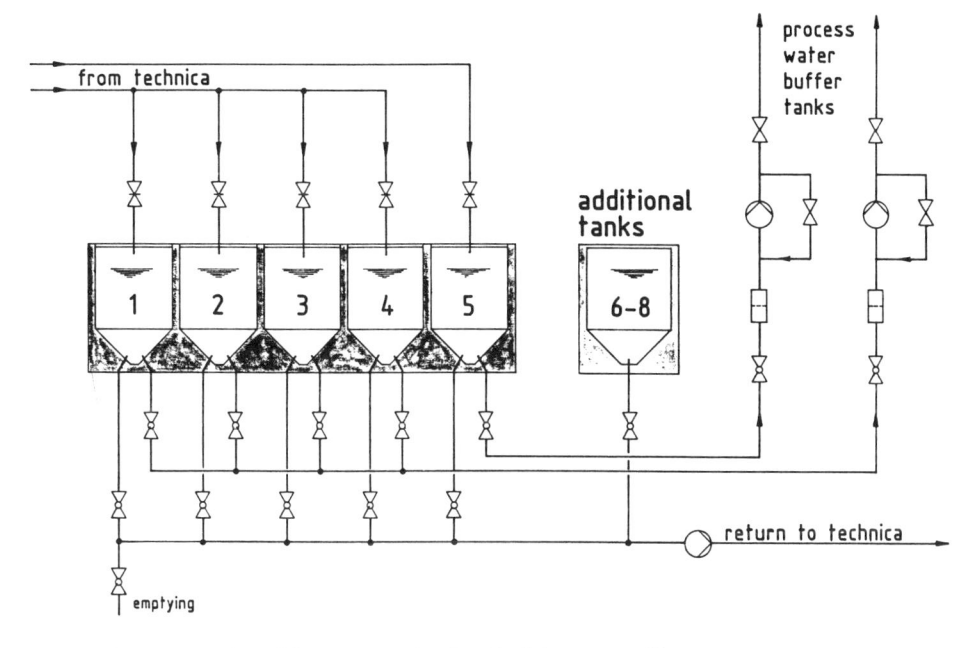

Figure 6. Treatment of special wastewater (1 - 4) and sulphate water (5)

tanks have a volume of 15 m³. There the wastewater will be treated, depending on its composition, with chemical oxidation, biologically with microorganisms, or with activated charcoal.

The special wastewater can be passed on via three different routes:
a) return to the technicum to recover the components
b) flow into the process water tanks
c) discharge to the special waste disposal facility

3.5 Sulphate Water

Composition. Sulphate-containing water consists of diluted sulphuric acid and sodium sulphate from the technicum. The sulphuric acid originates from the exchange of ions. Sodium sulphate serves as a drying agent in the synthesis of organics. The daily discharge amount is 2 m³/d with a concentration of 1 – 5% aqueous solution.

Treatment. The sulphate-containing water will be conducted directly into the fifth special wastewater tank which has a volume of 15 m³. From there it can be directed by means of a dosing pump to the three process water tanks. The sulphate-containing water can also be pumped back directly into one of the four technicas.

4 Construction and Costs

All of the tanks are constructed from steel concrete. The composition of the concrete was chosen so that it will not be vulnerable to attack by chemicals, in accordance with DIN 4030, and that the concrete is water resistant, in accordance with DIN 1045, Paragraph 6.5.7.2. In addition, the highest requirements for limiting the size of the cracks were met. All the tanks were constructed without joints.

The interior of all the tanks is coated. For the coating, different colors, tiles and stainless steel materials have been examined for their resistance and durability. The materials were tested with wastewater produced in the laboratory. The following coating was chosen:

a) For the settling tanks, sludge tanks and the process and pump water tanks, as well as for the existing plant: *epocid-tar-coating*.
   The coating was applied in the following manner:

   – sandblasting of the concrete surface, concrete moistness approx. 3% – priming
   – knifing the filler with mortar
   – three layers of top coat in alternating colours

   This coating is resistant to cyclohexane and diluted acids and lyes.

b) For the wastewater tanks: *furane resin coating reinforced with glass mats*.

   This coating is also applied in several stages and it is approximately 5 – 7 mm thick. It is resistant to sodium hydroxide, hydrochloric acid, methylene chloride and 1,1,1-trichloroethane and is stable for at least five days.

   The slides and valves of the inlet and outlet pipes will be controlled pneumatically to prevent the danger of an explosion.

   The total construction cost of the wastewater pretreatment plant amounts to 4 million DM. The technical part of construction including the coating amounts to approximately 2.4 million DM, and the technical and electrotechnical equipment come to 0.9 million and 0.6 million DM, respectively.

5 Final Checkpoint

The total amount of wastewater in the pretreatment plant flows into the final monitoring station. The highest level of discharge can only amount to 20 m^3/h and 328 m^3/d.

The pretreated wastewater that is discharged must have the following characteristics:

pH	: 6.5 – 10.0
suspended solids	: < 0.5 ml/l
COD (settled)	: < 1500 mg/l
AOX	: < 0.5 mg/l

No toxicity, no oil, no grease

The following analyses must be made and recorded in the operating diary: Twice a month, 2 hr. mixed samples from the effluent of the pretreatment tanks for the process water, the water containing polymer, and also from the final effluent must be analyzed for adsorbed organic halogen compounds (AOX) and pH. 4 hr. mixed samples must be taken from the final effluent monitoring plant and analyzed over a 24 hour period for the following parameters:

suspended solids
COD
BOD_5
toxicity by Toxiguard, continuous
discharge, continuous

6 Conclusion

The industrial wastewater from this dental-pharmaceutical company contains certain compounds which cannot be discharged into the municipal wastewater system without prior pretreatment without creating problems. In order to treat the different streams from the wastewater settling tanks, buffer and neutralization tanks are provided. If necessary, anti-foam agents can be added. Before the pretreated wastewater is discharged into the municipal wastewater system, it is monitored as to its composition. The plant is constructed in such a way that it gives the best possible protection to the ground water.

F. W. Günthert
Wasserwirtschaftsamt München
Praterinsel 2
8000 München 2
Fed. Rep. of Germany

P.-M. Hajek
Ing. Gesell. Michele/ Dr. Veits
Theresienstr. 140
8000 München 2
Fed. Rep. of Germany

Membrane Separation Processes for Industrial Effluent Treatment

P.S. Cartwright

Abstract

Industrial contaminants typically exist in one of the following forms: suspended solids, dissolved ionic or dissolved organic materials. Often, these contaminants represent a health or safety hazard. As a result, these discharges are becoming increasingly regulated with an eye towards eventual elimination.

The membrane processes of cross-flow microfiltration, ultrafiltration and reverse osmosis offer excellent potential for continuous removal of these contaminants. The selection of the optimum process is a function of the form of the contaminants present, as well as several other factors.

Membrane separation technologies cannot make the contaminants disappear; however, they are extremely effective at concentrating the contaminants or "dewatering" the stream in a pretreatment phase.

In many cases the contaminant is merely a component from an industrial manufacturing operation that "escapes" from the process during a rinsing operation, or is purposely removed.

This paper emphasizes the "point-of-source" concept of recycling or recovering specific components for reuse through the application of membrane separation technologies.

The fundamentals of these technologies are described, including membrane polymers and device configurations. In addition, complete system design considerations are presented.

Introduction

There are many treatment technologies available today either for removing contaminants from industrial streams or rendering the contaminants harmless.

The following table lists processes that are currently in use to remove contaminants from water supplies.

Figure 1 illustrates the relationship of some of these technologies to contaminant size range.

Membrane separation technologies offer the following advantages over competing processes:

— continuous processing (not batch)

	Dissolved BOD removal	Suspended Colloidal solids removal	Dissolved inorganic removal	Microorganism removal
Biological processes				
Activated sludge	×	×	-	×
Anaerobic digestion	×	×	-	-
Bio-filters	×	-	-	-
Extended aeration				
Bio-denitrification	L	-	-	-
Bio-nitrification	×	×	-	-
Pasveer oxidation ditch	×	×	-	×
Chemical processes				
Chemical oxidation				
Catalytic oxidation	×	×	-	×
Chlorination	×	×	-	×
Ozonation	L	-	-	×
Wet oxidation	×	×	-	×
Chemical precipitation	-	-	×	-
Chemical reduction	-	-	×	-
Coagulation				
Inorganic chemicals	×	×	-	×
Polyelectrolytes	×	×	-	×
Disinfection	-	-	-	×
Electrolytic processes				
Electrodialysis	-	-	×	-
Electrolysis	-	-	×	-
Extractions				
Ion exchange	-	-	×	-
Liquid-liquid (solvent)	-	-	×	-
Incineration				
Fluidized-bed	×	×	-	×
Physical processes				
Carbon adsorption				
Granular activated	×	×	-	-
Powdered	×	×	-	×
Distillation	×	×	×	×
Filtration				
Diatomaceous-earth filtration	-	×	-	×
Dual-media filtration	-	×	-	×
Micro-screening	-	×	-	×
Sand filtration	-	×	-	×
Flocculation-sedimentation	-	×	-	×
Foam separation	-	×	×	-
Freezing	×	-	×	-
Membrane processing				
Microfiltration	-	×	-	×
Ultrafiltration	×	×	-	×
Reverse osmosis	×	×	×	×
Stripping (air or steam)	×	×	-	-

L = Under specific conditions there will be limited effectiveness

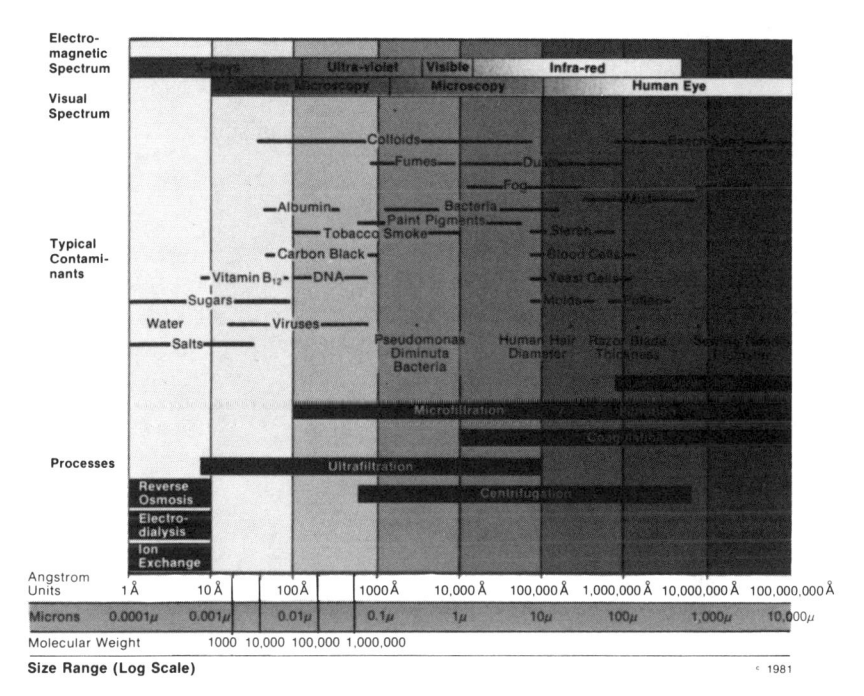

Figure 1. Contaminent removal capabilities of various separation processes

- low energy requirements (pumping only)
- modular construction
- simple maintenance demands
- positive barrier to contaminants
- ambient temperature operation

Listed below are the membrane processes recommended for the appropriate contaminant category.

Contaminant	Recommended membrane process
Suspended solids	
>0.1μ	Microfiltration
<0.1μ	Ultrafiltration
Dissolved solids	
Organic	Ultrafiltration/Reverse osmosis
Ionic	Reverse osmosis
Microorganisms	Microfiltration/Ultrafiltration

Background

All of the membrane processes utilize an engineering design known as "crossflow" or "tangential flow" filtration. In this mechanism, the bulk solution flows over and parallel to the membrane surface, and because the system is pressurized, water is forced through the membrane. The turbulent flow of the bulk solution across the surface minimizes the accumulation of particulate matter on the membrane and facilitates the continuous operation of the system.

Figure 2 illustrates both conventional and crossflow mechanisms.

Conventional filtration

Crossflow filtration

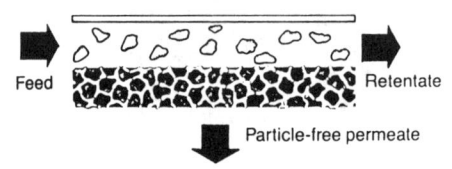

Figure 2. Filtration mechanisms

Figure 3 illustrates the mechanism of crossflow microfiltration. Microfiltration involves the removal of insoluble particulate materials ranging in size from 0.1 to 10.0 microns (1000 to 100,000 angstroms). Microfiltration membrane polymers include:

- Polycarbonate
- Polyester
- Mixed esters of cellulose
- Cellulose triacetate (CTA)
- PTFE (polytetrafluoroethylene)
- PVC (polyvinyl chloride)
- Thin film composite (TFC)
- Nylon

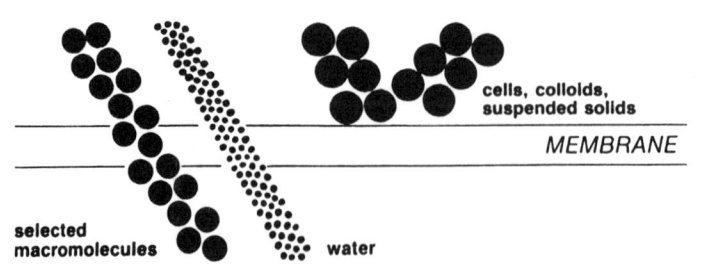

Figure 3. Microfiltration

Figure 4 depicts ultrafiltration, which is used to separate materials in the .001 to 0.1 micron range (10 to 1000 angstroms). Basically, ultrafiltration is used to remove dissolved materials whereas suspended solids are removed by microfiltration.

Typical polymers used include:
- Polysulfone
- Cellulose Acetate
- Polyamide

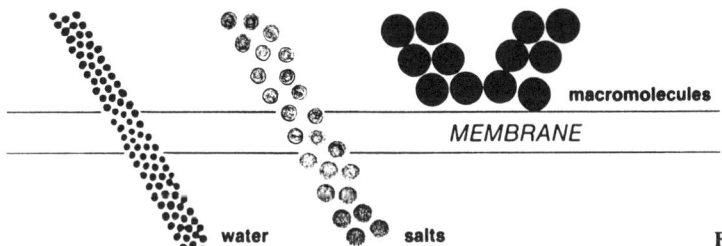

Figure 4. Ultrafiltration

Figure 5 illustrates reverse osmosis which typically separates materials less than .001 micron (10 angstroms in size). Reverse osmosis offers the added advantage of rejecting ionic materials which are normally small enough to pass through the pores of the membrane. As with ultrafiltration, reverse osmosis is used to remove dissolved materials.

Polymers used in the reserve osmosis membrane include:
- Polyamide
- Thin Film Composite (TFC)
- Cellulose Acetate
- Cellulose Triacetate

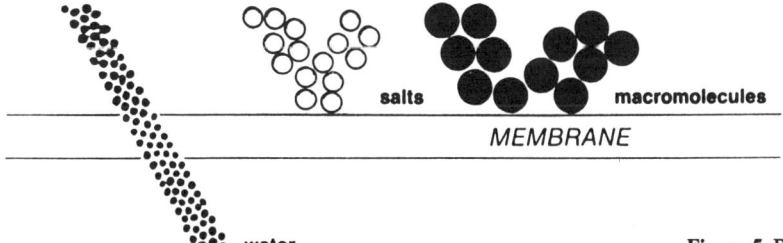

Figure 5. Reverse osmosis

Element Configurations

The membrane can be "packaged" in several element configurations, each offering particular advantage depending on the application.

Figure 6 illustrates the four basic configurations commercially available today.

Tabular. Manufactured from ceramic, carbon, or any number of porous plastics, these tubes have inside diameters ranging from 1/8 inch up to approximately 1 inch.

The membrane is typically coated on the inside of the tube, and the feed solution flows through the interior from one end to the other, with the "permeate" or "filtrate" passing through the wall to be collected on the outside of the tube.

Hollow Fiber. Similar to the tubular elements in design, hollow fibers are generally much smaller in diameter and require rigid support as is obtained from the "potting" of a bundle inside a cylinder. Feed flow is either down the interior of the fiber or around the outer diameter.

Spiral Wound. This device is constructed from an envelope of sheet membrane wound around a permeate tube that is perforated to allow collection of the permeate or filtrate.

Plate and Frame. This device incorporates sheet membrane that is stretched over a frame to separate the layers and facilitate collection of the permeate.

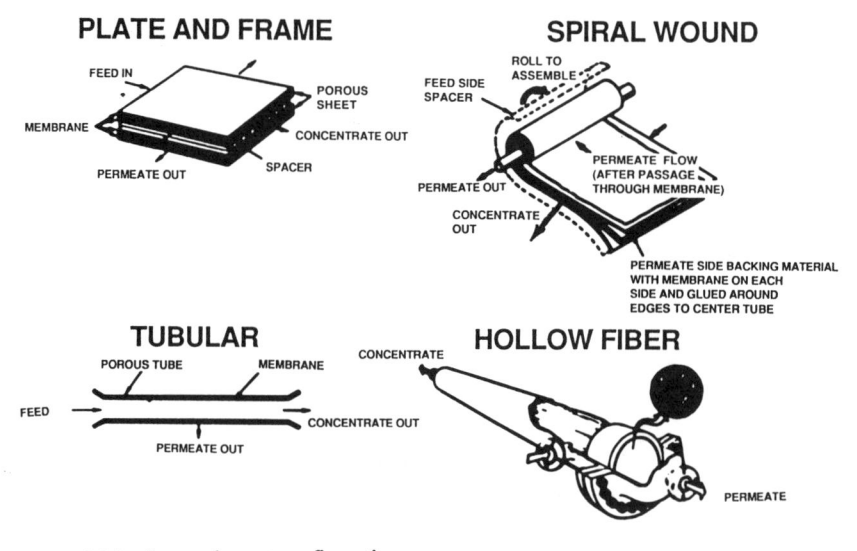

Figure 6. Membrane element configurations

The following table summarizes the important physical characteristics of the various membrane element device configurations available today:

Element configuration	Packing density *)	Suspended solids tolerance
Tubular	Low	High
Hollow fiber	Highest	Poor
Spiral wound	High	Fair
Plate and frame	Low	High

*) Membrane area per unit volume of space required

Because of the propensity of suspended or precipitated materials to settle out on the membrane surface and plug the membrane pores, turbulent flow conditions must be maintained (Reynolds numbers in excess of 4000).

System Design Considerations

In order to treat an effluent stream, it must be thoroughly analyzed for the following data:
- Total solids content
- Suspended (TSS)
- Dissolved organic (TOC)
- Dissolved inorganic (TDS)
- Specific chemical constituents
- Oxidizing chemicals
- Organic solvents
- pH
- Operating temperature

Usually the goal is to "dewater" the feed system as much as possible; that is, to remove solvent to facilitate either reuse or removal of the concentrated solute. Of secondary importance is the possible reuse of the purified solvent (usually water). These two considerations are significant in determining both the process and membrane device be to used.

Figure 7 depicts a general scheme for membrane processes. In these technologies the implication of increasing the dewatering process is described by the term "recovery", which is defined as the purified water volume divided by the incoming stream volume; in other words, percentage of the feed flow which is pumped through the membrane. Typically for effluent treatment applications, the recovery figure is at least 90%. As recovery is increased (to decrease concentrated solute volume), the concentration of solute and suspended solids in the concentrate stream increases.

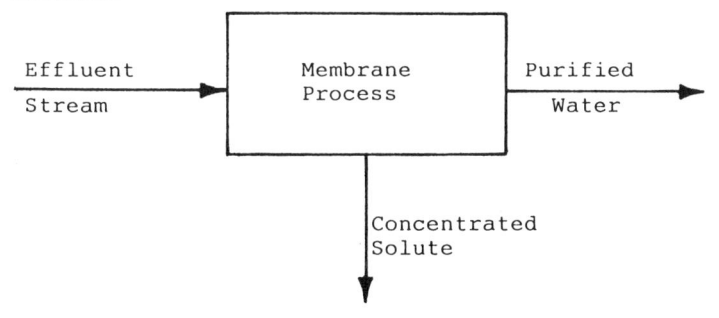

Figure 7. Membrane process schematic

No membrane is perfect in that it rejects 100% of the solute on the feed side; this solute leakage is known as "passage". Expressed as "percent passage", the actual quantity of solute which passes through the membrane is a function of the concentration of solute on the feed side. Under high recovery conditions, the concentration of solute on the feed side is increased and therefore the actual quantity of solute passing through the membrane also increases. Because most effluent applications demand that in addition to a minimum concentrate volume, the permeate quality be high enough to allow reuse or to meet discharge regulations, the "Catch-22" predicament of permeate quality decreasing as recovery is increased can impose design limitations.

Applications

Although membrane separation technologies have found applications in a number of specific areas, the potential for these unique processes has not yet begun to be tapped.

Following are descriptions of several specific applications and case histories illustrating where these processes have been utilized in industrial effluent treatment.

Metal Finishing Treatment

Point-of-source Recovery. In certain circumstances, it is possible to utilize reverse osmosis to effect a "zero discharge" electroplating rinse water recovery system. The rinse water from the first rinse is pumped to a reverse osmosis system that concentrates the salts and directs them back to the plating bath. The purified rinse water (permeate) is directed to the last rinse, and neither solute nor solvent is lost. In the United States there are approximately 150 reverse osmosis systems operating in this manner on nickel baths and 12 on acid copper. There are also a few installations operating on copper cyanide, hexavalant chrome and acid zinc.

Only those plating baths operating at relatively high temperatures (above 60 degrees Celsius) lend themselves to direct treatment by reverse osmosis. In most cases the cost of the system is recoverd by savings in plating salts within two years. Figure 8 illustrates this application.

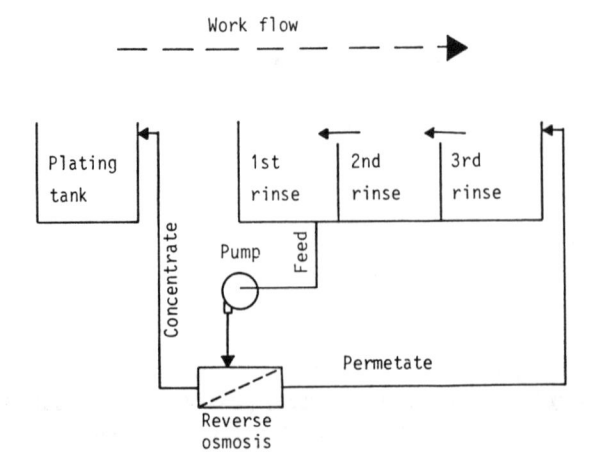

Figure 8. "Zero discharge" with reverse osmosis

"End of Pipe" Treatment. It is possible to use reverse osmosis and ultrafiltration to concentrate or "dewater" mixed effluent streams in order to reduce the hydraulic loading to downstream treatment processes. Typically at least 90% of the feed volume can be purified and often returned to the process, with the salts concentrated in the remaining 10%.

Conventional chemical treatment of metal finishing wastes will usually produce clarified effluent acceptable for discharge; however, in those applications where it is desirable or necessary to recover the clarified rinse water for reuse, the technologies are utilized to purify or "desalt" the effluent for reuse.

Microfiltration can be used to replace a clarifier in the chemical clarification of plating discharges. Compared to conventional equipment, it offers the advantages of continuous processing and significantly smaller space requirements. Figure 9 illustrates a microfiltration installation.

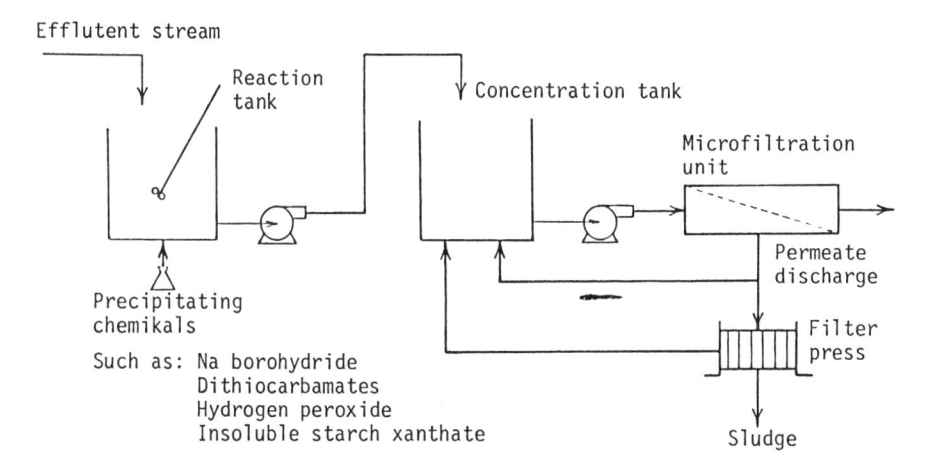

Figure 9. Microfiltration clarification

Oily Waste Treatment. Oil-water emulsion effluent streams are typically generated as a result of the following industrial activities:

- Metal cutting operations such as machining use oil-water emulsions for both lubrication and cooling.
- Metal forming operations use oil-water emulsions for lubrication.
- Hot and cold rollingo perations for steel and aluminium strip utilize oil-water emulsions for both lubrication and cooling.
- Heat treatment/quenching processes generate oil-water emulsions during the process of removing oily contaminants from metal parts.

Ultrafiltration can be utilized to separate the emulsion and dissolved oil from water. The specific ultrafiltration membrane polymer and pore size requirement are determined by the oil chemistry; however, the oil can typically be concentrated up to 60-80%, and in some cases, incinerated to recover energy in the form of heat. The permeate stream may be pure enough to be reused, or may require treatment with reverse osmosis prior to reuse.

Figure 10 illustrates the application of ultrafiltration to oily waste effluent treatment.

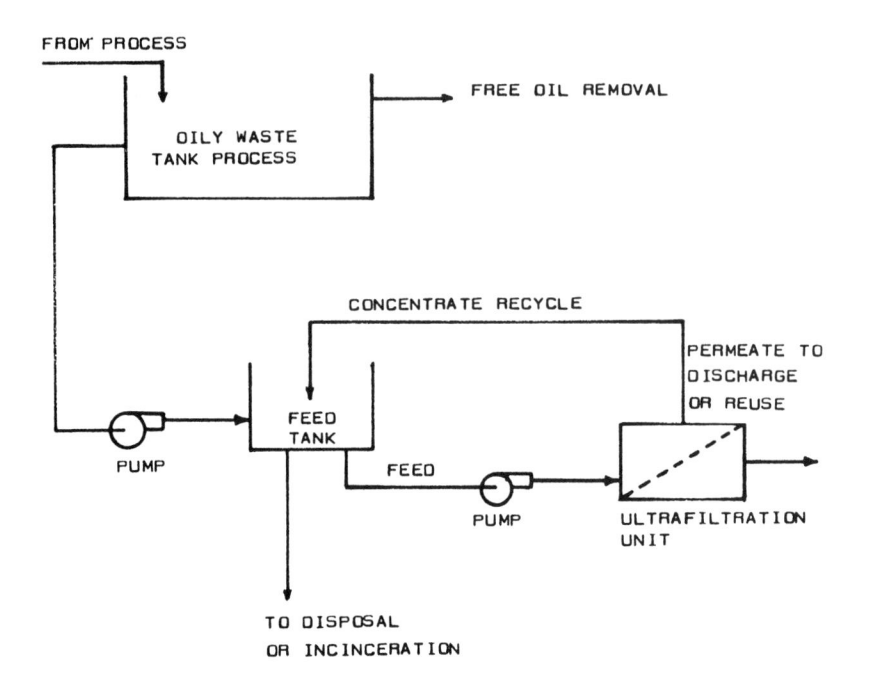

Figure 10. Ultrafiltration of oily wastes

Printed Circuit Effluent Treatment

Because printed circuit products require higher quality rinse waters than typical electroplated products, demineralized water is often used, and there is economic justification for recovery of this more expensive water.

The heavy metals used in printed circuit electroless plating (copper and nickel) are in chelated form (chemically "tied-up" in an organic matrix). The plating baths are more unstable than electroplating baths, thereby resulting in more frequent "dumping". As a result, waste treatment requirements in printed circuit manufacturing operations present special problems and opportunities for membrane separation processes.

Figure 11 illustrates a total printed circuit effluent treatment system utilizing reverse osmosis to recover purified water from mixed rinses and the airscrubber. Bath dumps and reverse osmosis concentrate are chemically treated, producing a sludge for landfilling and effluent suitable for discharge.

Membrane technologies can also be used in other parts of this total treatment system: microfiltration could be substituted for the clarifier (see Figure 9), and reverse osmosis could purify the clarified effluent for reuse.

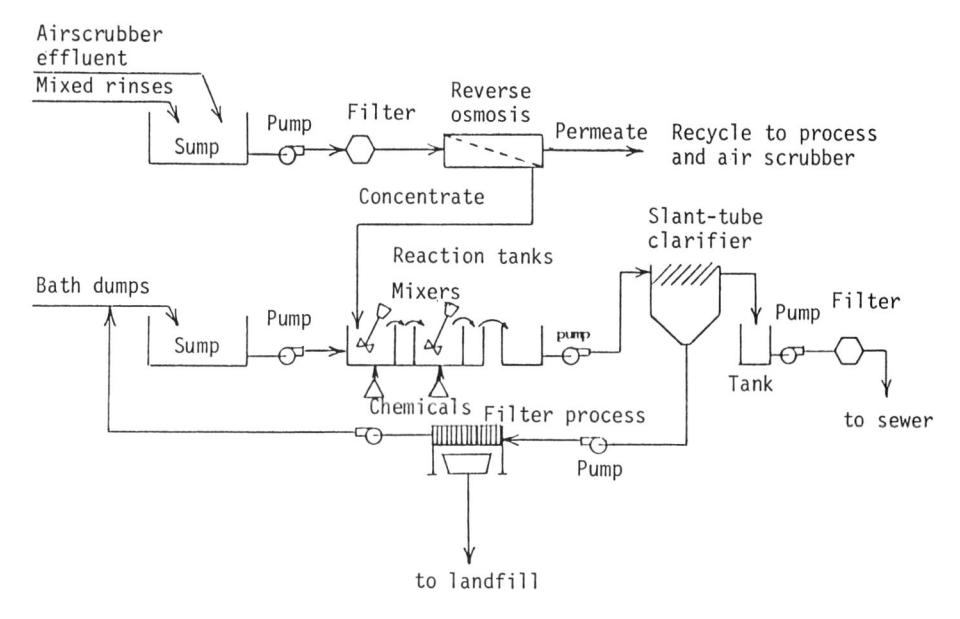

Figure 11. Printed circuit effluent treatment system

Semi-conductor Manufacturing Effluent Treatment

Because most industrial effluent streams contain a multitude of contaminants in many different forms, the greatest potential for membrane separation technologies is in combination with such other technologies as ion exchange, activated carbon adsorption, anaerobic digestion, etc.

An example of the integration of a number of these technologies to treat an effluent stream and recover high value rinse water is in a semiconductor manufacturing facility.

Figure 12 illustrates the basic design of a semiconductor rinse water reclamation facility which utilizes many of these technologies to reclaim approximately 90% of the contaminated rinse water and bring it back up to ultrapure water standards.

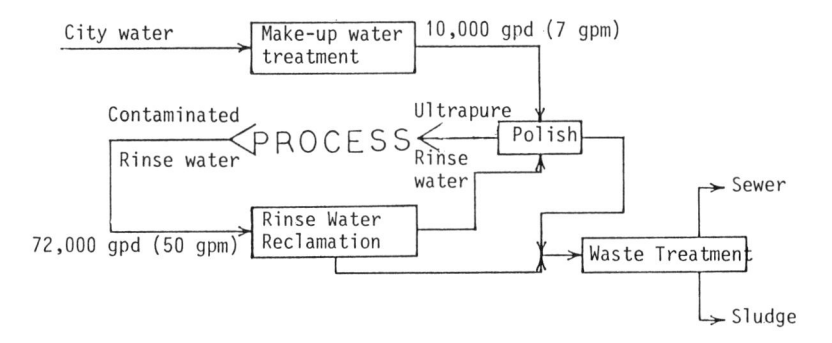

Figure 12. Semiconductor rinse water reclamation system

Figure 13 illustrates the specific components of the entire system. Because the manufacturing process involved heavy metals plating, the contaminated rinse water contained some hazardous materials, thereby requiring special waste treatment considerations.

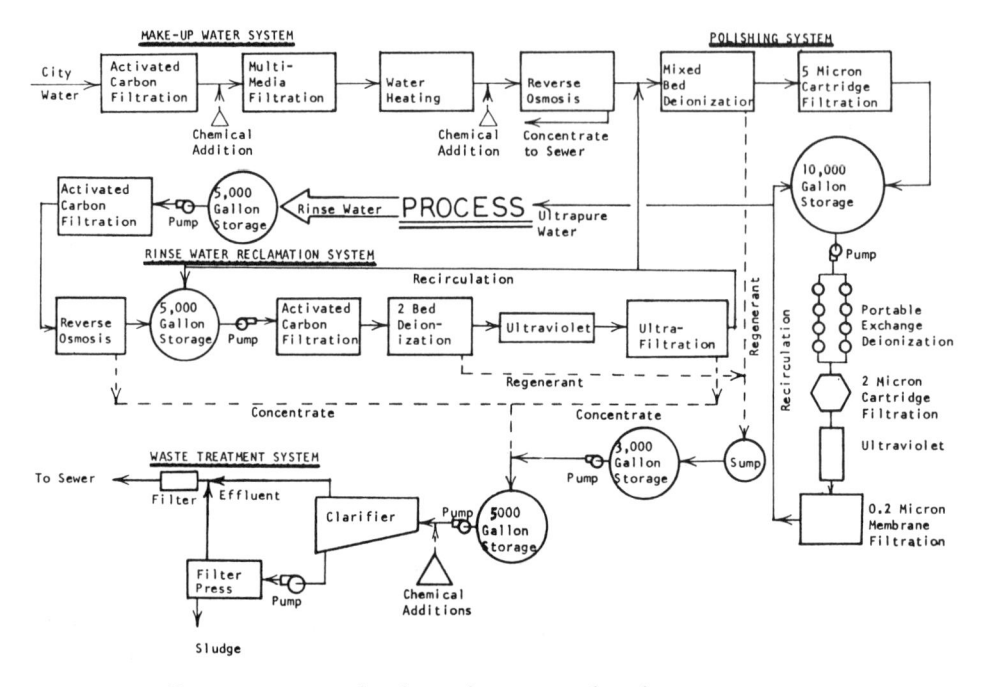

Figure 13. Specific components - semiconductor rinse water reclamation system

Conclusions

The membrane separation technologies of microfiltration, ultrafiltration and reverse osmosis offer many advantages in the treatment of industrial effluent streams. These advantages can only be realized through careful and deliberate integration into a total treatment system. Because these technologies are not well understood, it is hoped that through conference such as this, their outstanding potential will begin to be realized.

P.S. Cartwright
C3 International, Inc.
2019 West Cty Rd. C
St. Paul
MN 55113
USA

Alternative Treatment of De-icing Fluids from Airports

M. Boller

1 Introduction

In order to maintain safe air-traffic conditions during landing and take-off operations in airports at temperatures below freezing, deicing chemicals are used on the airplanes as well as on the runways. In most cases low molecular weight alcohols are used for deicing the airplanes, whereas for the runways alcoholic compounds are often combined with urea as a non-corrosive deicer. Accordingly, the run-off water from airplanes and runways contains high concentrations of the aforementioned chemicals. These in turn, cause tremendous heterotrophic growth and high NH_4^+-concentrations from the hydrolysis of urea in the receiving waters during winter time. At the Zürich airport, different ways of treating the deicing fluids were studied during the winter periods of 1985/86 and 1986/87. The purpose was to obtain alternative solutions for the removal of the organic substances and of urea and ammonia and to quantify the respective removal potentials.

In principle, two technical solutions were considered, namely (1) central deicing stations for the airplanes and recycling of the concentrated deicing fluids after treatment, for instance, by ultrafiltration and (2) aerobic or anaerobic treatment of the run-off water and biological hydrolysis and oxidation of the urea. Several alternatives of centralized deicing stations were considered. However, lack of space and high investment costs did not favour this solution. Furthermore, the run-off water from the runways would have to be treated separately, causing investment costs similar to those without central deicing stations. Therefore, only biological processes were considered as a feasible alternative for treating the combined deicer run-off. Since many process combinations of aerobic and anaerobic treatment are possible, different systems were investigated including the alternative of treating the run-off in a nearby municipal activated sludge plant.

Among the numerous possibilities, the process combinations shown in Table 1 were considered. Treating the deicing fluids with the activated sludge process (5 in Table 1) is known to cause excessive bulking sludge [1], and the anaerobic filter (6 in Table 1) does not promise a favourable energy balance because of low water temperatures and relatively low organic concentrations. Thus, only the first four process combinations were investigated in detail using pilot plant experiments.

Table 1. Possible process combinations for the biological treatment of deicer run-off water from airports

1) Anaerobic longterm storage / Discharge to the municipal activated sludge plant for degradation of residual organics and oxidation of ammonia (nitrification)
2) Aerobic (aerated) longterm storage with and without nutrient (phosphorous) addition / Discharge to the municipal plant for nitrification
3) Anaerobic storage / Rotating biological contactor for removal of organics / Discharge to the municipal plant for nitrification
4) Anaerobic storage / Rotating biological contactor for the oxidation of the organics and of ammonia (full separate treatment of the deicer water)
5) Anaerobic storage / Activated sludge treament of the stored water for the removal of organics, eventually with nitrification or discharge to municipal plant for nitrification
6) Anaerobic storage / Anaerobic filtration for reduction of organics / Discharge to the municipal plant for nitrification

2 Composition of the Deicer Wastewater

Depending on temperature and weather conditions, the type and dose of the applied deicing chemicals differ considerably from one event to another. In addition, rain and snowfall give rise to increased run-off from runways leading to unpredictable concentrations and load fluctuations. The deicer mixture, collected and stored over a longer period of time, contained various deicer substances, and the average concentrations of each are presented in Table 2.

Table 2. Concentrations and properties of deicer substances found in the airport run-off collected and stored over several weeks

Compound	Compound concentration mg/l	DOC concentration mg C/l	COD[1] concentration mg/l	COD/DOC
Ethylene glycol	1240	450	1570	3.33
Propylene glycol	1460	630	2280	3.56
Diethylene glycol	555	270	940	3.33
Isopropanol	520	270	1100	4.00
Urea	815	180	1965 [2]	10.67 [2]
Total	4950	1800	7855	---

[1] Theoretical COD for oxidation to CO_2 [2] Oxidation to NO_3^-

Compared to the nutrient requirements for biological growth, the deicing wastewater contains a surplus of nitrogen, but is characterized by a conspicuous lack of phosphorus. Unlimited growth conditions, and hence high degradation rates for the organic matter, are only possible if phosphorus is added before biological treatment. The effect of nutrient deficiency on deicer removal rates was shown experimentally and will be discussed later.

3 Characteristics of Wastewater Flow and Load

In order to gain more insight into the run-off behaviour of the deicing fluids under different weather conditions, continuous sampling and analysis of the wastewater was undertaken. That is, dissolved organic carbon (DOC) and urea- and ammonia-nitrogen were monitored with the help of flow meters, which were installed in three locations of the sewer system [2]. The following conclusions may be drawn from the data records of two winter periods:

1) Compared to the amount of deicer brought on to airplanes and runways, a considerable amount is lost in the embankments and the surrounding area by the spray effect during landing and take-off and by snow removal from the runways. Only 30–45% of the load which was applied is found in the sewer system.
2) The amount of deicer applied is strongly dependent on weather conditions. Therefore the run-off shows in addition to changing composition, enormous fluctuations in flow, concentration and load. The data summarized in Table 3 were measured during the observed winter periods.
3) The total amount of deicer used may differ from one winter to another and the distribution of the loads within a winter period is not predictable. Most of the chemicals are often applied within a period of only a few days (especially during icy rainfall).

Table 3. Monitored concentration range specific flow and loads of organic carbon and nitrogen in the airport run-off

	Flow	DOC		Urea + NH_4-N	
		conc.	load	conc.	load
	m^3/ha·d	g C/m^3	kg C/ha·d	g N/m^3	kg N/ha·d
Mean	21	290	6.6	38	0.5
Minimum	1.2	20	0.01	<1	0
Maximum	153	20 000	101	1000	16

An important conclusion for the design of the flow scheme is that for any biological treatment, storage and equilization facilities corresponding to hydraulic capacities of at least several weeks have to be considered in order to guarantee satisfactory and stable performance of the degradation processes. Longterm storage is therefore a prerequisite for all treatment alternatives studied.

4 Pilot Plants

An overview of the installed pilot plant facilities and the flow schemes are illustrated in Figure 1.

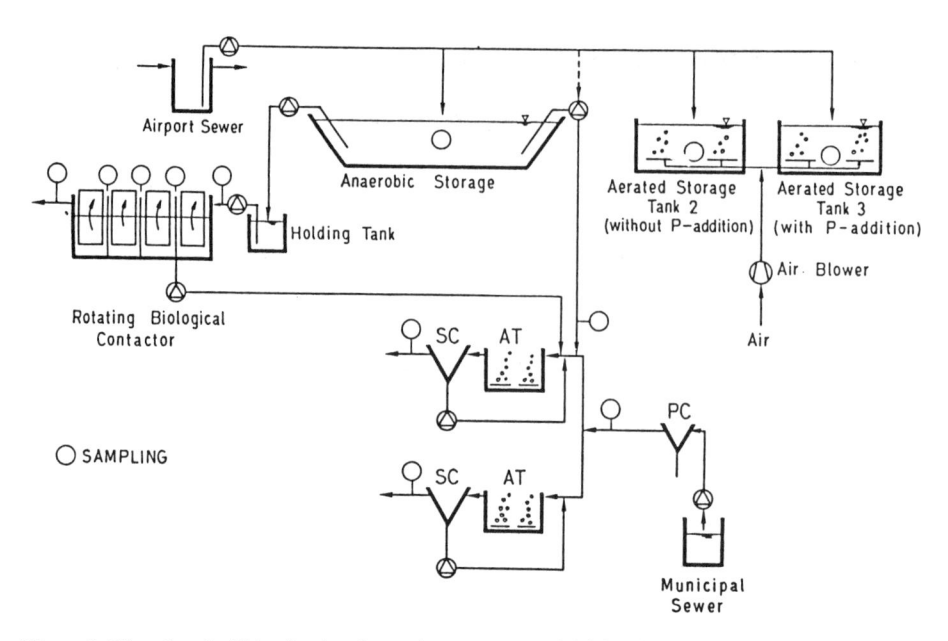

Figure 1. Pilot plant facilities for the alternative treatment of deicing fluids

4.1 Storage Facilities

A pump located at a water monitoring station and operated only above a desired DOC-level served to fill the different storage tanks. Three storage basins made it possible to investigate longterm storage behaviour under anaerobic and aerobic conditions. The geometry of the tanks is given in Table 4.

Table 4. Geometry and mode of operation of the equalization and storage tanks

	Storage volume m^3	Surface area m^2	Mean water depth m	Mode of operation
Storage tank 1	405	322	2.18	Anaerobic
Storage tank 2	137	96	1.42	Aerobic (aerated) without P-addition
Storage tank 3	126	105	1.18	Aerobic (aerated) with P-addition

4.2 Rotating Biological Contactor

The wastewater collected in the anaerobic storage tank was pumped to the rotating biological contactor (RBC) consisting of four rotating plastic units in series separated by walls. The contactors were designed such that the first unit would remove the organic compounds, and the subsequent units would serve as nitrification reactors. The flow was kept constant at 0.6 m³/d. The characteristics of one RBC unit may be summarized as follows: water volume 0.41 m³; diameter 1.2 m; internal surface area 134 m²; rotating speed 3.33 rpm.

4.3 Activated Sludge Plants

Two pilot plants were operated in parallel to simulate future conditions in an extended and nitrifying municipal activated sludge plant. One plant served as reference and was only supplied with domestic sewage, while the other was fed with 10% deicer wastewater either originating directly from the anaerobic storage tank or from the effluent of the first RBC unit. The operating data for the pilot plants are summarized in Table 5.

Table 5. Operating conditions of the activated sludge pilot plants

		Conventional plant	Plant with 10% deicer from storage 1	Plant with 10% deicer from RBC
Volume aeration tank	[m³]	0.25	0.25	0.25
Volume sec. clarifier	[m³]	0.05	0.05	0.05
Flow	[m³/d]	0.60	0.60	0.60
Sludge age	[d]	10.3	4.9	10.0
Sludge conc.	[kg/m³]	2.8	1.7	2.9
Temperature	[°C]	16.0	16.1	20.8

5 Results

5.1 Performance of Wastewater Equalization and Storage

Water quality and operating conditions in the storage tanks were monitored over a period of more than three months. The time history curves of the most important parameters in the three tanks are shown in Figures 2a – 2e.

The following conclusions may be drawn from the results:

Urea hydrolysis. Urea hydrolysis was very fast and was completed within the first four days of storage in all tanks. Urea hydrolysis is not a rate limiting step during deicer storage even at temperatures lower than the rather high temperatures of 7 – 8°C observed in this investigation. The hydrolysis rates measured in the three storage tanks are shown in Table 6. The hydrolysis of urea is an important process from the standpoint of the subsequent biological processes because it increases the alkalinity and thus the pH stability of the stored wastewater.

NH_4^+-incorporation and NH_3-desorption. The concentration time-history curves of ammonia in the storage tanks are strongly affected by NH_3-desorption from the

Table 6. Urea hydrolysis rates monitored within the first four days of storage in the three storage basins

	Urea hydrolysis rate g N/m^3 d	Temperature °C
Anaerobic storage 1	52.2	7.3
Aerobic storage 2 (without P-addition)	79.2	7.0
Aerobic storage 3 (with P-addition)	81.1	8.0

surface. The stripping rate depends on pH, temperature, and the gas transfer conditions at the water-air interface, which in turn are a function of the wind velocity above the anaerobic tank and of the aeration system in the aerobic basins. As can be seen from Figure 2e, the removal was highest in the open anaerobic storage tank, especially during the first period when high pH conditions prevailed, as a consequence of urea hydrolysis and low organic degradation rates. Figure 3 illustrates the influence of the different variables affecting the NH_4-concentration over a period of three months in anaerobic storage.

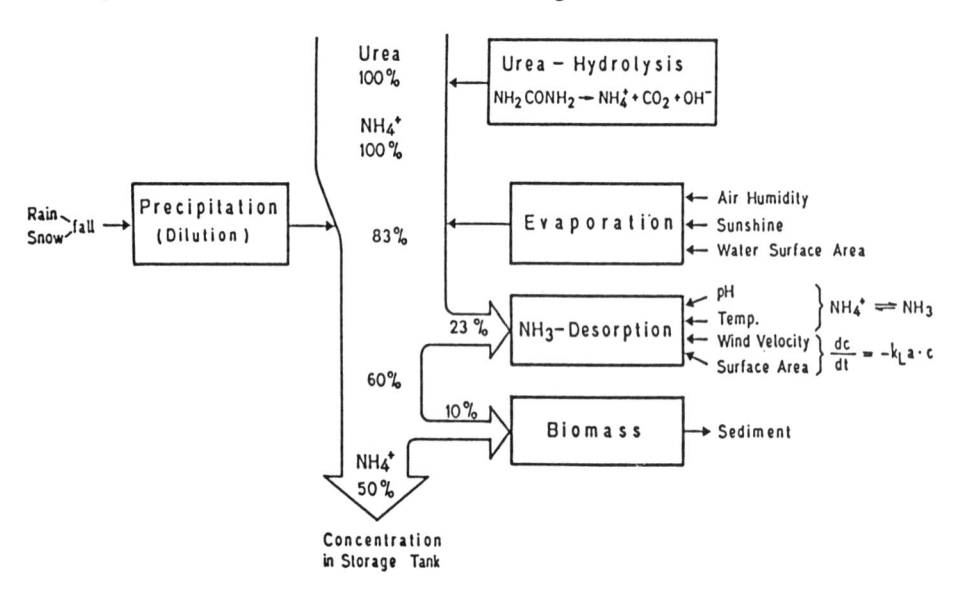

Figure 3. Effect of urea hydrolysis, dilution by rain, concentration by evaporation and desorption of NH_3 on the NH_4^+ concentration in the anaerobic storage tank

Removal of Organic Compounds. The faster removal of DOC achieved in the aerated storage tank with phosphorus-addition compared to the other two tanks without P-addition is apparent from Figures 2a–2d. Within a period of 40–50 days, the organic substances originally present are removed completely and the residual DOC reaches values in a range below 100 g C/m^3. Considerably lower degradation

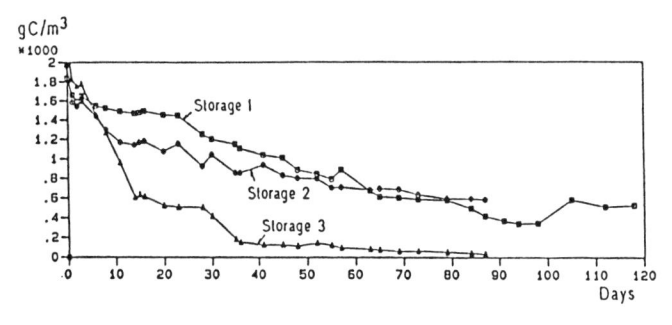

Alternative Treatment of De-icing Fluids from Airports

Figure 2a.
DOC concentrations in the three storage facilities

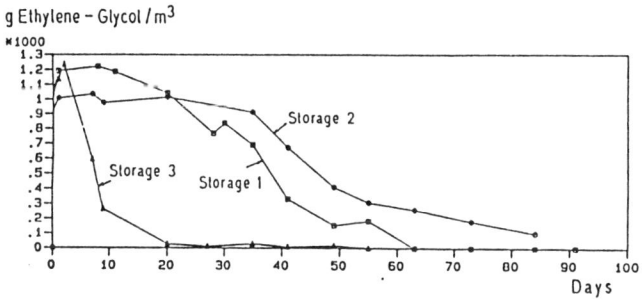

Figure 2b.
Concentrations of ethylene-glycol

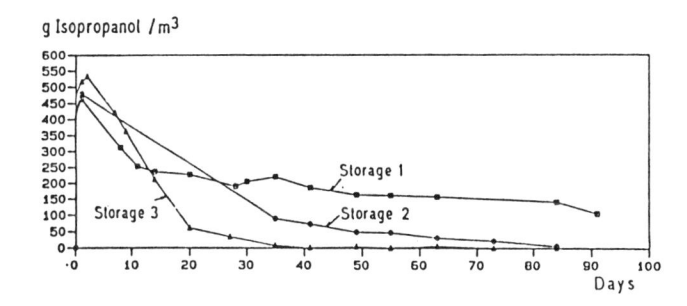

Figure 2c.
Concentrations of isopropanol

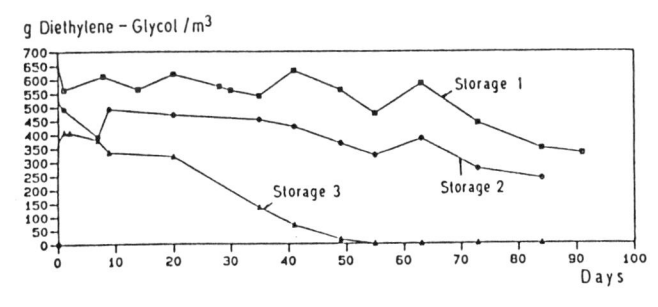

Figure 2d.
Concentrations of diethylene-glycol

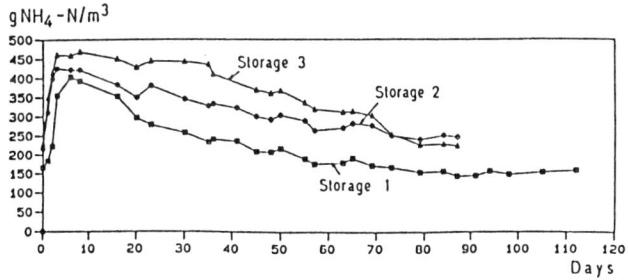

Figure 2e.
Concentrations of NH^+-N

rates were observed in the tank without P-addition and in the anaerobic basin. Based on the removal of single deicer substances as analysed by gas-chromatography, a classification with respect to biological degradability shows higher degradation rates for ethylene glycol and propylene glycol than for isopropanol and diethylene glycol in all tanks. The low rates for the latter two substances did not allow complete removal to occur in storage tanks 1 and 2 within the test period of 100 days. A summary of the removal rates of the DOC and the four deicer substances is given in Table 7.

Table 7. Removal rates of DOC and four deicer substances in the three storage tanks

	DOC removal rate $g C/m^3 d$	Degradation rates for deicer substances			
		Ethylene glycol $g C/m^3 d$	Propylene glycol $g C/m^3 d$	Isopropanol $g C/m^3 d$	Diethylene glycol $g C/m^3 d$
Anaerobic storage 1	9.5	8.1	9.7	2.0	0.56
Aerobic storage 2 (without P-addition)	10.4	4.6	6.9	3.4	1.45
Aerobic storage 3 (with P-addition)	15.7	22.6	30.0	10.3	3.8

Sludge production. Storage tanks are extensive biological reactors und therefore sludge production is low compared to conventional treatment systems. The sludge formed per m³ of stored water within 30 and 85 days and the corresponding yield coefficients are summarized in Table 8.

Table 8. Specific sludge production and yield coefficients at different storage periods in the three storage tanks

	Sludge production g dry sol./m³		Yield coefficient			
			g dry sol./g DOC rem		g dry sol./ g C rem	
	30 days	85 days	30 days	85 days	30 days	85 days
Anaerobic storage 1	---	110	---	0.13	---	0.05
Aerobic storage 2 (without P-addition)	135	30	0.25	0.04	0.10	0.015
Aerobic storage 3 (with P-addition)	1114	215	0.80	0.13	0.32	0.05

Oxygen Consumption in the Aerobic Storage Tanks. The O_2-consumption of the microorganisms is proportional to the oxidation rate of the organic substances. Corresponding to the rapid removal in the aerated storage tank 3 with P-addition, the O_2-consumption was as high as 1400 g $O_2/m^3 \cdot d$ during the initial period of operation. The O_2-input by the aerators was insufficient in the period lasting about 10 days, resulting in zero concentrations at that time. After removal of the major portion of DOC in storage tank 3, the O_2-consumption slowed down to about 100 g $O_2/m^3 \cdot d$ and less. In storage tank 2, the P-limited growth conditions only allowed O_2-consumption levels between $50-150$ g $O_2/m^3 \cdot d$ for the entire operating period.

Operation problems. In the anaerobic storage tank 1, heavy odor problems resulted from the formation of low molecular mass fatty acids, especially during pumping to the RBC plant. In the aerobic storage tanks 2 and 3, strong foam formation to a height of $50-100$ cm occured during the first 10 days of storage. The foam, which probably consisted of proteinaceous material, was degraded and disappeared completely in the course of further storage.

It was concluded that for further treatment of the stored water, aerobic and non-nutrient-limited storage can achieve a nearly quantitative removal of the organic substances within a period of about 6 weeks, and only nitrification is needed as final treatment step. With anaerobic storage, however, additional treatment processes are necessary for the removal of the organic substances even after long storage periods at higher temperatures.

5.2 Performance of the Rotating Biological Contactor

The water from the anaerobic storage tanks was treated by the RBC for further carbon removal and nitrification. Again, the DOC, the used deicer substances and the nitrogen compounds were analyzed in detail over a period of 5 months.

Removal of organic substances: In accordance with the DOC removed in the anaerobic storage tank, the DOC concentrations in the influent to the RBC decreased with time (see Figure 4a). The time-history curves of the DOC and the deicer concentrations in the effluent of the first two RBC units in Figures 4b and 4c clearly show the limited removal during the first period under P-limited conditions as well as the expected high removal rates, especially in the first RBC, during the period with P-addition. As a design aid, the observed removal rates in the first RBC unit can be plotted versus the organic carbon surface load as shown in Figure 5. The curves show that for attaining high removal rates, rather low loads have to be applied. The deicer substances isopropanol and diethylene glycol were not removed completely by the total number of RBC units during P-limitation. However, no deicers were present after the second unit when phosphorus was dosed continuously. The break-through of deicers in the first RBC unit is probably due to partial clogging of the RBC element.

Removal and nitrification of NH_4^+: Part of the removal of NH_4^+ was due to biological incorporation in the course of heterotrophic growth in the first RBC cell at the beginning, accompanied by a slight desorption of NH_3. The removal of $60-70$ g NH_4^+-N/m^3 could be attributed to the production of surplus sludge. The RBC units 2 to 4 were designed for biological nitrification. Nitrification started in the second unit as soon as the organic load was below 1.6 g $C/m^2 \cdot d$. However, the production of H^+ by the nitrifiers led to a considerable decrease of the pH and consequently to inhibition of the nitrification process. The oxidized nitrogen components showed the typical start up pattern of NO_2^--production, resulting in high

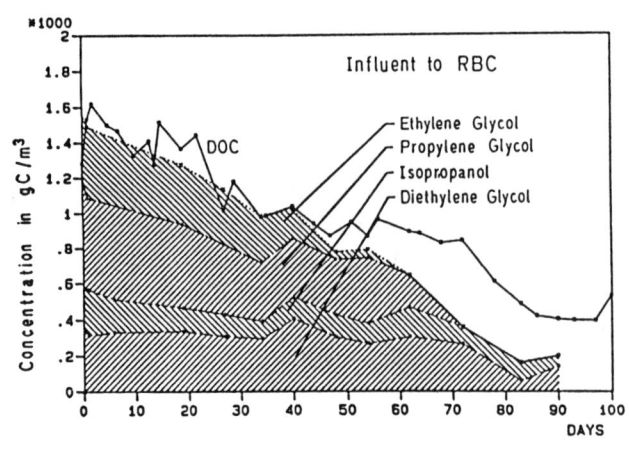

Figure 4a. Doc and deicer concentrations in the influent to the first RBC unit

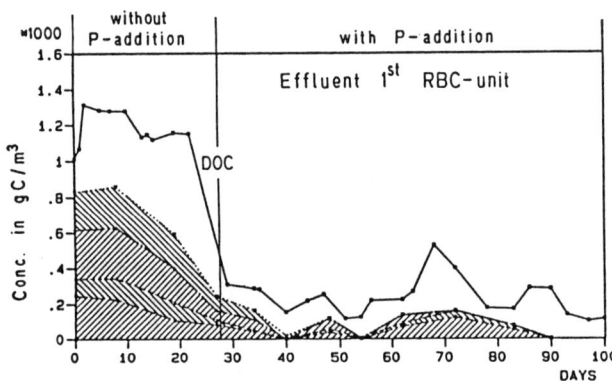

Figure 4b. DOC and deicer concentrations in the effluent of the first RBC unit during periods without and with P-addition

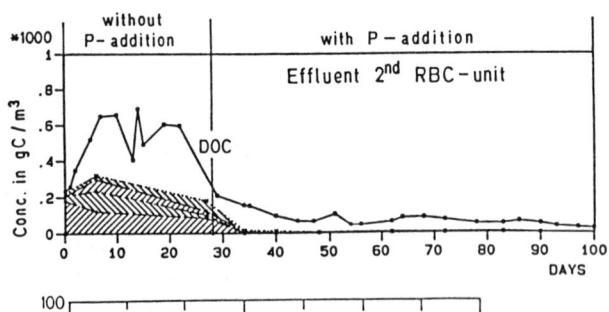

Figure 4c. DOC and deicer concentrations in the effluent of the second RBC unit during periods without and with P-addition

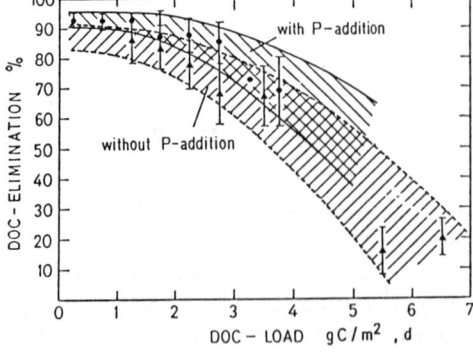

Figure 5. DOC-removal in the first and second RBC unit as a function of the specific carbon surface load

NO_2^--concentrations, followed by a steady increase in the nitrification rates and the production of NO_3^-. Nitrification increased while the pH decreased to about 6.3. In the range of pH = 6.0 the oxidation step of NO_2^- to NO_3^- was strongly inhibited, resulting again in high NO_2^--concentrations reaching values of more than 100 g NO_2^--N/m^3 (see Figure 6). High nitrification rates can only be maintained when neutralizing chemicals, e.g. $Ca(OH)_2$, are added continuously, leading to additional operational costs. Considering that the nitrification step is inhibited and unstable and that phosphorus and neutralizing chemicals are an additional cost, the most appropriate solution is to transfer the pretreated deicer wastewater to a municipal treatment plant for nitrification where sufficient alkalinity is present.

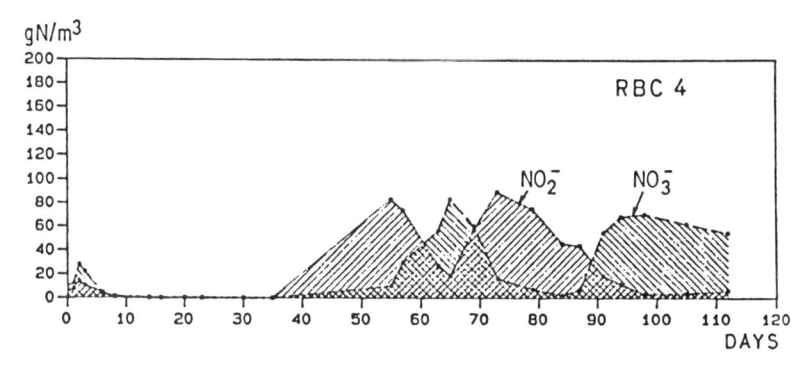

Figure 6. NO_2^- and NO_3^- concentrations in the effluent of the nitrifying RBC units showing inhibited nitrification

Sludge production and sludge characteristics: The greatest amount of sludge production occured in the first RBC unit, because of heterotrophic growth, at a level of approximately 4.4 g dry solids/$m^2 \cdot d$. A carbon-mass balance indicates that about half of the organic substances were incorporated into biomass and half were oxidized to CO_2. The sludge produced in the first unit was further oxidized in the subsequent nitrifying RBC units by endogenous respiration leading to a decrease in total sludge production. The sludge produced by the RBC showed very bad settling and dewatering characteristics and the supernatant of the settled sludge was extremely turbid. The settling velocity of 1 m/h at concentrations of 1.6 kg dry solids/m^3 was about four times lower than that of conventional activated sludges. The high turbidity of the supernatant was due to high concentrations of single non-flocculating microorganisms. The treatment of the effluent by flocculating chemicals is possible. However, jar tests revealed the need for an extremely high dose on the order of 100 g Al/m^3 of the most efficient chemical (poly-Al-chloride) investigated. This is considered not to be economical.

5.3 Performance of the Activated Sludge Plants

In order to simulate future conditions in the nearby municipal treatment plant, both activated sludge pilot plants were operated with domestic sewage with a sludge age of 10 days for safe nitrification. In the first experimental period, plant 2 received deicer fluids from the anaerobic storage tank on the order of 10% of the flow, and plant 1 served as reference. In the second period, 10% of the effluent from the first RBC unit was added to reference plant (1), and in the other plant (2) recovery from the deicer load was monitored until reference conditions were reestablished. The analyzed parameters, their concentrations and the flow schemes are depicted in Figures 7a – 7c.

TOC	120	20-160
DOC	85	23.5
TSS	90	2-500
NH_4-N	33	15.0
NO_2-N	~0	1.0
NO_3-N	~0	14.6

a) Activated sludge plant with 10% deicer fluid from the anaerobic storage tanks

TOC	60	12.0
DOC	33	9.0
TSS	80	7.2
NH_4-N	24	0.42
NO_2-N	~0	0.07
NO_3-N	~0	20.3

b) Activated sludge plant with 10% pretreated deicer fluid from the RBC plant

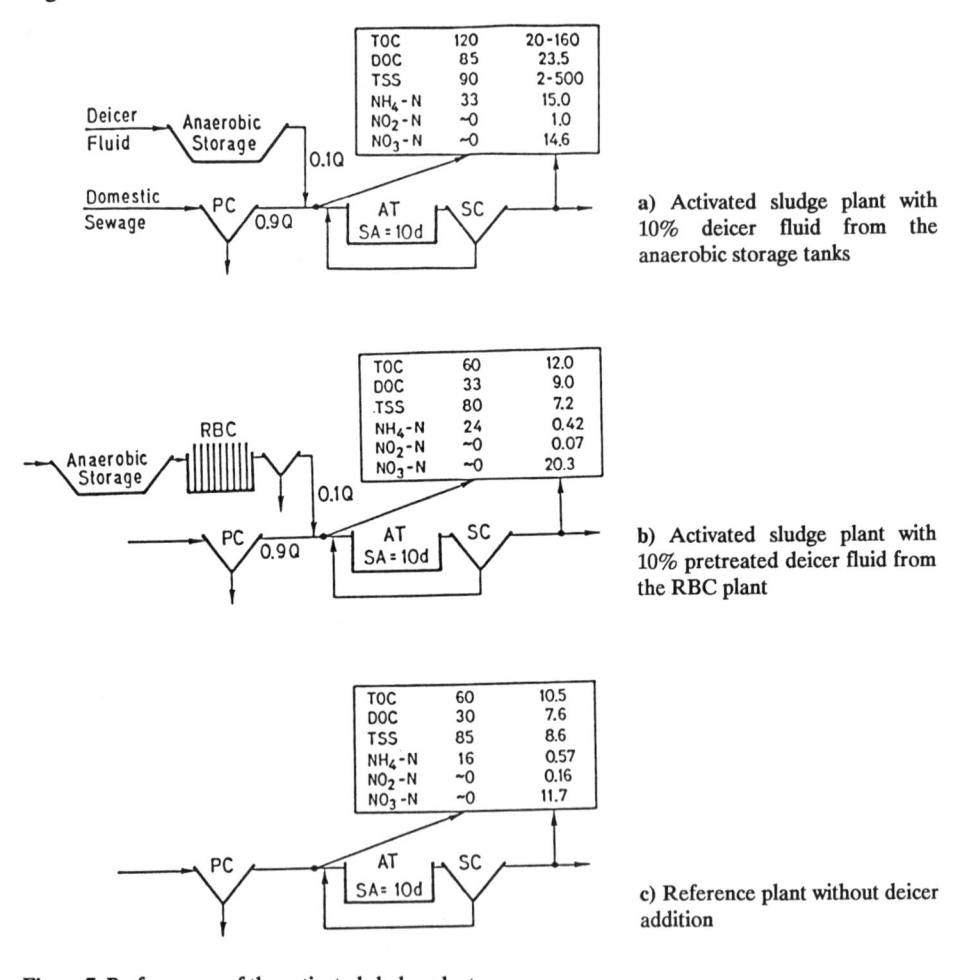

TOC	60	10.5
DOC	30	7.6
TSS	85	8.6
NH_4-N	16	0.57
NO_2-N	~0	0.16
NO_3-N	~0	11.7

c) Reference plant without deicer addition

Figure 7. Performance of the activated sludge plants

The results clearly indicate that the desired operating conditions (sludge age 10 days, 3.0 kg dry solids/m^3) could not be maintained in the plant with 10% deicer. Consequently, unacceptably low removal of organics and nitrification rates were observed. The reason for the bad performance was the formation of bulking sludge resulting in extremely high SVI (>1000 l/kg) and very low settling velocities (<0.2 m/h) Accordingly, most of the activated sludge was lost in the clarifier effluent, making the operation of the plant under the desired conditions impossible. On the other hand, the addition of already treated deicer wastewater from the RBC effluent did not cause any negative effect on process performance and operation of the plant. Sufficient and stable removal of organics as well as full nitrification was observed during the entire period. Some of the measured operational parameters illustrating the behaviour of the activated sludge plants during the two aforementioned experimental phases are shown in Figures 8a – 8c.

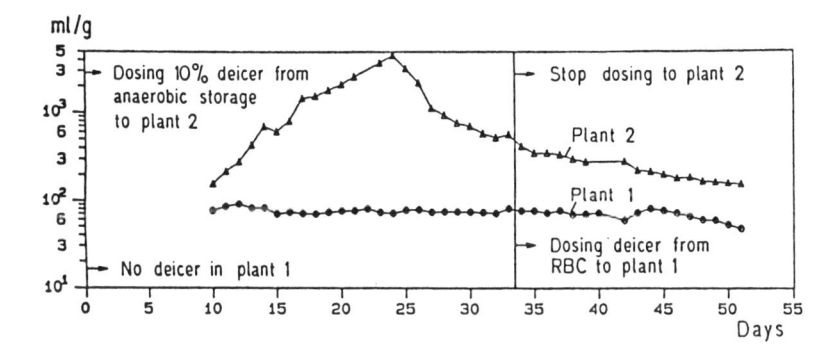

a) Sludge volume index

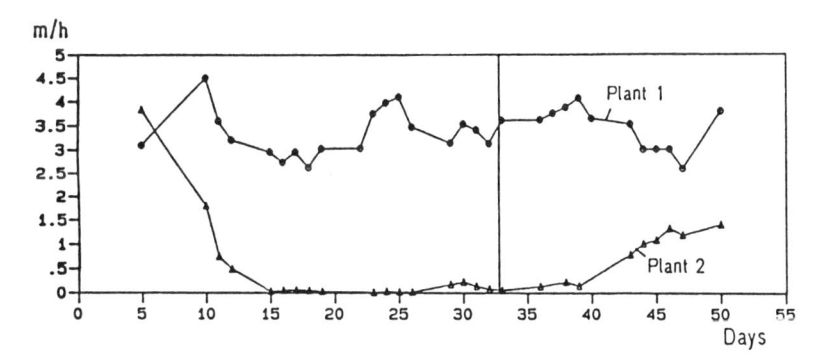

b) Sludge settling velocity

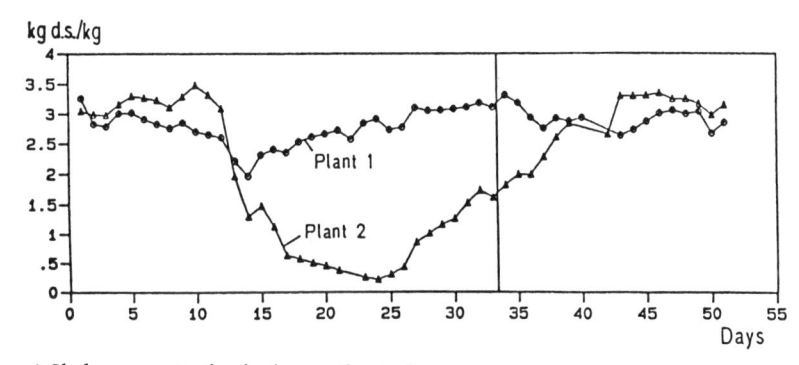

c) Sludge concentration in the aeration tank

Figure 8. Time history curves of some operational variables in the activated sludge plants with and without addition of deicer

6 Conclusions

The following conclusions may be drawn from the experimental findings:

1) The enormous fluctuations of the concentrations, loads and composition of the deicer wastewater require longterm (several weeks) equalization and storage facilities.
2) The treatment of the anaerobically stored or the fresh deicer water combined with municipal sewage in a communal plant with existing settling tanks is not feasible because of the formation of an extreme amount of bulking sludge, even if the deicer water makes up for only 10% of the flow.
3) The degradation rates of the deicer substances, especially of isopropanol and diethylene glycol, are low during anaerobic storage. This water has to be treated, preferably by fixed biofilm processes or activated sludge plants with flotation in order to remove the organic substances before transferring it to a municipal plant or discharging it to a receiving water.
4) The treatment of anaerobically stored deicer water in fixed biofilm reactors such as RBCs leads to comparatively large amounts of surplus sludge with poor settling and dewatering characteristics. To prevent RBCs from clogging, the surface area of the biomass carrier must not exceed approximately 150 m^2/m^3.
5) Full nitrification and thus full treatment of the deicer water in a separate treatment plant is only possible if neutralizing chemicals are added to overcome alkalinity limitations.
6) Storage in aerated tanks leads to full degradation of all deicer substances and to low DOC residuals after a storage period of 1.5 to 2 months. However, the addition of an appropriate dose of phosphorus to enhance biological growth is a necessary prerequisite. The water quality after the aforementioned storage period allows direct discharge to a municipal plant for nitrification. In the case studied, the alkalinity of the domestic sewage is high enough to guarantee full nitrification without the addition of chemicals. An additional benefit of aerobic degradation is a temperature increase of approximately $3-4°C$ during the initial period of high biomass activity, stimulating the biological degradation processes.

The evaluation of the advantages and disadvantages of the investigated treatment schemes and a cost analysis led to the conclusion that longterm aerated storage facilities, preferably at different locations at the airport, and subsequent nitrification in a municipal treatment plant promise to be the most feasible solution for the removal of the deicer fluids.

References

[1] Jank, B.E., Guo, H.M., Cairns, V.W.: Biological Treatment of Airport Wastewater Containing Aircraft Deicing Fluids. Report EPS 4-WP-73-6, Canada Environmental Proction Service, July 1973

[2] Ins & Knecht, Eng. Office Zürich: Flughafen Zürich - Abwasseruntersuchungen 1985/86 and 1986/87

M. Boller
Swiss Federal Institute for Water
Resources and Water Pollution Control
(EAWAG)
8600 Duebendorf
Switzerland

Separators and Emulsion Separation Systems for Petroleum, Oil, and Lubricants

H. Nöh

Only about 3% of the enormous amounts of water existing on earth are suitable for drinking. The remaining 97% must presently be discounted as drinking water due to the high content of salt. Nevertheless, about 1 million cubic meters of water are available for every person on earth, not counting animals and plants.

The amount remains the same. But what about the quality?

Unfortunately, many people still treat water as something without any real value, just like a quite ordinary liquid. In reality water is an unusual liquid, maybe even the most unusual one. Surely it is the most important one, because water means life.

The most significant attribute of water is the fact that it absorbs other substances, meaning, it dissolves them. If this were not the case, there would be no life on earth. This is a positive feature of water.

A negative aspect is that water also dissolves substances that we cannot use or that are even harmful in higher concentrations, including all the waste that man produces in ever-increasing amounts due to industrialization and the affluence of our society. This leads to a notable change in the original make-up of the water. Past environmental catastrophies have shown the tremendous impact on plants, animals, and man.

Environmental problems with which we are confronted daily, stem, among other things, from the production, transportation, storage, and processing of petroleum, oil, and lubricants, and also from the disposal of waste oil.

Mineral oils present a high pollution factor for water. For instance, a concentration of 1 g of oil/l of water is generally toxic to fish, and 5% of this concentration is classified as hazardous.

An important instrument for protecting water is the POL separator.

Separators have been known for about 100 years. Their primary function in those days was not the protection of waters, but rather to prevent the formation of explosive mixtures in sewer systems. Considering today's criteria, neither the design nor the required efficiency rates were very restrictive.

For instance, a German standard that was still in effect 15 years ago required a minimum efficiency of 95% for a separator. That meant that 5 l of the 100 l of fuel oil introduced into a separator could leave the same together with the effluent. The result: 5 l of oil, or 5 million liters of water made undrinkable.

Consequently, the minimum degree of efficiency was raised to 97%. But even 97%, which is equal to a residual concentration of oil of approx. 124 mg/l of water, is too much. The legislature therefore reacted by imposing considerably tighter restrictions on discharge entering sewer systems and public waters.

German law states that, effective 1987, any effluent discharged into the sewer system shall not contain more than 20 mg of residual oil per liter of water.

For discharge into lakes and rivers, this figure is even lower with a maximum concentration of 10 mg/l of water. Further restrictions are anticipated for the near future. A proposed ruling intends to limit the residual oil to 5 mg/l of water for direct discharge.

How then is it possible to achieve such high efficiency and/or low residual concentrations of oil?

The basic principle of POL separation, namely, to use the uplift forces in the gravitational field to separate liquids with different density, is well known. The problem is how to retain all of the POL particles in the effluent in the separator. This can certainly not be done in gravity separators.

The most important data for designing optimum efficiency separators can be gained in tests using original size separators. It is obvious that tests of that nature go hand in hand with enormous development and test expenditures.

We have noted that tests with model separators can at best establish tendencies. Naturally, model tests furnish important data, for instance about droplet size, uplift behavior as compared to the density, water temperature, and the degree of purity of water and petroleum, oil, and lubricants. However, model tests do not show that in the original size separator, turbulence and flow velocities have totally different effects, mostly negative ones.

The diagram of a gravity separator is well known and is often used even today to explain how POL separators work. See Figure 1.

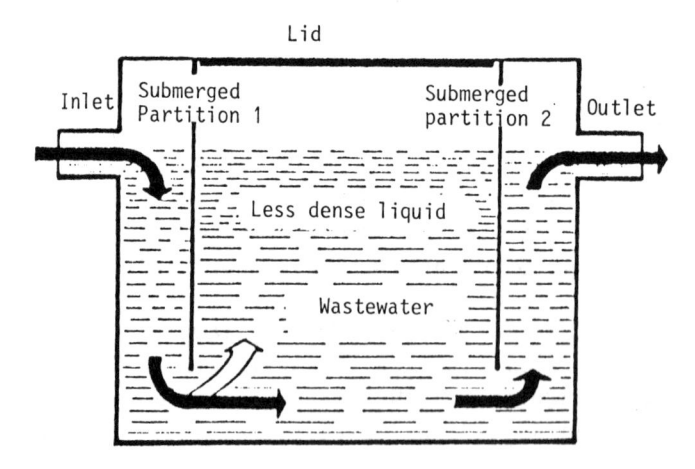

Figure 1

Even though oil is separated from water in the separators operating on these principles, the amount of oil that does not settle between the stilling partitions, but reaches the outlet, is much too high and would never satisfy today's requirements.

Why is That?

1) The design of the inlet is most unfavorable. Together with the influent water, the oil is forced down, although it is supposed to rise to the top.
2) Turbulence that breaks up larger oil droplets into smaller ones is created between the separator wall on the inlet side and the stilling partition. Minute droplets cannot rise since their upward force is not sufficient to get past the frictional resistance of the heavier water. They are transported to the outlet together with the water flow.

This effect can be greatly reduced by allowing longer settling times in the separator; however, this renders the separator less economical.

If unfavorable inlet conditions exist, the dimensions of the separator must be notably larger than those of a separator with a favorably designed inlet, to achieve equal efficiency. See Figure 2.

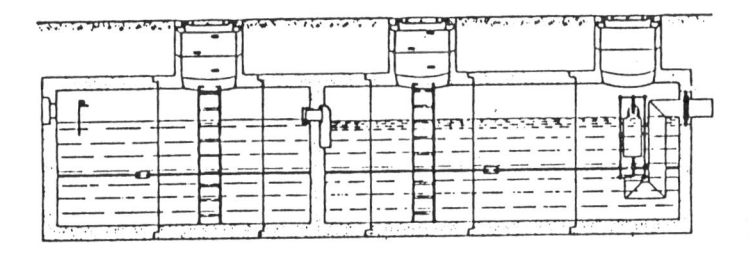

Figure 2

Figure 2 shows a system with an inlet that is located directly at the head wall and that leads the incoming water/oil mixture vertically to the bottom. Systems of this type must have a minimum efficiency of 95%, meaning 212.5 mg of oil/l of water were allowed to leave the separator.

A simple structural change of the inlet led to a reduction of the residual oil in the discharge water by at least 40%. The efficiency could thus be increased from 95% to 97%, which equals a reduction to a maximum of 127.5 mg of residual oil/l of water. At the same time, it was possible to raise the flow rate from 80 l/sec to 100 l/sec, for example. See Figure 3.

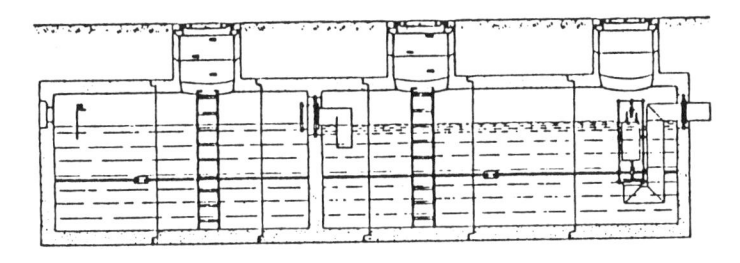

Figure 3

The purpose of the inlet is to have the water/oil mixture enter as closely as possible to the surface of the water, to achieve good utilization of the surface by return against the separator head plates, and to lengthen the flow distance to the outlet.

It is vital, however, that only a small portion of the minute droplets reach the bottom third of the separator; in this way the majority of the droplets stay outside the effective range of the outlet opening.

Separators of this type generally have an efficiency of 98 to 99.5%, meaning 85 to 21.25 mg oil/l of effluent.

Figure 4 shows the influent principle of the so-called box inlet which permits optimum hydraulic conditions to be obtained as described above. This inlet system is standard for all the different BUDERUS separators.

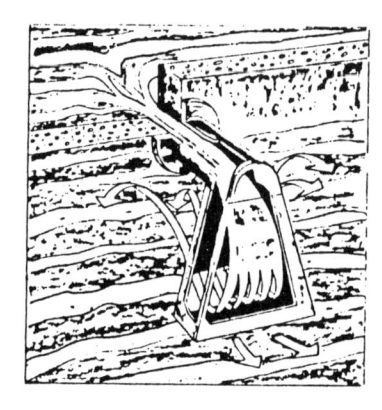

Figure 4

An important step was taken in the development of large capacity separators favorable to the environment and, at the same time, less costly. It became apparent, however, that the capacity of gravity separators was limited, not only from a technical but also from an economic point of view.

Through intensified development and test efforts, it was possible to construct systems consisting of a single pipe with a length of no more than 4 meters and a capacity of 100 l/sec.

In the meantime, however, the regulations for discharge into sewer systems and lakes and rivers has become more restrictive.

The maximum value permissible for discharge into the sewer system was no more than 20 mg of oil/l of water, and into lakes and rivers no more than 10 mg of oil/l of water.

The diagram in Figure 5 clearly shows the difference between a separator with an efficiency of 97% and a residual concentration of oil of 20 or 10 mg/l of water.

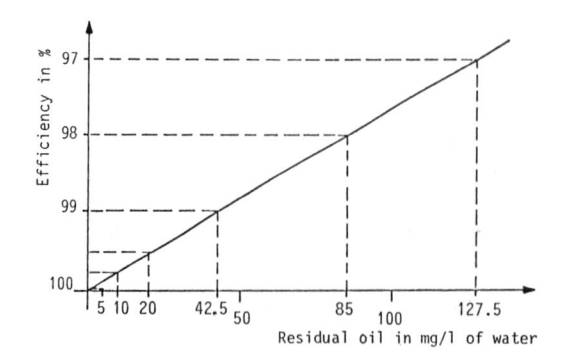

Figure 5

Frequently, very low concentrations of residual oil can only be obtained through further treatment. Since these oil particles are predominantly minute droplets that reach the outlet with the flow of effluent, additional treatment is required in front of the outlet, so that not only the larger droplets that rise easily, but also the dispersed, microscopically small droplets are removed.

Therefore, the physical processes of adsorption and coalescence are used in addition to gravity in the so-called adsorption-coalescence separators, and residual oil concentrations of less than 5 mg/l of water are the result. A prerequisite is a water/oil mixture without wetting agents, meaning without wash-active substances.

An adsorption/coalescence process can be carried out using an insert filled with tube granules of a specific size. Figure 6 serves to demonstrate this.

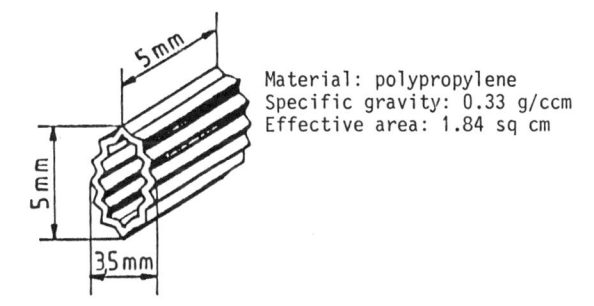

Material: polypropylene
Specific gravity: 0.33 g/ccm
Effective area: 1.84 sq cm

Figure 6

Each of these tubes has an effective surface area of 1.84 cm². 1 m³ contains 8,101,650 tubules with a total surface area of 1,490.7 m². About 48 l of coalescent with approx. 72 m² surface area are available for a flow capacity of 1 l/sec.

Minute oil droplets adhere to this adsorption-coalescence material (adsorption) and finally combine into one large drop (coalescence) that can rise to the surface. See Figure 7.

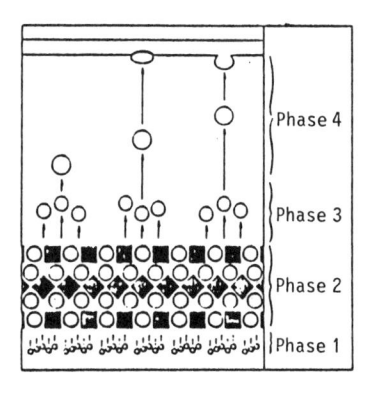

Figure 7

Figure 8 shows the system in a complete separator system, in this case one with a rated size of 100.

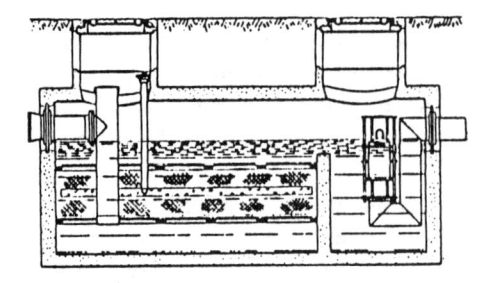

Figure 8

In the future, the concentrations of residual oil obtained in such a system may not, depending on the criteria, exceed 5 or 10 mg/l of water.

The disadvantages of this concept are
– variable utilization of the adsorption/coalescence surface
– increased pollution hazard
– charging the system with the entire oil volume introduced into the separator
– comparatively maintenance intensive

To remedy this situation, a pre-separation device retaining at least 90% of the influent oil from the adsorption-coalescence stage was established. See Figure 9.

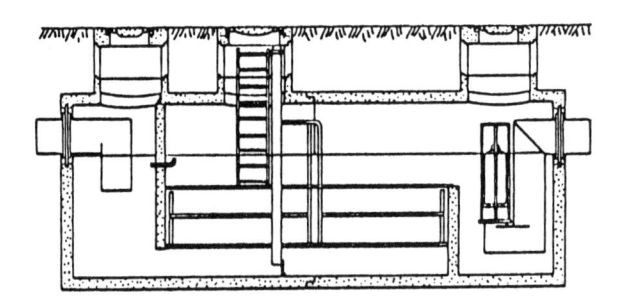

Figure 9

The water/oil mixture is pre-separated in the inlet chamber. The oil droplets that reach the adsorption-coalescence insert together with the water combine with others and slowly rise to the surface. The resulting concentration of residual oil in the effluent is less than 5 mg/l of water.

In order to prevent too much of the separated oil from collecting in the pre-chamber, there is a transfer into the main chamber by way of an overflow pipe. This is also required to ensure the effectiveness of the automatic shut-off.

For construction and maintenance reasons, the system was turned almost 180 degrees in another series of separators. See Figure 10.

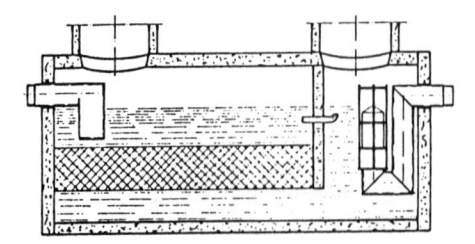

Figure 10

The water/oil mixture is no longer forced from the bottom to the top through the adsorption-coalescence stage, but rather the advantage of the box inlet that guides the oil mostly into the upper layers of the liquid is utilized.

This means that very good pre-separation takes place in the upper third of the influent liquid.

The minute oil droplets that are transported to the bottom together with the flow in spite of this pre-separation, accumulate mostly in the adsorption-coalescence stage and rise back to the surface from there. The concentration of residual oil in the effluent is at least as low as that shown for the system in Figure 8.

All additional systems in a separator are costly, both in terms of labor and materials, and result in increased maintenance efforts by the user. These considerations led to the construction of a separator that works without the adsorption-coalescence insert. See Figure 11.

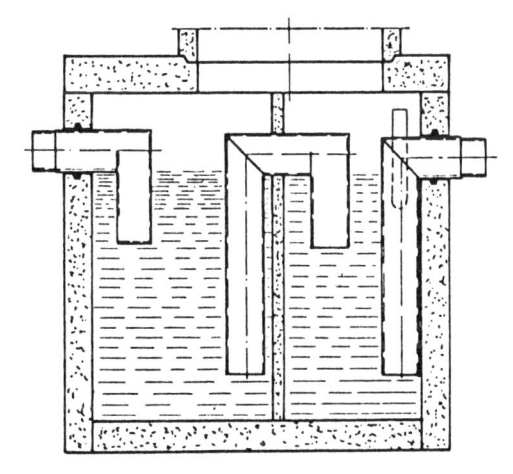

Figure 11

The separator consists of 2 chambers, namely the one in front where the main separation is performed, and the one in back, where the separation of the minute droplets takes place.

The system works as follows: most of the oil is removed in the pre-chamber. Those few droplets that reach the riser pipe together with the water combine into larger droplets that have the ability to rise. The most minute droplets are also forced into the top third of the fine separation chamber. This process can be repeated as often as required. See Figure 12.

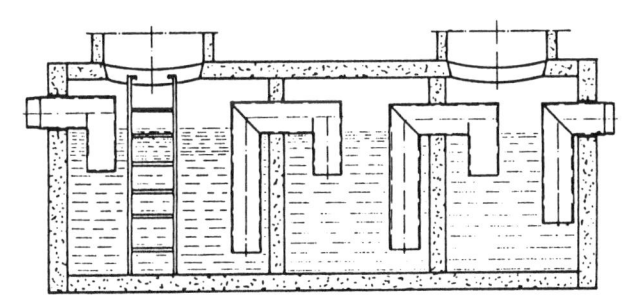

Figure 12

To reach residual oil concentrations that are as low as those in the separators in the adsorption-coalescence stage, the separator load in l/sec must be reduced by about 30%, if the volume is the same, or the system must be enlarged accordingly.

The kinds of systems described above have the advantage of not being maintenance intensive.

The separators described so far can only be used for treating effluents in which the dispersion of the petroleum, oil, and lubricants was not stabilized by emulsifiers or detergents, which are chemicals that reduce the surface tension.

Effluents containing substances that impair POL separation or act as emulsifiers cannot be treated in gravity separators. These kind of effluents have to undergo special treatment, for instance in emulsion treatment systems.

A system where the hydrocarbon content does not exceed 5 mg/l of water, to be used in connection with separators or by itself, was developed specifically for car washes, motor vehicle service stations that perform partial cleaning and remove protective coatings, and also for machine shops, railroad and aircraft maintenance facilities, and maritime operators who have to dispose of the bilge water.

The difference in comparison to other known methods is that the effluent containing emulsions is continuously treated in the flow process method. See Figure 13.

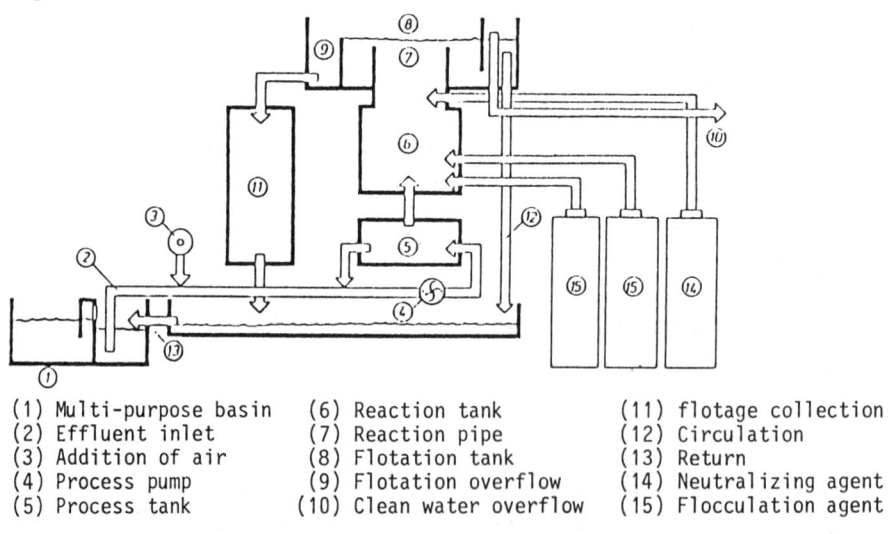

(1) Multi-purpose basin
(2) Effluent inlet
(3) Addition of air
(4) Process pump
(5) Process tank
(6) Reaction tank
(7) Reaction pipe
(8) Flotation tank
(9) Flotation overflow
(10) Clean water overflow
(11) flotage collection
(12) Circulation
(13) Return
(14) Neutralizing agent
(15) Flocculation agent

Figure 13

Following separation of the free floating oil, the emulsified effluent is led from the multi-purpose tank (pre-separator and storage tank for the effluent to be treated) to the emulsion separation system.

Prior to entering the circulation, additional load carriers are formed by adding air. Then the treated effluent is fed into a tank by means of the process pump, and is again mixed vigorously. During the rising process, the water reaches the so-called reaction tank via a slide control device. Here microflocs form after the addition of a cathodic electrolyte. These microflocs rise in the reaction pipe together with the water. To speed up the process, an anionic polyelectrolyte is added, and the microflocs combine instantaneously into macroflocs. The trapped air bubbles provide these flocs with enough updrift to separate them from the water. The flotage enters a self-draining collecting tank by way of an overflow lip. The purified water can be discharged into the sewer system or reused, as required.

The efficiency of the emulsion separation system is shown in Figure 14.

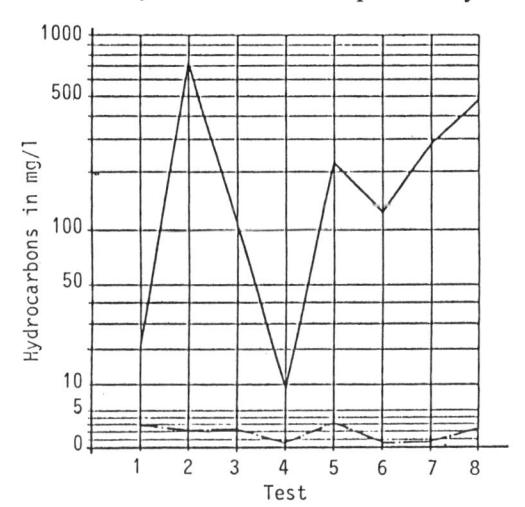

——— Hydrocarbons in mg/l before emulsion separation
—·— Hydrocarbons in mg/l after emulsion separation

Figure 14

Prior to treatment in this separation system, wastewaters with values ranging from 10 mg hydrocarbons/l of water to 700 mg hydrocarbons/l of water were determined for the emulsions. Following treatment in the separation system, the measured peak concentration was no more than 3 mg of hydrocarbons, meaning the system works with a safety factor of at least 3.3 according to the influent criteria presently in effect.

Finally, let me add that the proper function of the separation system is ensured only if the pH of the effluent to be treated is in the neutral range. pH can be adjusted using acids or bases. The other advantage of this method is that the effluent does not require further treatment before discharge into the sewer system or the receiving water.

H. Nöh
Buderus Bau- und Abwassertechnik GmbH
Postfach 1220
6330 Wetzlar 1
Fed. Rep. of Germany

Chemical Treatment of Flue Gas Washing Liquids

A.N. Grohmann, L. Bauch and H.-P. Scheerer

1 Introduction

Sources of heavy metal pollution in the aquatic environment include the iron and steel manufacturing industries, the metal-working industry, the glass industry, and phosphate fertilizing, but also tire abrasion and the burning of refuse and coal. In the latter two fields, pollution due to heavy metals can be carried over from the air into the water through rinsing of the flue gas. By means of chemical precipitation, destabilization of the precipitated products and their flocculation and separation, the heavy metals are incorporated into a disposable sludge.

It is necessary to test whether this can also be done when the chloride concentration of the circulating rinse water increases to more than 35 g/l. Chloride complexes, e.g. with cadmium, are known to be difficult to precipitate.

Thanks to the co-precipitation technology which uses metal-hydroxo-oxides and destabilization of the precipitated products, it should be possible to precipitate even chloride complexes, thus making a significant contribution to removing heavy metals from the aquatic environment.

2 Basis

2.1 Precipitation

Heavy metals form oxides, carbonates and sulfides which are poorly soluble. Cadmium is an example of this, see Figure 1. For wastewater from the electroplating industry, a residual concentration of cadmium in the treated water, which corresponded approximately to the equilibrium concentrations and was at times even less [1], could be achieved using carbonate precipitation. Figure 1 also shows:

— the range from pH 8 to 10 is particularly suitable
— sulfides allow particularly low residual concentrations to be achieved.

These conclusions are also valid for other metals and their complexes which form poorly soluble sulfides (cf. Table 1).

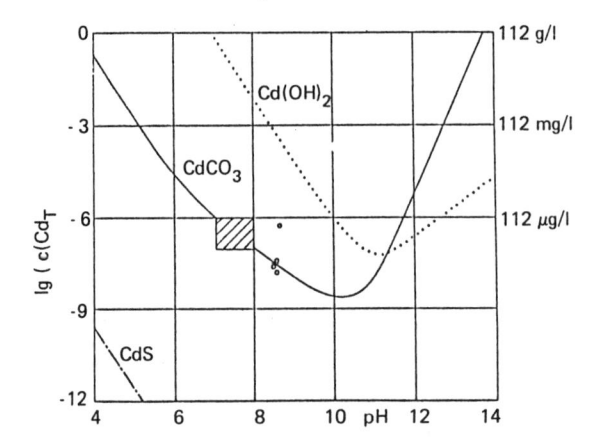

Figure 1. Residual concentration for Cd elimination by flocculation and filtration in comparison to concentration in accordance with the solubility equilibrium in a carbonate (0.01 mol/l C_T) or sulfide system (10^{-5} mol/l S_T)

Table 1. Solubility equilibria for hydroxides, carbonates, and sulfides at 25 °C and I = 0 [2]

$$Me(OH)_n + n\,H^+ = Me + n\,H_2O$$
$$MeL_s = Me + L\ (L = CO_3^{2-}\ \text{or}\ S^{2-})$$

Me	lg K for		
	OH	CO_3	S
Cd	+ 13.65	-11.7	-27.0
Cu	+ 7.65	-2.0 [1)]	-36.1
Hg	+ 2.56		-52.7
Mn	+ 15.2	-9.3	-13.5 (green)
Ni			-26.6
Pb	+ 13.1	-13.13	-27.5
Zn	+ 12.45	-9.8 [2)]	-24.7 (alpha)
Mg	+ 16.84	-7.46	
Ca	+ 22.8	-8.42	

[1)] $Cu(OH)(CO_3)_{0.5}$ [2)] $Zn(OH)_{1.2}(CO_3)_{0.4}$

2.2 Kinetics

The thermodynamical calculations according to Figure 1, however, do not yet say anything about the actual chemical reactions which take place, nor do they say whether or not the precipitation is effective.

If crystallization results from the precipitation, as is the case with calcium carbonate, then the process is slow and can be disturbed easily with inhibitors. This is because the molecules or ions must orient themselves in particular, preferred positions on the nucleus of the crystal in order to allow the crystals to grow. These positions can be blocked physically by adhering polyphosphates, humic acids and organic acids.

If, on the other hand, chemisorption of the metal complexes to the active sites of the particles is involved in the precipitation step, the process can proceed extremely rapidly. The hydroxo groups of isopoly bases, such as those formed in water by iron or aluminium salts during hydrolysis, are particularly suited for this.

Table 2. Scheme of the specific adsorption of substances to hydroxide groups of iron or aluminium hydroxo complexes

Ion exchange reaction
$\equiv MeOH + HPO_4^{2-} = \equiv MeHPO_4^- + OH^-$

Condensation reaction
$\equiv MeOH + HO^-Cd^+ = \equiv Me\text{-}O\text{-}Cd^+ + H_2O$

$Me = Fe$ or Al

The background for these reactions was given in Stumm [3a, 3b], who described them as a form of ligand exchange for specific adsorption. It does not matter whether the reactive hydroxo groups are bound on the surface of the oxides or on amorphous, newly precipitated oxide-hydrates or on still soluble, polymeric hydroxo complexes, approximately with a polymerization level $n = 2000$ for $Fe_n(OOH)_n$ [4]. The so-called "co-precipitation" during the hydrolysis of iron or aluminium salts and the ligand exchange process on the surface of oxides are thus special forms of one and the same mechanism, as was also shown by the investigations by Leckie [5]. This process, which can also be understood as a condensation of various species onto the hydroxo groups (cf. Table 2), presumably proceeds just as quickly as the polymerization of the hydrolyzing iron hydroxo complexes. It could be shown, for example, that the precipitation of phosphates can be completed in less than 0.1 s [6].

The specific adsorption model explains in simple terms why water decreases in the species of a substance more than the thermodynamic equilibrium suggests; in other words, why measured values are possible below the equilibrium curve in Figure 1. Thus it must be expected that in the course of aging to the degree that the amorphous Fe or Al-oxyhydroxide together with the adsorbed species are converted into thermodynamically stable modifications, the species concentration in water increases.

The model thus requires:
- rapid mixing of the precipitation chemicals
- addition of Fe or Al salts
- short residence time of the water in the unit
- subsequent dosing of Fe or Al salts ahead of a filter in order to form reactive hydroxo groups.

The metal species will probably react with the hydroxo groups of the iron or aluminium at different rates. No previous experience is available on this subject, but problems with cadmium precipitation in water with a high chloride content might be explained by the fact that the chloride complexes of heavy metals react more slowly with Fe-OH than the hydroxo complexes.

2.3 Destabilization and Aggregation

After particulate precipitation products are formed, these must aggregate into flocs which can be removed. This does not occur readily. For example, sulfide precipitation is very effective with respect to the residual concentrations according to solubility equilibria in Figure 1 and Table 1. However, this technique is complicated and is used only reluctantly in practice. It is customary to overdose and subsequently to precipitate remaining sulfide with bivalent iron. This results in negatively charged suspensions, however, which are difficult to floc and to filter.

The logical suggestion, therefore, is to add sulfide in sub-stoichiometric quantities, because a portion of the heavy metals is already present in the suspension as oxides or carbonates, and to destabilize the negatively charged particulate precipitation products by adding Me(III) salts again. The subsequent aggregation and flocculation is achieved using conventional methods.

3 Description of the Raw Water from the Flue Gas Rinse

In a refuse incineration plant with a capacity of 6 t/h, 290 m^3 of wastewater with pH values from 0.3 to 1.0 and a composition shown in Table 3 is produced daily in the flue gas rinser which follows an electrostatic filter [7].

Table 3. Composition of flue gas rinse water from a refuse incineration plant and from a hard coal plant and achieved residual concentrations

	Cl g/l	Ca g/l	Mg g/l	Cd mg/l	Cu mg/l	Hg mg/l	Mn mg/l	Ni mg/l	Pb mg/l	Zn mg/l
Refuse incineration	14			1	2	10		2	19	37
Flue gas desulphurization	36	20	4	1	1	0.1	100	4	0.6	50
Filtrate	pH 9									
Carbonate precip.	36			0.5	0.07				0.03	0.2
TMT ppt.	36			0.5	0.07					0.1
Sulfide	36			0.001	0.001	0.001	2	0.2	.001	

For a two-step flue gas purification, the flue gas is first pre-treated with water, whereby hydrogen chlorine and hydrogen fluorine dissolve in the water. In the second step, the exhaust gas is desulphurized by limestone washing, whereby in some cases water from the cooling tower is used, which is already preloaded with heavy metals. The wastewaters are combined and result in 36 m^3/h wastewater for a 750 MW$_{EL}$ power plant block; the composition of the wastewater is shown in Table 3 [8]. When low-grade coal having 21 MJ/kg calorific value and 0.3% Cl is used, a volumetric flow rate of 50 to 100 m^3/h wastewater results, depending on the mode of operation. If, in contrast, coal having 28 MJ/kg and 0.1% Cl is burned, the volumetric flow rate of the wastewater is reduced to 15 m^3/h in a 750 MW$_{EL}$ power plant [9].

When governmental regulations concerning the quality of the flue gas rinse water permitted for release into rivers or into the sewer system are published and a wastewater output tax is determined, the operators of flue gas rinsers will be motivated by economic factors to keep the wastewater volumetric flow rate and the mass flow of heavy metals as low as possible. From the technical point of view, the progressively increasing concentration of pollutants in the rinse water is reasonable, because the purification makes possible constant, low effluent concentrations regardless of the concentration of pollutants in the raw water [1]. When the permissible effluent values are determined in this way, the pollutant loads removed, in this case the load of heavy metals, become larger with increasing influent concentrations. However, high effluent concentrations may disturb the precipitation process. On the one hand, the solubility of the heavy metal hydroxides and carbonates increases due to the increasing ionic strength; on the other hand, there is the possibility of further complex formation, e.g., due to the high chloride concentration.

For these reasons, the precipitation experiments were carried out not only with the usual concentrations found in flue gas rinse water, but also with concentrated flue gas rinse water with higher concentrations. Tests for sulfide dose, precipitation and filtration were carried out using only the concentrated chloride solutions.

4 Methods

In order to determine the effect of the chloride concentration on precipitation, jar tests [10] were carried out using modified stators in jars [13] and with paddle mixers. The rapid mixing phase lasts 10 s with 300 rpm and the slow mixing phase 1200 s with 30 rpm. This corresponds to an average velocity gradient (G value) of 800 and 40 s^{-1}, respectively.

The test solutions or the flue gas rinse water were brought to 10 mmol/l alkalinity (pH 4.3) using $NaHCO_3$ and adjusted to the desired pH value with HNO_3 and NaOH.

Flocculation was carried out using 0.2 mmol/l $FeCl_3$. At the end of the rapid mixing phase, the pH value was determined and 0.3 mg/l weakly anionic polyacrylamide (flocculation aid) was added.

The flocculated and sedimented sample was filtered and acidified to pH 2 with HNO_3. For the voltametric determination of the heavy metals, the sample was also treated with H_2O_2 and disintegrated by UV radiation [11].

Sulfide precipitation was carried out in a laboratory apparatus as follows: flocculation in pipes with turbulent flow and filtration over gravel filters, which were described in [12]. The technical specifications of this apparatus are as follows:

Clarity measurement of effluent from the sand filter

Volumetric flow rate	100 l/h
1st pipe (microflocculation)	d = 10 mm l = 20 m G = 200 s^{-1}
2nd pipe (macroflocculation)	d = 16 mm l = 20 m G = 85 s^{-1}
Sedimentation lamella separation	4-edged pipe 130 x 130 x 1000 mm
Angle of inclination	60°
Sand filter	d = 40 mm h = 500 mm F = 1590 mm^2
Sand packing	d = 2 mm

The pH value of the raw water was adjusted with HNO_3 and NaOH. Sulfide precipitation was carried out with 0.2 mmol/l Na_2S and flocculation with 0.2 mmol/l $FeCl_3$. 0.6 mg/l flocculation aid was added prior to macroflocculation.

5 Results

The results are shown in Figures 2 through 5 and in Table 3. Chloride seems to have a negative effect only in the case of cadmium. The cadmium-chloride complexes account for 90% of the total cadmium concentration when $c(Cl^-) = 0.7$ mol/l [13]. Accordingly, the solubility of $Cd(OH)_2$ or $CdCO_3$ increases by a factor of 10 in the transition from $c(ClH^-) = 0.0025$ to 0.7 mol/l. The solubility is, however, partly compensated by the effect of co-precipitation of the cadmium species with $Fe_n(OOH)_n$ [13].

Only the residual concentrations of Zn are lower than the calculated equilibrium concentrations, especially at high pH values. For this reason, addition of carbonate seems to be of interest only for pH buffering.

The addition of sulfide is to be interpreted differently, though. Together with satisfactory destabilization, easily separable precipitation products result, and only very minimal residual concentrations of heavy metals remain in the filtered rinse water (cf. Table 3).

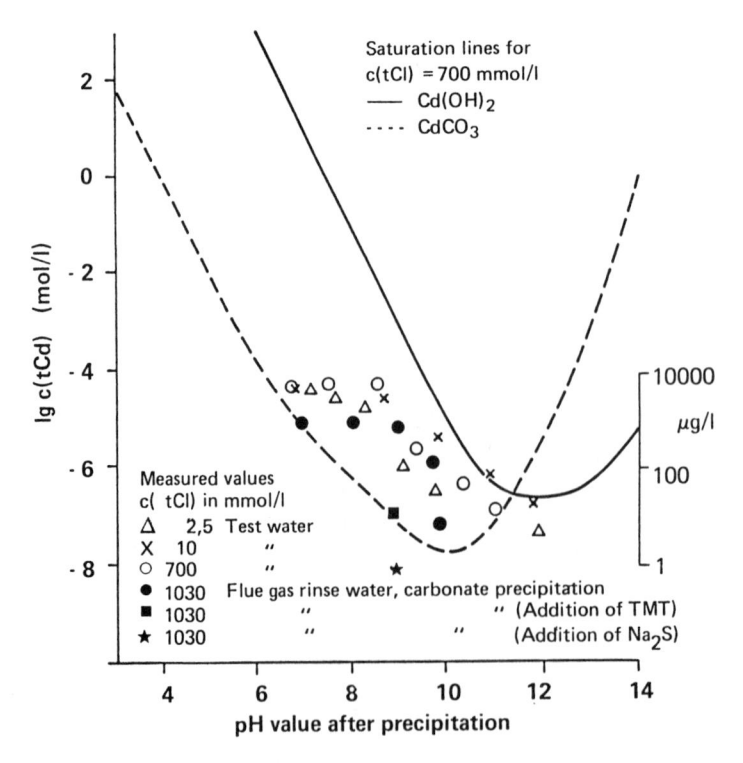

Figure 2. Residual concentrations of cadmium for various levels of total chloride (tCl) in the water and $c(tC) = 10$ mmol/l (inorg. C)

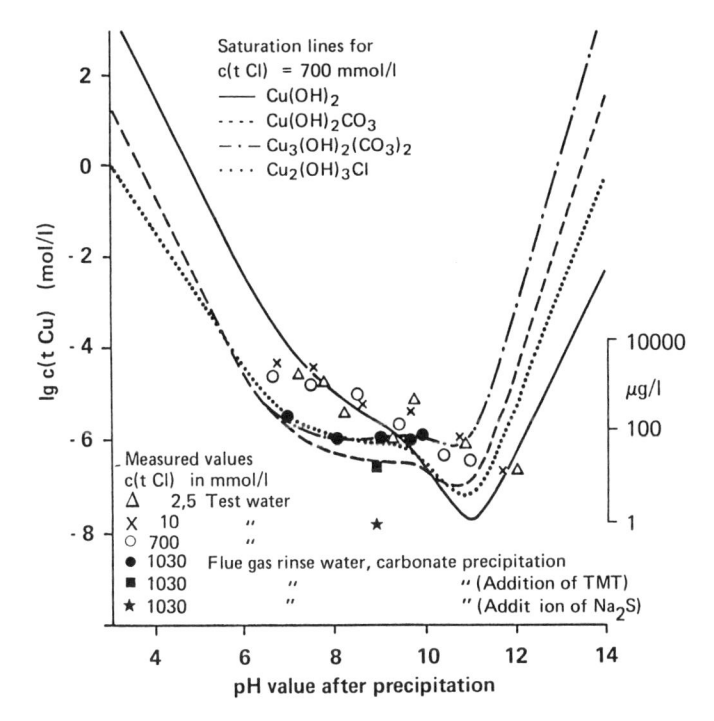

Chemical Treatment of Flue Gas Washing Liquids

Figure 3. Residual concentration of copper for various levels to total chloride (tCl) in the water and $c(tC) = 10$ mmol/l (inorg. C)

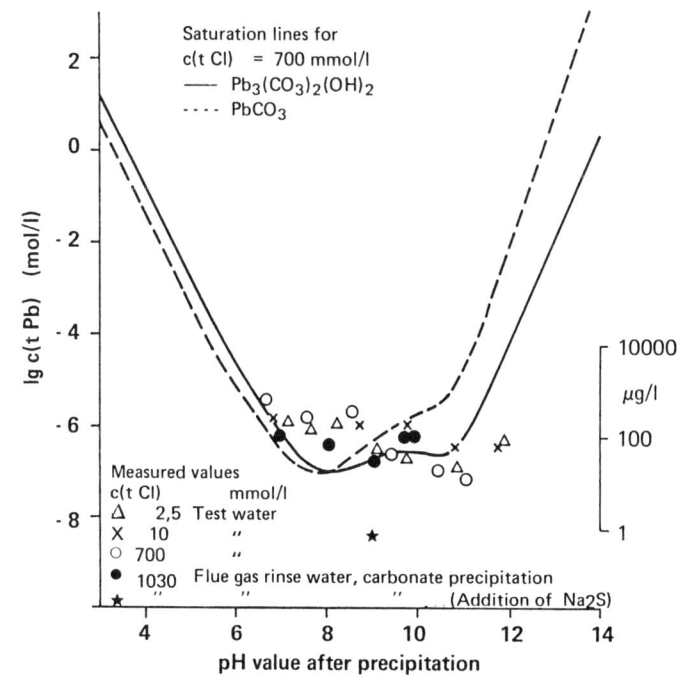

Figure 4. Residual concentrations of lead for various levels of total chloride (tCl) in the water and $c(tC) = 10$ mmol/l (inorg. C)

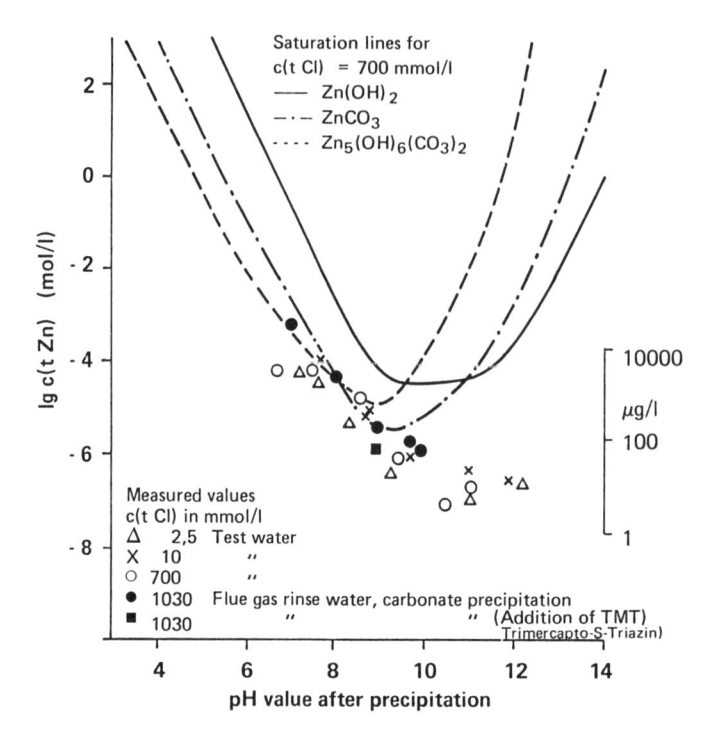

Figure 5. Residual concentrations of zinc for various levels of total chloride (tCl) in the water and $c(tC) = 10$ mmol/l (inorg. C)

6 Calculation of Improvement to the Aquatic Environment

The concentrations in Table 3 refer to a demonstration plant for refuse incineration with a capacity of 6 t/h and a rinse water effluent of 2 m^3/h [7]. From this the load per ton of refuse can be calculated. Using 30 million tons/year as a figure for household garbage in the Federal Republic of Germany, the load of heavy metals which is found in the flue gas can be estimated for complete refuse incineration. If the wet flue gas rinse is introduced simultaneously with the refuse incineration, then precipitation prevents heavy metal pollutants from being introduced into the aquatic environment.

A similar calculation can be performed for power plants which use hard coal. A 750 MW$_{EL}$ plant has a rinse water flow rate of 40 m^3/h with a chloride content of 4 g/l [8]. The data in Table 3 refer to rinse water that was circulated and concentrated 10 times.

Thus, 4 m^3/h of a concentrated flue gas rinse water should be considered. Then the annual load per MW can be determined. For 45,000 MW installed capacity for hard coal and brown coal powerplants in the FRG, the total load, and therefore the improvement to the environment which can be expected, can be estimated when the rinse water is treated with a combination of sulfide- carbonate precipitation.

Table 4. Environmental pollution due to flue gas rinse water and improvement to the environment when this water is treated

	Cd	Cr	Cu	Hg	Mn	Ni	Pb	Zn
Load per ton of refuse for refuse incineration [g/t]	2.5	1.4	4	20		4	38	74
Annual load without treatment [t/yr]	7.5	42	120	600		120	1140	2220
with treatment [t/yr]	0.08	0.04	0.12	0.06		18	0.1	0.6
Load per MW for desulphurization of flue gas [g/(MW yr)]	40	40	4	8000		400	20	2000
Annual load without treatment [t/yr]	1.8		1.8	0.2	360	18	0.9	90
with treatment [t/yr]	0.002		0.002	0.002	7.2	0.9	0.001	0.2

7 Summary

When the flue gases of refuse incineration are wet-purified, pollutants from the refuse are transferred from the air into the water. The heavy metals can be removed from the water completely (up to 0.001 mg/l) through a combined carbonate-sulfide precipitation process, even though the pH value of 9 is not exceeded so that precipitation of magnesium is avoided. The rinse water from the flue gas desulphurization process can be treated in the same way.

This allows a significant improvement of the environment with respect to heavy metals. Large quantities of mercury (up to 500 t/yr), lead (1000 t/yr), zinc (2000 t/yr), cadmium (60 t/yr) and copper (100 t/yr) accummulate particularly during refuse incineration, and using this method they can be concentrated in precipitated sludge, dewatered and disposed of or recycled.

References

[1] Grohmann, A., Horstmann, B., Sollfrank, U. (1982): Ein einfaches Konzept zur vollständigen Eliminierung von Schwermetallen am Beispiel Cd, Cr, Cu und Zn. Vom Wasser 58, 269 - 289

[2] Smith, R.M., Martell, A.E. (1976): Critical Stability Constants. Plenum Press, New York/London

[3a] Stumm, W., Morgan, J.J. (1982): Aquatic Chemistry. John Wiley & Sons, New York, 2nd edn., pp. 599

[3b] Stumm, W., Sigg, L. (1974): Kolloidchemische Grundlagen der Phosphorelimination. Fällung, Flockung und Filtration. Z. Wasser Abw. Forsch. 12, 37 - 46

[4] Sillen, L.G. (1954): On Equilibria in Systems with Polynuclear Complex Formation. II. Testing Simple Mechanisms which give "Core+Links" Complexes of Composition $B(A,B)_n$. Acta Chem. Scand. 8, 318 - 335

[5] Leckie, J.O., Benjamin, M.M., Hayes, K., Kaufman, G., Altmann, S. (1980): Adsorption/Coprecipitation of Trace Elements from Water with Iron Oxyhydroxide. Electric Power Research Inst., CS 1513 Res. Proj. 910-1

[6] Grohmann, A., Althoff, H.W., Koerfer, P. (1984): Geschwindigkeit der Phosphatfällung mit Eisen(III)-Salzen und der Einfluß von Calciumionen. Vom Wasser 62, 171 - 189

[7] Reimann, D.O. (1983): Schwermetallreduzierung im Müllheizkraftwerk Bamberg. Müllverbrennung und Rauchgasreinigung. Thome-Kozmiensky (Hrsg.), Freitag
[8] Sieth, J. (1984): Abwasser aus Rauchgasendreinigungsanlagen. Haus der Technik 490, Vulkan, Essen
[9] Frick, B.R. (1985): Rauchgasreinigung. Vom Wasser 65, 145 - 156
[10] Hudson, H.E. Jr., Wagner, E.G. (1981): Conduct and Uses of Jar Tests. J. AWWA 49, 1414 - 1424
[11] Philar, B., Valenta, P., Golimowski, J., Nürnberg, H.W. (1980): Z. Wasser Abw. Forsch. 13, 130 - 138
[12] Grohmann, A. (1985): Flocculation in Pipes, Design and Operation. In: Chemical Water and Wastewater Treatment. Grohmann, A.N., Hahn, H.H., Klute, R. (Hrsg.), Fischer, Stuttgart
[13] Scheerer, H.-P. (1987: Untersuchungen des Einflusses hoher Chloridkonzentrationen auf die Fällung von Schwermetallen. Diplomarbeit, TU Berlin

A.N. Grohmann
Institute for Water,
Soil and Air Hygiene
Federal Health Office
1000 Berlin
Fed. Rep. of Germany

L. Bauch, H.-P. Scheerer
Dep. of Water Purifikation
Inst. for Env. Technology
Technical University of Berlin
1000 Berlin
Fed. Rep. of Germany

Section III

Wastewater and Sludge

Hydrogen Sulphide Control in Municipal Sewers

T. Hvitved-Jacobsen, B. Jütte, P.H. Nielsen and N.Aa. Jensen

Abstract

Increased concern for the impact of hydrogen sulphide produced under anaerobic conditions in wastewater has lead to the need for pretreatment of the sewage.

Intensive field investigations on hydrogen sulphide formation, impacts and control in sewerage systems were carried out during the period 1985–1987. Emphasis was given to determining the importance of sulphate, organic matter, temperature and anaerobic residence time on sulphide formation in a 3.9 km pressure main.

Based on these studies an empirical model for the sulphide buildup in sewers was proposed and evaluated. Variation of the diurnal sewerage composition and flow and the importance of the type of sewage (municipal/industrial) on sulphide formation was also taken into account.

Furthermore, full scale experiments in which iron sulphate was added to the sewage were carried out under fluctuating conditions in order to optimize precipitation of the sulphide produced and reduce the impact of the hydrogen sulphide. A system for control and adjustment of the chemicals dosed was developed based on the proposed model and an evaluation of the importance of the external parameters and conditions.

Introduction

During recent years widespread occurrence of hydrogen sulphide related problems in municipal sewers has been observed in Denmark. This fact is associated with increased centralized wastewater treatment which has caused a corresponding need for pressure mains to transport sewage over long distances. In these full flowing sanitary sewers, strictly anaerobic conditions may often prevail for several hours. Therefore, sulphate reducing bacteria which develop in the wall slimes (biofilms) on the pipes are given favourable conditions for growth and may therefore cause extensive hydrogen sulphide formation and related effects [1, 2].

Among the problems observed, until now special attention has been paid to the occurrence of hydrogen sulphide in the atmosphere in pumping stations and gravity sewers located downstream of a pressure main, Figure 1. Odor problems, health aspects and corrosive effects on metals and concrete are all well known examples to be taken into account [3, 4]. Furthermore, because wastewater treatment plants with sensitive biological processes like nitrification and biological phosphorus removal are going to increase, the influx of excessive amounts of sulphide should be avoided. The entire complex of hydrogen sulphide problems requires that the pretreatment concept for wastewater be more seriously considered in the future in order to reduce undesired effects.

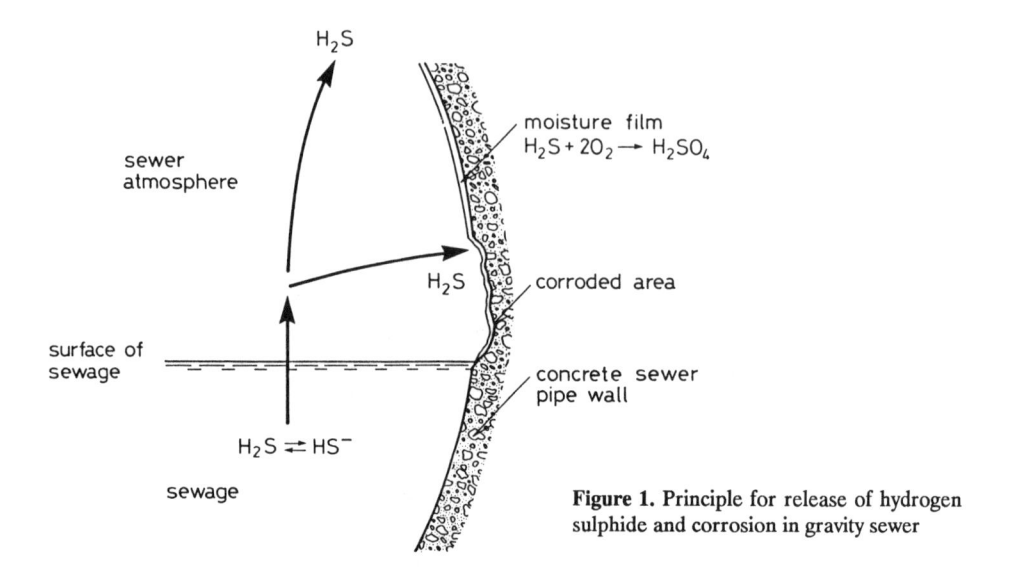

Figure 1. Principle for release of hydrogen sulphide and corrosion in gravity sewer

Impact of Sulphides on Wastewater Treatment Processes

The activated sludge wastewater treatment processes are affected to different degrees by the influx of hydrogen sulphide.

Sulphide is often responsible for bulking in sewage treatment plants caused by filamentous microorganism. Especially the sulphide-oxidizing *Thiothrix* and *Thiothrix*-like filamentous bacteria are found in bulking activated sludge in both the USA and Europe including Denmark [5, 6, 7]. Reduced sulphur compounds also affect wastewater treatment by increasing the need for aeration in the pretreatment stages, e.g. the grit chamber, mg S^{2-}/mg O_2 = 0.5. Furthermore, sulphides are known to inhibit the nitrification process [8]. Especially in areas with a relatively cold climate, i.e. winter temperatures in the sewage below 5 – 8°C, the toxic effect of sulphides may be particularly harmful to the nitrogen removal processes.

In order to overcome the sulphide problems in the sewers and the wastewater treatment plants, the addition of the chemical agent must follow the daily variation of the main variables which determine the formation of hydrogen sulphide. Therefore, the development of a scheme for an appropriate pretreatment of sulphide-containing sewage must fulfill the following demands: a model for prediction of hydrogen sulphide formation in full flowing sewers must be available and a corresponding procedure for optimal dosage of a chemical agent should be developed and designed.

The objective of this paper is to quantify both hydrogen sulphide formation and its control in sewers during dry weather conditions.

Sewer System Overview

The sewer system under investigation is located in the northern part of Jutland, Denmark. Sampling and measurements took place in two pumping stations located upstream (Vejby) and downstream (Boerglum), respectively, of a 3.9 km pressure main.

The sewer catchment area upstream of the site amounts to 163 ha of which 97 ha are combined sewers. During dry weather periods the daily flow varies between 25 and 90 m^3/h. For rainfall runoff the design peak flow is 500 m^3/h corresponding to a pipe velocity of 1.25 m/s.

Each pumping station is operated by three sewage pumps. During dry weather conditions one pump in operation results in a pipe flow of 260 m^3/h with a pipe velocity of 0.65 m/s. Therefore, there is intermittent flow with one pump in action for approximately 5 to 20 minutes per hour. Each active pumping period lasts for 3 – 5 minutes depending on the flow to the wet well.

Methodology

Field studies were conducted during two periods: July-November 1985 and September – October 1986. The objectives were:

- Determination of the rate of hydrogen sulphide formation in the pressure main in order to set up an empirical model. Procedures for measurements, sampling and analyses are described in [1] and [2].
- Determination of a daily variation of dosage of iron sulphate as a chemical agent for suppressing the impact of hydrogen sulphide.

Addition of iron sulphate to the sewage took place at the inlet into the upstream pumping station, Vejby. A solution at around 0.14 kg FeSO$_4 \cdot$7H$_2$O/l was pumped from a 4 m^3 container.

Two different principles for adding iron to the sewage were in operation during the investigation period:

The fundamental principle for the original procedure which was implemented by the municipality was based on adding iron sulphate to the wet well. Addition of the chemical agent took place only while the sewage pumps were in operation. During these pumping periods iron was added at intermittent time intervals.

The alternative procedure which was studied worked in the following way: Based on the model developed for hydrogen sulphide formation in the pressure main, the daily variation for continuous dosage of iron sulphate to the wet well in the upstream pumping station, Vejby, was calculated.

In the practical situation addition of iron sulphate was regulated day and night by an electronic control device on 10 different levels. The dosage pump was chosen so that the addition of iron sulphate on each level was as close to continuous addition as possible. Required efficiency of the dosage was evaluated by continuous monitoring of the hydrogen sulphide concentration in the sewer atmosphere in the downstream pumping station, Boerglum.

Design Equation for Hydrogen Sulphide Suppression in Sewers

As a part of the entire investigation program an empirical model for prediction of the sulphide flux from a biofilm surface in a pressure main was developed [1, 9]:

$$r_a = a\,(COD_{sol} - b)^c \tag{1}$$

r_a Rate of sulphide formation (sulphide flux from the biofilm surface) [g S/m² hr]
COD_{sol} Chemical oxygen demand for filtered sample [g O_2/m³]
a, b and c are parameters

In a pressure main where anaerobic conditions exist, the following empirical equation for predicting the increase in the hydrogen sulphide concentration of the wastewater flow was developed ($b = 50, c = 0.5$):

$$\Delta C_s = a\,\sqrt{COD_{sol} - 50}\;\; 1.07^{t-20}\,t_h\frac{A}{V} \tag{2}$$

ΔC_s Formation of total sulphide (determined as soluble + metal-associated) in a pressure main, [g S/m³]
t Temperature [°C]
t_h Anaerobic residence time in the pressure main [hr]
A Inside area of the pipe [m²]
V Total volume of the pipe [m³]

According to both laboratory and field investigations the following choice of the parameter a was proposed in order to simulate hydrogen sulphide formation, Table 1.

Table 1. Proposed values of the parameter a in equation (2)

Model type	a	Remarks
A	1.5×10^{-3}	Typical Danish domestic sewage with little or no content of industrial sewage
B	3×10^{-3}	Sewage from both municipal and industrial (foodstuff) sources
C	6×10^{-3}	Sewage from foodstuff industries

Based on Equation 1 and Table 1, Figure 2 shows the hydrogen sulphide formation rate versus the concentration of soluble COD. As indicated Models A and B should be restricted to COD_{sol} values less than 500 and 1000, respectively.

Parameters important for the formation of hydrogen sulphide under anaerobic conditions in pressure mains are discussed in [1] and [9]. From Equation 2 it is evident that soluble COD, temperature, anaerobic residence time and the sewer dimensions are the governing parameters. Not until the concentration of sulphate is less than 5 – 10 mg SO_4-S/l does it affect the rate of sulphide formation [9].

It should be emphasized that Equation 2 represents a valuable tool both for the prediction of hydrogen sulphide formation and for the dosage of the chemical agent in question.

Hydrogen Sulphide Control in Municipal Sewers 243

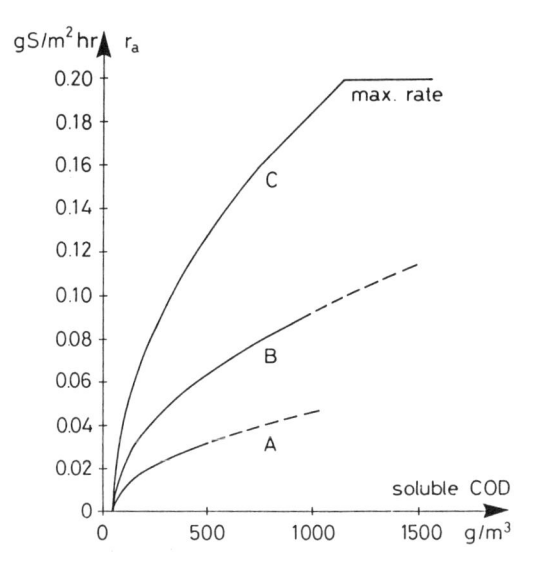

Figure 2. The hydrogen sulphide formation rate versus the sewage concentration of soluble COD; cf. text

Especially the parameters COD_{sol} and anaerobic residence time, i.e. flow, are subjected to daily fluctuations in a sewer pipe. In order to control the variation of the required amount of iron sulphate added to the sewage, these fluctuations must be known. Technically, it is possible - although still difficult under intermittent flow conditions – to monitor the flow variations. However, now there is still no reliable method for real time control of COD_{sol}. Therefore, the daily variation in both COD_{sol} and flow must be predicted in order to calculate (Eq. 2) the fluctuations in the downstream concentration of hydrogen sulphide as the basis for dosage of iron sulphate. An example of the daily fluctuation in COD_{sol} and flow is shown in Figure 3.

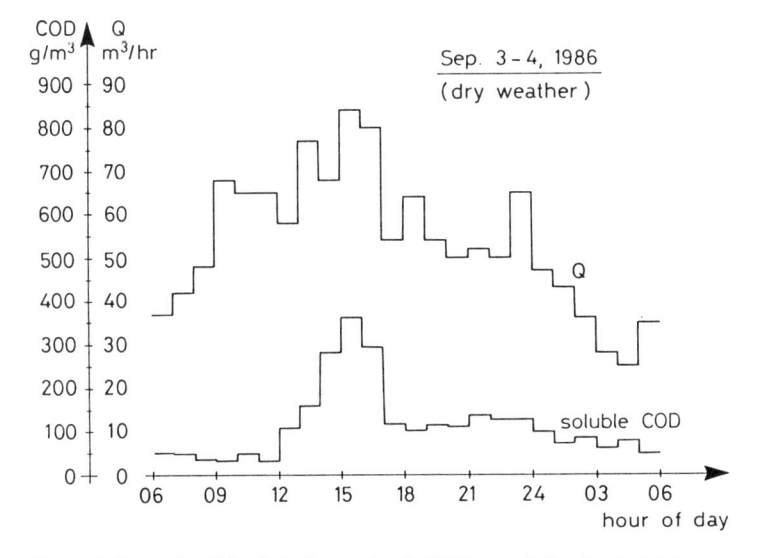

Figure 3. Example of the daily fluctuation in COD_{sol} and flow in the 3.9 km pressure main

In addition, the variation in the downstream concentration of hydrogen sulphide was monitored during the same period of time as shown in Figure 3. The result is represented in Figure 4 and shows the need for a regulated addition of a chemical agent.

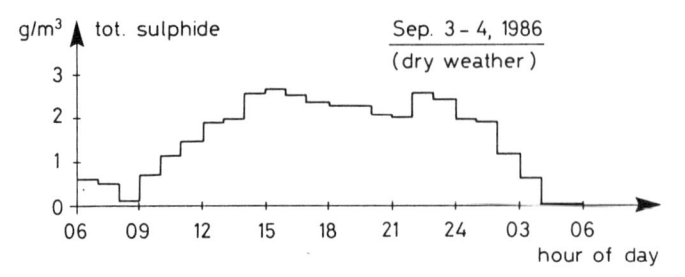

Figure 4. The daily variation in total sulphide in the 3.9 km pressure main

For the pressure main in question, Model B was chosen as the best alternative for predicting hydrogen sulphide formation. This occurred because the type of sewage was partly domestic, partly industrial (a slaughter-house). The calculated dosages of iron sulphate for precipitation of sulphide show the importance of the industrial source since the weekday demand is high compared with Saturday/Sunday consumption, Figure 5.

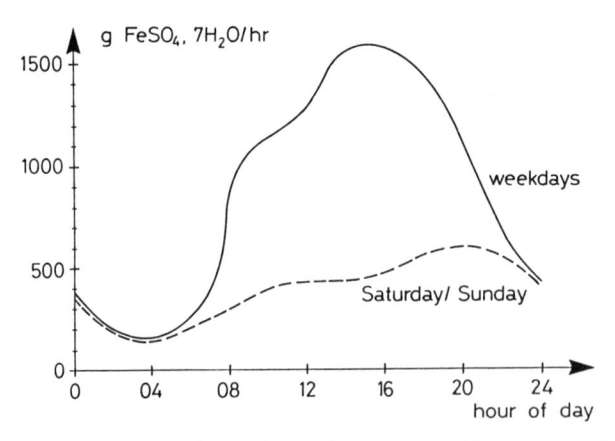

Figure 5. Calculated curve for continous dosage of iron sulphate for precipitation of sulphide in a pressure main

Optimal Procedure for Dosage of Iron Sulphate to Sulphide-Containing Sewage

Iron salt is a well known chemical agent which precipitates insoluble iron sulphide in wastewater and thus prevents effects due to soluble sulphide [10]. In Denmark iron sulphate is the most commonly used chemical agent for hydrogen sulphide control. It is found to be both cheap compared with agents like oxygen and nitrate, easy to handle and efficient as it reacts almost stoichiometrically with sulphide to form insoluble iron monosulphide. Furthermore, it is of technical importance that iron sulphate could be added to the sewage before sulphide is present and still react efficiently when sulphide is produced. This means that quantity of iron sulphate based on a model prediction could be added to an upstream point in a sewer line in order to suppress hydrogen sulphide at any point downstream.

As previously described in this paper two different principles of adding chemicals agents to the sewage were studied.

Using the simple principle of adding iron sulphate only while the sewage pumps were working, it was found necessary to use $100 - 140$ kg $FeSO_4 \cdot 7H_2O$ per day on weekdays under dry weather conditions in order to precipitate the sulphide and maintain low H_2S-concentrations in the atmosphere of the wet well in the downstream pumping station, Boerglum.

By using this simple dosing principle the concentration of iron sulphate in the sewage pumped into the pressure main during one pumping period varied by a factor of $2 - 2.5$. This was due to the variation of the volume in the wet well and only dosing when emptying the wet well. Except for this variation of the iron sulphate concentration during each pumping period there was no change in the concentration following the daily variation in the flow and the COD concentration in the sewage.

Calculations showed that it would be necessary to use only $25 - 30$ kg $FeSO_4 \cdot 7H_2O$ per day in order to precipitate the sulphide produced in the pressure main from Vejby to Boerglum. This result was confirmed by implementing the principle of adding $30 - 35$ kg $FeSO_4 \cdot 7H_2O$ per day to the sewage entering the pressure main, at any time of the day sufficient to precipitate the sulphide produced later in the pipe. From the measurements on the system the daily weekday flow and COD variation were known. Based on these values the variation in sulphide production was calculated using the previously described empirical model.

In order to introduce a simple system for controlling the addition of iron sulphate into the pumping station Vejby, the addition was kept constant for periods of $2 - 4$ hours, using different amounts of iron sulphate per hour, following the diurnal variation in sulphide production, Figure 6.

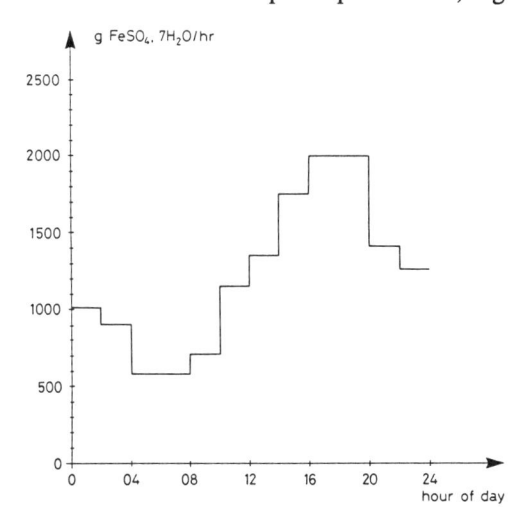

Figure 6. Different dosages of iron sulphate to precipitate sulphide in a pressure main

From the measurements of H$_2$S in the sewage atmosphere in the downstream pumping station Boerglum, it was possible to determine any changes to the actual pattern for the addition of iron sulphate that might be required. By means of the daily variation of residence time in the pressure main, Figure 7, it was easy to find those periods where increase or decrease in the added amount of iron sulphate was required.

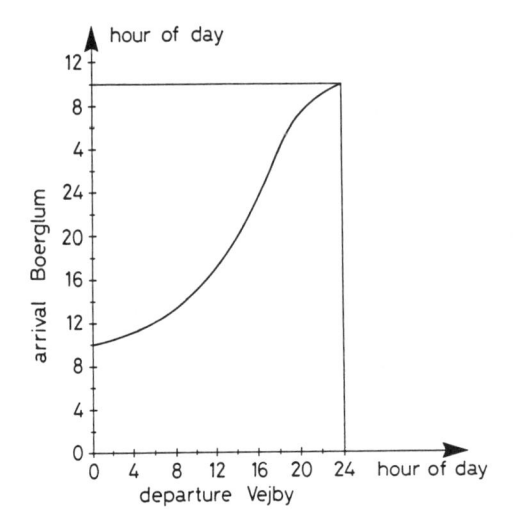

Figure 7. Determination of residence time for the sewage between Vejby and Boerglum. Results are based on average weekday and dry weather conditions

When adding the chemical agent almost continuously to the pumping station Vejby, the concentration of iron sulphate in the sewage in the pressure main only varied with the dosage. This meant that it was possible to control the amount of iron sulphate in every part of the sewage in the pressure main, which was necessary in order to make it react efficiently with the sulphide.

Finally, it should be emphasized that the procedure for determining optimal dosage of a chemical agent is of general applicability, e.g. it seems well suited for a chemical agent like nitrate for suppressing hydrogen sulphide formation.

Conclusions

An empirical model was presented for simulating increases in the hydrogen sulphide concentration in domestic sewage passing through a pressure main. The model could be used for the prediction of hydrogen sulphide formation with variations in soluble COD, temperature, flow and sewer dimensions. Furthermore, the type of the sewage (municipal/industrial) is considered in the model.

The empirical equation was shown to be a valuable tool for determining optimal dosage of a chemical agent (iron sulphate) to sewage in order to suppress undesired hydrogen sulphide concentrations in both sewers and wastewater treatment plants. A procedure for determining optimal dosage corresponding to the daily fluctuation in the sewage parameters — soluble COD and flow — was proposed and evaluated.

Acknowledgements

The authors wish to acknowledge the financial support of the Danish National Agency of Environment Protection and the Danish Technical Research Council.

References

[1] Hvitved-Jacobsen, T., Nielsen, P.H. (1986): TNO Committee on Hydrological Research. Proceedings No. 36, 39-52, Wageningen
[2] Postgate, J.R. (1984): The Sulphate Reducing Bacteria, 2nd edn., Cambridge University Press, Cambridge
[3] Thistlethwayte, D.K.B. (ed.) (1972): The Control of Sulphides in Sewerage Systems. Butterworths, Sidney Melbourne Brisbane
[4] Pomeroy, R.D. (1976): The Problem of Hydrogen Sulphide in Sewers. Clay Pipe Development Association Ltd., London
[5] Strom, P.F., Jenkins, D. (1984) J. Wat. Pollut. Control Fed. *56*, 449-459
[6] Farquhar, G.J., Boyle, W.C. (1971) J. Wat. Pollut. Control Fed. *43*, 779-798
[7] Nielsen, P.H. (1984) Wat. Sci. Techn. *17*, 167-181
[8] Beccari, M. et al. (1980) Environm. Technol. Lett. *1*, 245-252
[9] Nielsen, P.H., Hvitved-Jacobsen, T. (1988) J. Wat. Pollut. Control Fed. (in press)
[10] Pomeroy, R., Bowlus, F.D. (1946) J. Sewage Works *18*, 598-640

T. Hvitved-Jacobsen, B. Jütte,
P.H. Nielsen, N.Aa. Jensen
Environmental Engineering Laboratory
University of Aalborg
Sohngaardsholmsvej 57
9000 Aalborg
Denmark

Coagulation as the First Step in Wastewater Treatment

H. Ødegaard

Abstract

In this paper contamination in municipal wastewater is characterized in terms of particle size, and the fate of particles in biological treatment is discussed. Based on this, it is concluded that coagulation, used as the first step in wastewater treatment, is rational. To demonstrate what is achievable by coagulation of raw wastewater, treatment results from over 100 Norwegian chemical treatment plants are presented.

Introduction

Wastewater treatment is normally carried out in steps, traditionally with primary settling as the first step, biological treatment as the second and often chemical treatment as the third step. It is this author's opinion that this procedure seems more to be based on historical development than on the composition of municipal wastewater. It is the intention of this paper to discuss the composition of raw municipal wastewater and from that viewpoint, determine what kind of process should be used as the first step in wastewater treatment.

Classification of Contaminants in Wastewater

Surprisingly enough, very few studies have been performed in order to characterize the contaminants in municipal wastewater in terms of particle size. Very early, however, Imhoff [1] presented in his classical pocket-book, tables showing the distribution of settleable and non-settleable suspended matter and soluble matter in wastewater, demonstrating that more than half of the organic matter was associated with the suspended solids.

In some early American studies (Balmat [2], Heukelekian and Balmat [3] and Richert and Hunter [4]), the organic contaminants in wastewater were separated into four size fractions by successive sedimentation, centrifugation and filtration. The fractions were classified by size range as settleable, supracolloidal, colloidal and soluble (Table 1).

It was demonstrated that only 25–30% of the COD/TOC was found to be truly soluble according to the definition used. Furthermore it was shown that the soluble

Table 1. Composition of organic materials in wastewater [2, 3, 4]

Size Range	Classification			
	Soluble	Colloidal	Supra-colloidal	Settleable
	<0.08 μm	0.08-1.0 μm	1-100 μm	>100 μm
COD (% of total)	25	15	26	34
TOC (% of total)	31	14	24	31
Org. constit. (% of tot. solids)				
Grease	12	51	24	19
Protein	4	25	45	25
Carbohydrates	58	7	11	24
Biochemical oxid rate, k, d^{-1}	0.39	0.22	0.09	0.08

organic matter smaller than 1 μm was found to degrade biochemically at a much more rapid rate than the particulate organics larger than 1 m.

Based on this, Balmat [2] actually suggested that the larger particles in sewage should be grinded to smaller ones in order to increase the capacity of biological treatment plants. Another conclusion might be that improved particle separation prior to biological treatment would enhance the biochemical oxidation.

A later study by Munck et al. [5] (Table 2) confirmed the results of the earlier investigations, even though Munck used a slightly different definition of the size ranges and used only filtration to separate the fractions.

Table 2. Classification of contaminants in wastewater [5]

Size Range	Classification			
	Soluble	Colloidal	Supra-colloidal	Settleable
	<0.025 μm	0.025-3 μm	3-106 μm	>106 μm
BOD_5 (% of total)	17	16	46	21
COD (% of total)	12	15	30	43
TOC (% of total)	22	6	36	36
Tot P (% of total)	63	3	12	22
Org. N (% of total)	27	15	38	20

As shown in Table 2, Muck et al. [5] also investigated the distribution of total phosphorus and organic nitrogen. Even if most of the phosphorous in wastewater appeared as soluble phosphates, the amount of phosphorous associated with particles was found to be significant, and even more so was the amount of organic nitrogen.

It is often thought that effluents from food-processing industries contain mainly soluble organics. Ødegaard [6] showed, however, that in such wastewaters, as much as 40–70% of the total COD may be associated with the suspended solids (Table 3).

Table 3. Content of suspended COD in food processing effluents [6]

Type of industry	COD_{Total} g/m^3	$COD_{Filtered}$* g/m^3	%$COD_{Suspended}$
Cheese processing	3036	1452	52
Potato processing	6735	1575	76
Slaughterhouse	1833	718	61
Vegetable processing	1588	977	47
Dairy	971	539	45

* 1 µm glassfibre filter

Other contaminants such as metals, bacteria and viruses and organic micropollutants are also strongly associated with particles. Table 4 shows that the percentage of total metal associated with suspended matter in the raw water of a Swedish wastewater treatment plant was in the range of 50 – 75%, with the exception of nickel.

Table 4. Percentage of total metal associated with suspended solids in the raw water at the Eskilstuna wastewater treatment plant in Sweden [7]

Metal	Zn	Cu	Ni	Cr	Pb	Cd
%-suspended	51	48	13	71	71	82

Bacteria and viruses are particles in the size range of about 0.01 – 10 µm, and even if some of the viruses are very small (0.01 – 0.1 µm), several investigations [8, 9, 10] have demonstrated that 60 – 100% of the viruses in sewage are adsorbed to particles.

It is also well known that some of the organic micropollutants such as PCB and PAH have a high affinity for particulate matter [7, 11].

The goal of the first part of this paper has been to demonstrate that a very significant part of the contaminants in wastewater is associated with particles, and that consequently a significant reduction in contaminants may be expected as a result of direct particle separation. By optimizing particle separation at an early stage in the process train, considerable savings could be made in the magnitude of the subsequent processes.

In order to be able to select and optimize particle separation processes, however, much more knowledge about the characterization of particles and particle distributions in wastewater is needed.

Techniques for Particle Size Characterization

Contaminants of interest in wastewater range in size from less than 0.001 μm to well over 100 μm. The size ranges of typical organic contaminants characteristic of settled municipal sewage are presented in Figure 1 [12].

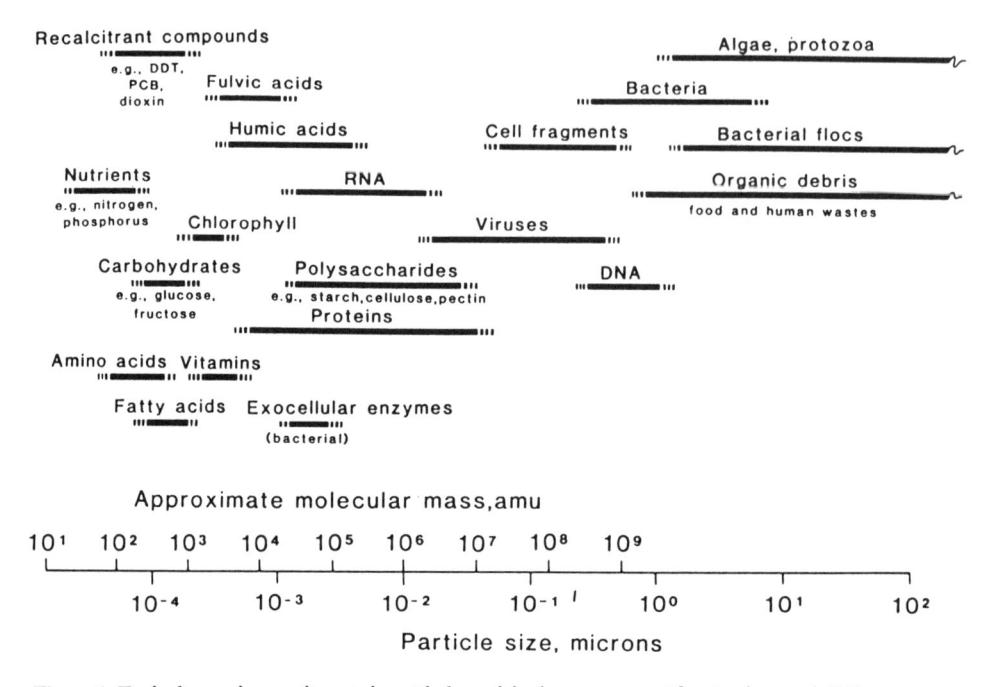

Figure 1. Typical organic constituents in settled municipal wastewater. After Levine et al. [12]

The technique still most commonly used for the characterization of particulate matter in wastewater is the analysis of suspended solids. The particles are separated from the water by filtering through a glass fibre filter with an effective pore size of about 1.2 μm. Levine et al. [12] suggested, however, that the use of a filter with a pore size of 0.1 μm would be more logical for characterizing suspended material in wastewater. This view was based on the review of size ranges of organic matter in wastewater (Figure 1), the available analytical techniques and their finding that 63 – 70% of the organic material measured by the TOC-test was associated with particles larger than 0.1 μm, while virtually no organic matter was detected in the size range between 0.01 – 0.1 μm (Figure 2).

Levine et al. [12] performed a thorough study of particle size distribution in several wastewaters of different quality and concluded that the organic contaminants in municipal wastewater were effectively classified as either greater than or less than 0.1 μm.

One should always analyze contaminants as they enter and leave the different steps in a treatment plant for both unfiltered and filtered samples in order to evaluate whether the contaminants are associated with particles or not. In day to day operation much information is gathered by using the traditional GF/C-filter (pore size about 1 μm), but in optimization and choice of particle separation methods, one should consider the use of serial filtration based on a series of polycarbonae membrane filters.

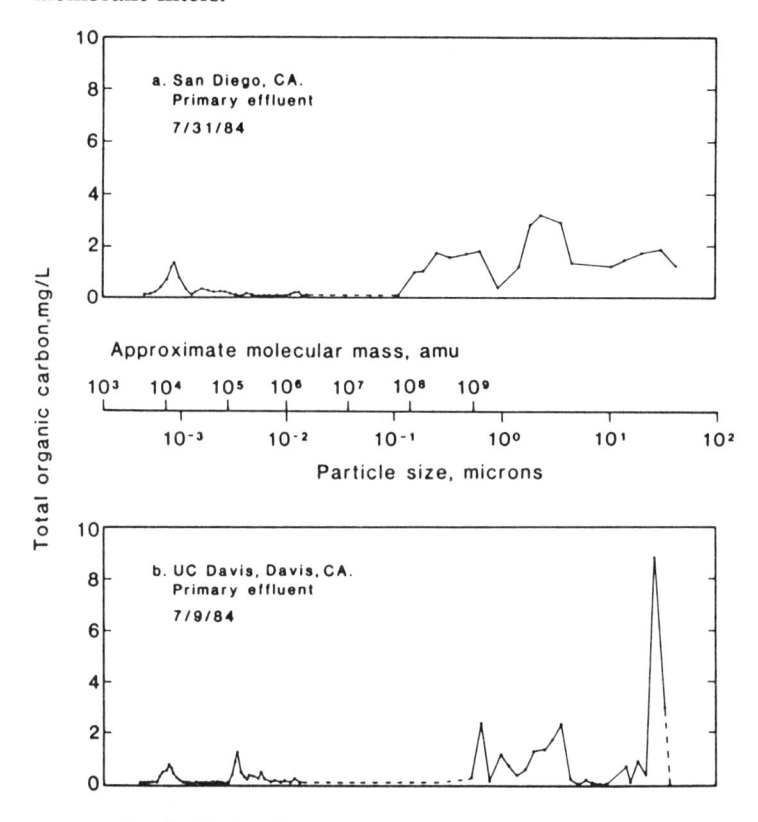

Figure 2. Size distribution of the organic constituents in primary effluent from two locations [12]

The Fate of Particles in Biological Treatment

Principally a biological treatment process consists of a bioreactor and a particle separation reactor. In the bioreactor soluble material (organics, ammonia etc.) is changed from its soluble state to a particulate state (e.g. bacteria, protozoa, algae, etc.), and the particles have to be separated from the water in the separation reactor. Both reactors are equally important for the total treatment result. In spite of this, much more work has been performed in order to understand and describe what is happening in the bioreactor than in the separation reactor. Likewise, much more effort has been expended in studying the fate of the soluble matter than the fate of the particulate matter in the bioreactor.

The bioreactor itself is a captor of particles. The aeration tank of an activated sludge plant, for instance, has a tremendous floc volume fraction, and the flocculation increases performance with increasing floc volume fraction. In practice we make use of the flocculation capability of activated sludge in the contact stabilization process and in the practice of returning excess sludge to the primary clarifier in order to promote clarification efficiency.

The Influence of Particles on Biodegradation

Even if it can be demonstrated that the activated sludge process has a great potential for separating particles from the water by flocculation, the question may be raised whether it is economically sound to use the bioreactor as a captor of particles.

Gujer [13] demonstrated that particulate organic material in primary effluents is rapidly taken up into the activated sludge, but its degradation is very slow. In addition, as shown in Table 1 of this paper, the lower the biochemical oxidation rate, the larger the particles [2]. Likewise, there are strong indications from the work of Särner [14, 15, 16] and Rusten [17, 18], that particulate and dissolved organics interact on biofilms during the removal process, and that particle adsorption on biofilm surface decreases the removal rate of dissolved organics.

In a plug-flow biofilm reactor, the biofilm in the last part of the reactor is loaded with biofilm-particles from the first part, also leading to a lower removal rate of the dissolved organic substances in the last part. This effect may explain why some authors [19] have found that after the depth of a trickling filter exceeds a certain value, there is little change in treatment efficiency, contrary to the many models based on soluble BOD kinetics that illustrate that efficiency should increase with increasing depth.

Särner [20] proposed that the negative effect of adsorbed organic particles on the removal of dissolved organics may be caused by a local oxygen shortage inside the biofilm that develops from the degradation activity of the adsorbed particles, see Figure 3.

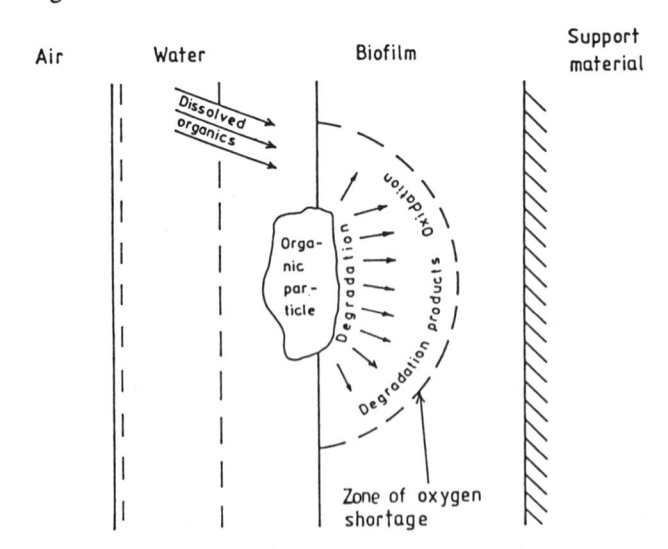

Figure 3. Possible effect of an adsorbed organic particle on the biofilm surface [20]

Based on the findings that particulate matter has a negative effect on the degradation rate both in activated sludge and in biofilm processes, it seems to be logical to remove the particles ahead of the bioreactors, both in order to reduce the overall load and to promote rapid biodegradation.

Coagulation of Wastewater

Since such a large amount of the organic matter and other contaminants in wastewater are associated with particles, and since it seems that particles have a negative effect on biodegradation in biological processes, it seems reasonable to start the wastewater treatment processes train with a good particle separation process.

Sedimentation is, of course, the most widespread sewage separation unit process, and is used both for pretreatment and in later stages in the process train. Sedimentation is, however, only effective for the removal of particles larger than about 50 μm on a practical basis. Based on Stokes law and a particle overflow rate in the range of $1-2$ m/h, the minimum particle size that would be removed by sedimentation is in the range of $50-70$ μm, assuming an average density of particles of 1.2 g/m^3 and a temperature of 15°C. A major part of the particles larger than 0.1 μm may, however, be separated by sedimentation subsequent to coagulation/flocculation.

When used in wastewater treatment, aluminium salts, iron salts or lime are used as coagulants resulting in phosphate precipitation as well as particle coagulation. The process, often referred to as chemical treatment, consists of a reaction step where particles are coagulated and soluble substances, like phosphates, are precipitated, followed by a flocculation step where the coagulated/precipitated particles are aggregated to separable flocs, to be removed by some floc separation process, normally sedimentation.

The reaction step of the process, namely coagulation/precipitation, is very rapid and is terminated within seconds while the flocculation and floc separation processes are far slower. Investment cost is consequently almost entirely linked to the particle separation step. Both the reaction and the particle separation steps are, however, equally important for the total treatment result. Optimization of particle removal by the different unit processes is dealt with thoroughly elsewhere [21, 22, 23]. Experiences, both in research and in practice, show that optimized chemical treatment of raw water removes most particles resulting in a BOD$_7$ reduction in the range of $60-80\%$ and a reduction in SS in the range of $85-95\%$.

Norwegian Experiences with Chemical Treatment of Wastewater

Chemical treatment may be applied to raw wastewater as shown in Figure 4.

In Norway we have well over 100 such treatment plants. Earlier these plants were built as conventional secondary precipitation plants (see Figure 4), but since it became evident that almost equally good results were obtained with primary precipitation, the latter is now more often chosen because of lower cost.

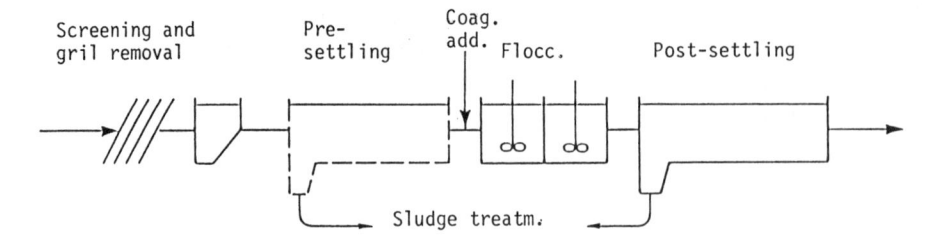

Figure 4. Schematic of non-biological, chemical treatments plants

In many cases, the treatment results obtained in these plants are acceptable for the receiving water in question. The treatment scheme shown in Figure 4 may thus be regarded as final treatment in some cases and as pretreatment before biological processes in others. Table 5 shows the 1985 yearly average treatment results obtained in 56 randomly selected chemical treatment plants in Norway [21].

Table 5. 1985 yearly average treatment results obtained in 56 chemical, non biological treatment plants in Norway, ranging in size from 2000 - 750,000 PE [21]

Process	Number of plants	BOD				COD			
		$n^{3)}$	In	Out	%	$n^{3)}$	In	Out	%
PP[1)]	23	210	216	42	80.6	303	313	92	70.6
SP[2)]	33	287	238	36	84.9	322	404	74	81.6
PP & SP	56	497	229	39	83.0	625	360	83	76.9

Process	Number of plants	SS				Tot. P			
		$n^{3)}$	In	Out	%	$n^{3)}$	In	Out	%
PP[1)]	23	294	172	27	84.3	338	5.5	0.54	90.2
SP[2)]	33	371	218	22	89.9	419	7.1	0.50	93.0
PP & SP	56	665	387	24	87.6	757	6.2	0.52	91,6

1) Primary precipitation   2) Secondary precipitation   3) Number of samples used in the average

Table 5 demonstrates that good particle removal and generally good treatment results (in terms of suspended solids, organics and phosphates) may be obtained by means of this simple treatment. The removal of organic matter in many of the plants is surprisingly high, and higher than what is normally experienced in the pre-precipitation step of continental wastewater treatment plants. There are probably several reasons for this. The sewerage system leading to many of the Norwegian plants is quite short and the wastewater temperature relatively low. A long detention time in the sewers and a high wastewater temperature may result in particle degradation caused by hydraulic shear forces, solubilization and enzymatic hydrolysis, and therefore to a smaller fraction separable by coagulation/flocculation.

Influents heavily influenced by low molecular weight organic substances originating from food-processing industries, discard water from septage treatment and so on, will negatively influence the treatment results. In pre-precipitation plants in other countries, where the coagulation step is followed by a biological treatment step, the chemicals are often added directly to the inlet of the presettling tank or to the grit chamber, and the overflow rate is high since the presettling tank normally is designed for mechanical treatment only.

The Norwegian plants mentioned previously are designed for optimum particle separation since chemical treatment is the only treatment. They are therefore normally equipped with a mechanical flocculation facility, normally with 3 or 4 chambers with a total residence time in the flocculator chamber of about 30 min and with a typical overflow rate in the settling unit of 0.8 – 1.0 m/h. These results demonstrate what may be achieved when optimization of particle separation is taken into account.

In a recent, more extensive investigation, more detailed information was obtained concerning treatment results and operational experiences from Norwegian wastewater treatment plants [24]. This investigation included all the chemical treatment plants in the country, out of which 43 plants used primary precipitation and 64 plants used secondary precipitation. Figure 5 shows how the mean effluent quality, based on samples taken throughout the year, with respect to total phosphorous, was distributed among the plants.

It was shown that 45% of all the primary precipitation plants and 60% of the secondary precipitation plants had a mean effluent concentration of equal to or less than 0.5 mg P/l. About 90% of the total number of plants met a value of 1.0 mg P/l.

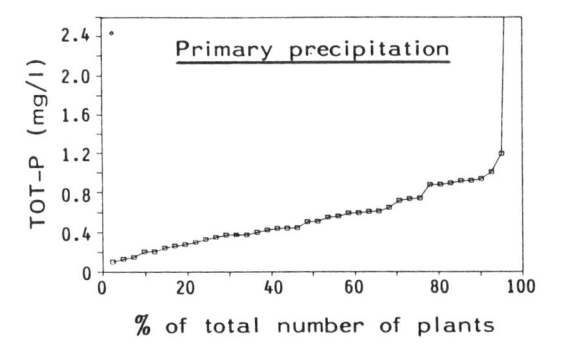

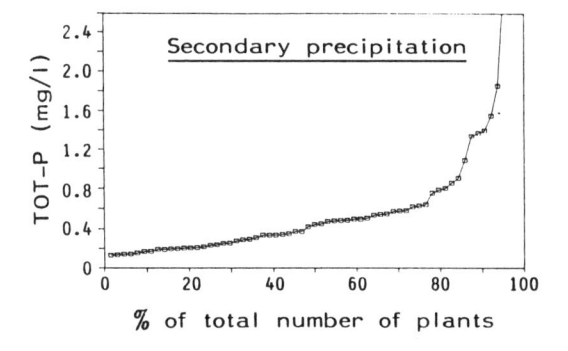

Figure 5. Mean concentration of total phosphorous in the effluent for 41 primary precipitation and 63 secondary precipitation plants in Norway, 1985 [24]

Both of these figures demonstrate that very good treatment results may be obtained with either process, and that the variation in stability of the process is small.

Several of the Norwegian plants use COD or TOC as a measure of organic matter instead of BOD_7, so the results shown in Figure 6 are for a fewer number of plants than those where total P was monitored [24].

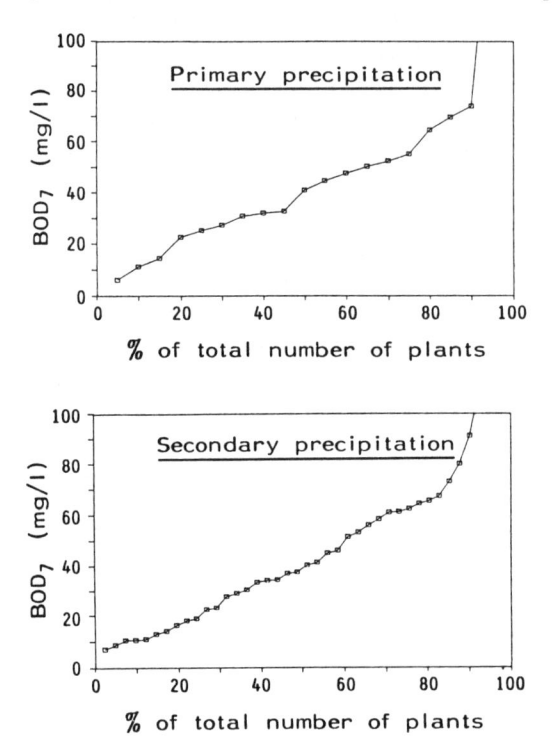

Figure 6. Mean effluent concentration of BOD_7 for 19 primary and 40 secondary precipitation plants in Norway, 1985 [24].

It is evident, however, that about 50% of both the primary and the secondary precipitation plants had a mean BOD_7 concentration in the effluent equal to or lower than 40 mg BOD_7/l. Almost 90% of the total number of plants met a value of 70 mg BOD_7/l.

The fact that the distribution curves for the effluent BOD_7 concentrations are steeper than those for total phosphorous demonstrates that there is variability among the plants with respect to the size of the fraction of the raw water BOD_7 that is particulate, and therefore able to coagulate.

The most frequent cause of process disturbance was found to be poor quality of the sewerage system resulting in hydraulic overloading and poor floc separation in snow melting periods.

Conclusions

The following conclusions may be drawn from this work:

1) Since the major part of the contaminants in raw wastewater is associated with particles, direct particle separation is an effective way of lowering raw water contaminant levels.
2) Particles have a negative effect on the biochemical oxidation rate both in activated sludge and in biofilm processes, and should therefore be removed prior to biological treatment in order to reduce the organic load but also to promote rapid biodegradation.
3) Chemical treatment (coagulation followed by floc separation) removes most of the particles in raw wastewater and is consequently probably the most cost-effective process that can be used as the first treatment step.
4) The results from Norwegian chemical treatment plants show that a very considerable reduction of contaminants and a very stable effluent quality is obtained by coagulation of raw wastewater.

References

[1] Imhoff, K. (1939): Taschenbuch der Stadtentwässerung. 8. Auflage, Oldenbourg, München
[2] Balmat, J.L. (1957): Biochemical Oxidation of Various Particulate Fractions of Sewage. Sew. and Ind. Wastes 29, 7, p. 757
[3] Heukelekian, H., Balmat, J.L. (1959): Chemical Composition of the Particulate Fraction of Domestic Sewage. Sew. and Ind. Wastes 31, 4, p. 413
[4] Richert, D.A., Hunter, J.V. (1971): General Nature of Soluble and Particulate Organics in Sewage and Secondary Effluent. Water Research 5, 7, p. 421
[5] Munch, R., Hwang, C.P., Lackie, T.H. (1980): Wastewater Fractions Add to Total Treatment Picture. Water and Sew. Works, Dec., p. 49
[6] Ødegaard, H. (1985): Treatment of Effluents from the Foodprocessing Industry. NTNF-Ind.Rens-project, Report No. 7 (in Norwegian)
[7] SWEP, Sewage Works Evalution Project 1985: Specifika föroreningar vid kommunal avloppsrening. Rapport SNU PM 1964, Naturvårdsverket, Stockholm (in Swedish)
[8] Lewis, G.O., Austin, F.J., Loubit, M.W., Sharples (1986): Enterovirus Removal from Sewage — The Effectiveness of Four Different Treatment Plants. Water Research 20, 10, p. 1291
[9] Wellings, F.M., Lewis, A.L., Moutain, C.W. (1976): Distribution of Solid Associated Viruses in Wastewater. Appl. Envir. Microbiol. 31, p. 354
[10] Vasl, R.J., Kott, R. (1982): Differential Adsorption Rate of Enteroviruses Onto Sewage Solids. Abstracts of the Annual Meeting of the American Society of Microbiology, p. 211
[11] Luin, A.B. van, Starkenburg, W. van (1984): Hazardous Substances in Wastewater. Wat. Sci. Tech. 17, p. 843
[12] Levine, A.D., Tchobanoglous, G., Asano, T. (1985): Characterization of the Size Distribution of Contaminants in Wastewater: Treatment and Reuse Implications. Journ. WFCF 57, 2, p. 805
[13] Gujer, W. (1980): The Effect of Particulate Organic Material on Activated Sludge Yield and Oxygen Requirements. Prog. Wat. Tech. 12, p. 79
[14] Särner, E. (1980): Plastic-Packed Trickling Filters. Ann Arbor Science Publishers, Ann Arbor, Mich.
[15] Särner, E. (1981): Removal of Dissolved and Particulate Organic Matter in High-Rate Trickling Filters. Water Res. 15, p. 671

[16] Särner, E., Marklund, S. (1985): Influence of Particulate Organics on the Removal of Dissolved Organics in Fixed-Film Biological Reactors. Wat. Sci. Tech. *17*, p. 15
[17] Rusten, B. (1984): Wastewater Treatment with Aerated Submerged, Biological Filters. Journ. WPCF *54*, p. 424
[18] Rusten, B. (1983): Purification of Domestic Wastewater in Aerobic Submerged, Biological Filters. SINTEF-report STF 21, A83076, Norwegian Institute of Technology, Trondheim (in Norwegian)
[19] Bruce, A.M. (1971): Some Factors Affecting the Efficiency of High-Rate Biological Filters. 5th Int. Wat. Poll. Res. Conf., Pergamon Press, Oxford
[20] Särner, E. (1986): Removal of Particulate and Dissolved Organics in Aerobic Fixed-Film Biological Processes. Journ. WPCF *58*, 2, p. 165
[21] Ødegaard, H. (1987): Particle Separation in Wastewater Treatment. Dokumentation, 7. EWPCA-Symposium, Munich 19. — 22. May
[22] Grohmann, A. Hahn, H.H., Klute, R. (eds.) (1985): Chemical Water and Wastewater Treatment. Fischer, Stuttgart/New York
[23] Hahn, H.H., Klute, R., Balmér, P. (1986): Recycling in Chemical Water and Wastewater Treatment. Schriftenreihe des ISWW Karlsruhe *50*
[24] Storhaug, R. (1988): Treatment Results at Norwegian Wastewater Treatment Plants in 1985. Statens Forurensningstilsyn (in Norwegian)

H. Ødegaard
The Norwegian Institute of Technology
7034 Trondheim-NTH
Norway

Pre-precipitation for Improvement of Nitrogen Removal in Biological Wastewater Treatment

I. Karlsson

Abstract

Chemically coagulated sewage water gives an effluent low in both suspended matter and organics. A fixed biosystem can be a natural consequence for treating this water. Adapted to an existing sewage plant this volume saving process affords possibilities for nitrogen removal without tank expansion. Pilot plant and full scale results are discussed. The precipitated sludge contains 75% of the organic matter and represents a valuable internal carbon source. Some economical aspects concerning conventional biological treatment and the pre- precipitation process are taken into consideration.

Introduction

In Sweden we have had considerable experience in the area of chemical precipitation as phosphorus reduction demands were introduced in the country some 15 – 20 years ago. Figure 1 shows treatment developments in Sweden up to the present day. For phosphorus reduction the mechanical and biological treatment processes have mostly been supplemented by the addition of a third stage – a chemical post-precipi-

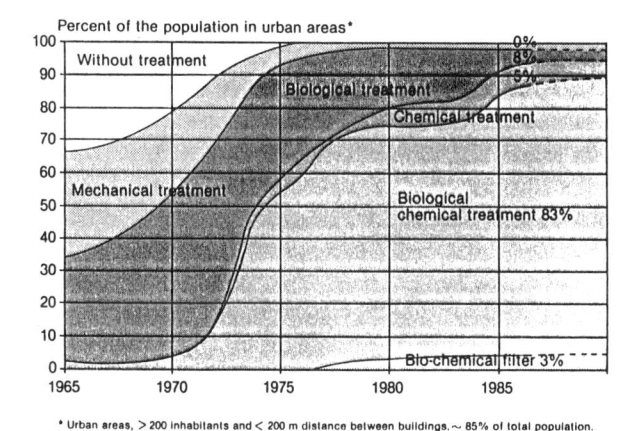

Figure 1. Sewage treatment in Sweden 1965 — 1986

tation stage – which has made it possible to achieve total phosphorus effluent values of 0.5 mg/l and below. Biological phosphorus removal has had no great success in Sweden because it is difficult to achieve total values below 2 mg of phosphorus per litre of water. On the other hand, chemical post-precipitation has not achieved great success outside the Scandinavian countries because of the high investment cost involved. Simultaneous precipitation processes are common when phosphorus requirements of 1 mg of total phosphorus per litre are required in the effluent water. In Sweden the five year old practice of dosing precipitation chemicals in the beginning of the treatment process, pre-precipitation, has achieved increasing importance. Initially pre-precipitation was used as an optimizing process, primarily for financial reasons, but operating advantages and better results were also obtained [1]. Today, the energy saving effect and a possibility for nitrification and denitrification has drawn attention to the process even outside Scandinavia.

Chemical Sewage Treatment

It is generally known that trivalent metal salts are good at removing suspended solids from a turbid water. In the process used to treat surface water, to produce potable water, aluminium salts are often used to destabilize and flocculate the pollutants. Sometimes, highly charged metal ion complexes such as polyaluminium sulphate or polyaluminium chloride are used because they are more effective in this process than a salt with a lower valency or charge. Also the combination of an inorganic salt and an organic compound can be and is used. In the dosing point mixing has to take place as quickly as possible and with a high energy input and be followed by a floc formation stage. The same types of chemicals are also used for sewage treatment, both for direct precipitation, pre-precipitation and also for post-precipitation.

The organic matter not removed by gravity in the pre- sedimentation tank is in stable suspension, consisting of particles smaller than 0.1 mm. This stable suspension may be destabilized by using the low molecular, highly charged metal salts mentioned earlier. These salts are also able to form fluffy hydroxide flocs that can arrest the destabilized particles through general sweep coagulation. Large aggregates capable of being sedimentated are formed. Dissolved compounds like phosphorus are precipitated at the same time. Figure 2 shows that with an effective precipitant and with good initial mixing and floc formation, particles down to 0.1 may be destabilized, coagulated and removed from the waste water [2]. Measured in terms of organic substance these correspond to a total reduction of some 75% or even more for normal municipal sewage. Phosphorus reduction then often exceeds 90%. A good example of a direct precipitation plant ist the large VEAS treatment plant outside Oslo, see Figure 3, which has been designed to handle about 400,000 m^3 a day. The retention time in this plant is only 3 hours. In Norway with direct precipitation more than 50 sewage plants show a reduction efficiency of more than 75% for BOD and more than 90% for total phosphorus [3], Figure 4.

COD:

```
         65%                    35%
   Biological treatment    Mechanical
                           treatment
```

0.0001µ 0.001µ 0.01µ 0.1µ 1µ 0.01 mm 0.1 mm 1 mm 10 mm

```
  Biological          Chemical treatment
  treatment
     25%                    75%
```
COD:

Figure 2

VEAS

Me^{3+}

$V = 3 \times Qh$

$Q = 300.000$ m³/d
Dosage:
430 m mol Me^{3+}/m^3

- 6 l/m³ 5% TS
+ 30g CaO/m³
570 g/m³ 37% TS

	Influent mg/l	Effluent mg/l
BOD	140	30
SS	110	10
P tot	3.5	0.2
N tot	15	11

Figure 3. Direct Precipitation

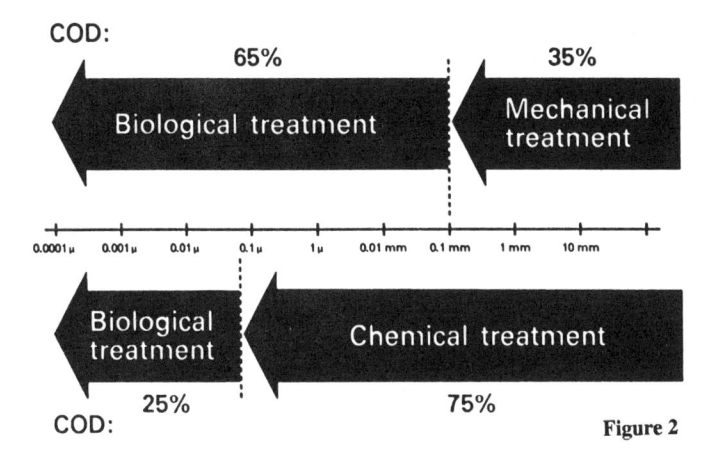

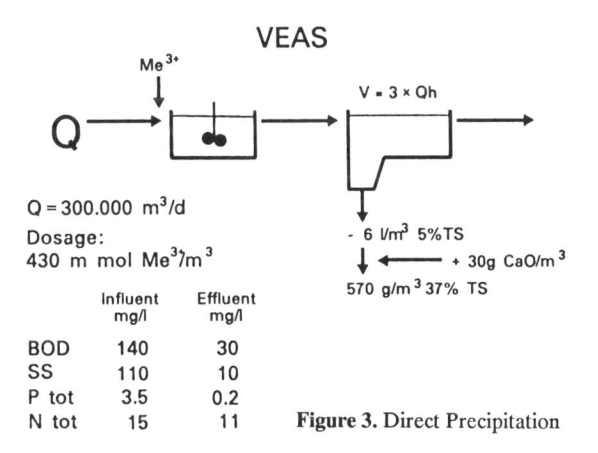

Process	Number of plants	SS			BOD			COD			TOT.P		
		in	out	%	in	out	%	in	out	%	in	out	%
PP	23	172	27	84.3	216	42	80.6	313	92	70.6	5.5	0.54	90.2
SP	33	218	22	89.9	238	36	84.9	404	74	81.6	7.1	0.50	93.0
PP & SP	56	387	24	87.6	229	39	83.0	360	83	76.9	6.2	0.52	91.6

Figure 4. 1985 yearly average treatment results obtained in 56 chemical, non-biological treatment plants in Norway, ranging in size from 2,000 — 750,000 PE

Nitrification

In Scandinavia we are now facing a new situation concerning nitrogen. Our coastal line waters in northern Europe show very serious problems with alga blooms and anaerobic bottom conditions. We must decrease the discharge of nitrogen to our seas. For our sewage plants this is a problem because most of them have insufficient biological volumes and retention times to achieve nitrification and denitrification in existing volumes. We have to increase the volumes according to today's technical knowledge. Here chemical pre-treatment and unloading with the pre-precipitation technique is very favourable. To build up a nitrifying system it is necessary to increase the sludge age enough so that the slow growing autotrophs are not washed out of the biological system. Especially during winter with our low water temperatures, often below 10°C, it is necessary to have a high sludge age. With a trickling filter the nitrifiers start to grow when most of the organic matter is consumed and the number of the fast growing heterotrophs start to decrease. Large volumes and surfaces are necessary if the BOD/N-ratio is high. The lower the BOD/N-ratio, the smaller the surface in the trickling filter required to achieve nitrification [4], Figure 5.

Number of nitrifying organisms

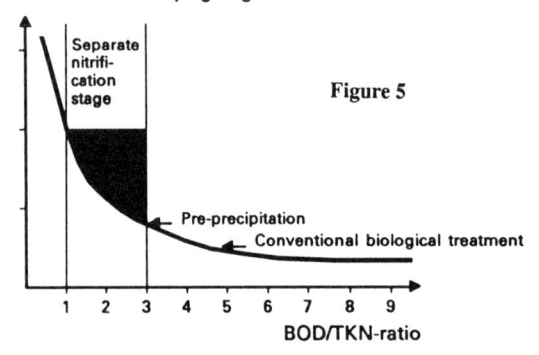

Figure 5

Separate nitrification stage

Pre-precipitation
Conventional biological treatment

BOD/TKN-ratio

For an activated sludge process the sludge age is one of the most important parameters for nitrification. For cool water a very high sludge age is necessary.

An increased removal of organic matter and suspended solids over the primary tank up to 100%, which can be expected when pre-precipitation is introduced, will also double the sludge age if the same tank volume and sludge concentration is used. Of course, simultaneous precipitation decreases the sludge age because of the sludge production of the chemical in the biological stage.

Sewage water from chemical treatment normally contains just about 25% of the original organic content. The organic matter is mainly in a dissolved readily degradable form. Figure 6 shows the chemical oxidation rate of different fractions of the sewage water [5].

The fraction remaining after chemical treatment shows the highest oxidation rate, about four times higher compared to some portions of the effluent from primary treatment.

To use the chemically-treated water in a compact biological process with a biological fixed film system can be of great interest, as this combination can be volume saving, Figure 7.

When sewage water passes the fixed biological film soluble organics and nutrients diffuse into the biofilm together with oxygen. In this oxidation process it is

Figure 6

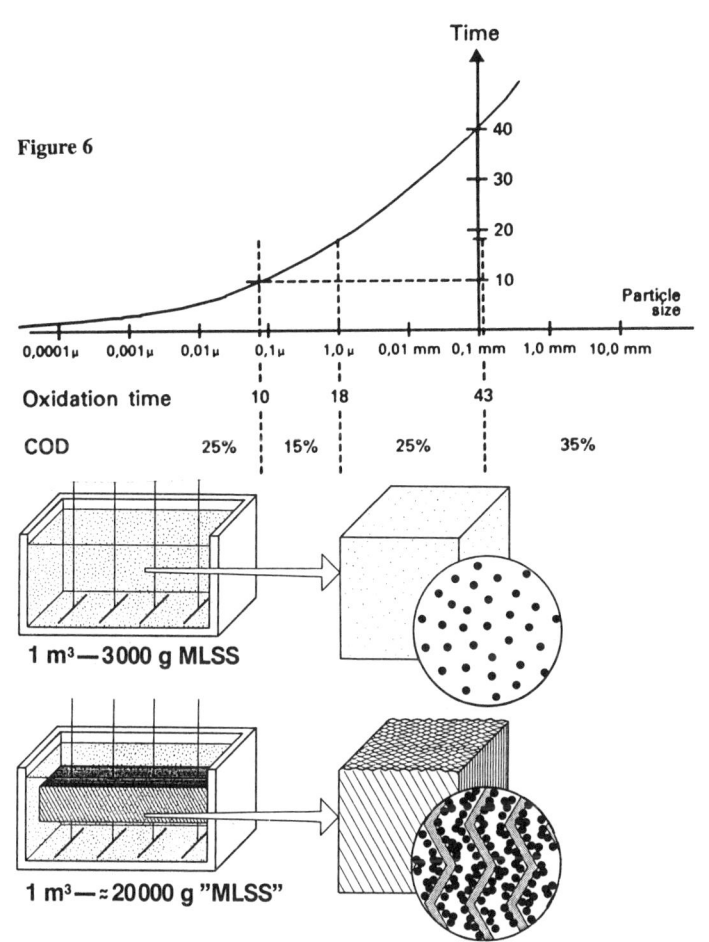

Figure 7

an advantage if the organic matter is in a soluble form, because it can penetrate the bacterial membrane directly.

Organic matter in the form of colloids and larger particulates is adsorbed on the biofilm's surface and solubilized by enzymatic activity before it can penetrate the bacterial membrane and serve as an electron donor.

Pre-precipitation produces water with a very low content of particles which should be an advantage in combination with a fixed film biosystem. The possibility of increasing the sludge content together with a high oxidation rate of the organic matter is volume saving and increases the sludge age compared to conventional biological treatment.

Results

The pilot plant test to study the fixed film system was done at Helsingborg's wastewater plant. Two separate lines were pursued: The first without chemical pretreatment, just primary settling tanks, and the second with chemical treatment in the primary settling tanks. Two identical biological stages were followed by the pre-treatment stage. Each stage has three aerated tanks with submerged biological blocks followed by a sedimentation tank. Each tank was filled to 65% with filter medium, Plasdek, with a specific surface area of 100 m^2/m^3. The hydraulic load was the same in both approaches. After more than a month the chemical treatment oxidized all ammonia. A good degree of nitrification was shown already after the second tank. After another month nitrification also began in the other line of treatment. The retention time in each tank was 4 hours. The second primary settling tank in this line of treatment did not show any nitrification until the BOD load was halved by decreasing the flow, so that now each tank in this line of treatment has 8 hours' retention time.

The ammonia removal rates for the chemical treatment was a little higher than 1 g of ammonia nitrogen per m^2 and day and the settled water showed about half that rate. The BOD load was about 2 g/m2·d.

Influent BOD   200 g/m^3
Primary effluent BOD   100 g/m^3
Pre-precipitation BOD   40 g/m^3

When return sludge pumping, 100% of the influent flow, was started, almost all the ammonia was oxidized already after the first aeration tank. The flow was increased and the retention time decreased to 2.5 hours for the pre-precipitation line of treatment, but for the conventional treatment 5 hours. In the third tank very little sludge was produced on the submerged material so that it was shut off and taken out of order. The following result was found. For two tanks when the BOD load was 2.25 $g/m^3 \cdot d$, 1.75 g H_4N-N was oxidized per $m^2 \cdot d$ in the chemical treatment. The oxidation rates are probably much higher, maybe double, since almost all nitrogen was in the form of nitrate already in the first tank's effluent. With pre-precipitation it is possible to operate the plant with 100% nitrification at twice the hydraulic load compared to the unprecipitated line of treatment. The sludge showed very good characteristics with good settleability, with the sludge volume index below 100 in both lines of treatment, Figures 8, 9 and 10.

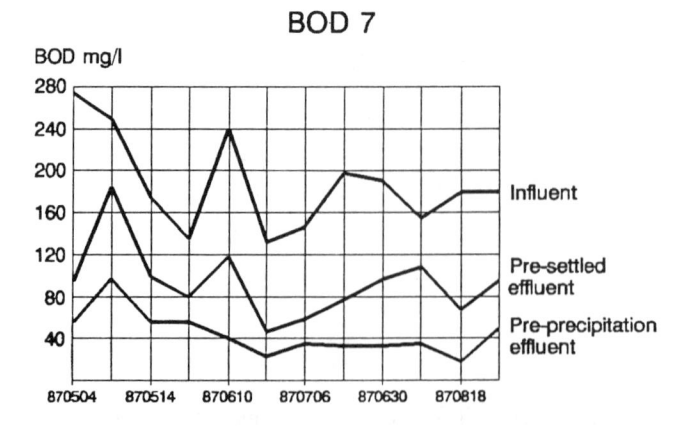

Figure 8. Helsingborg Pilot

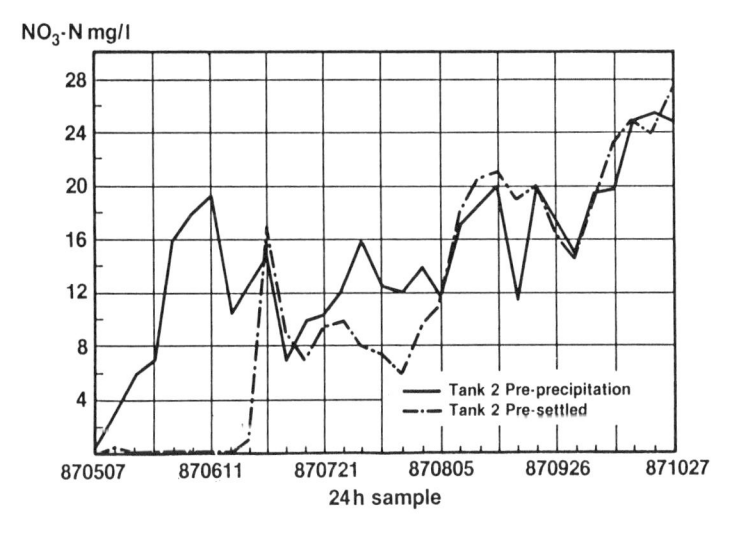

Figure 9

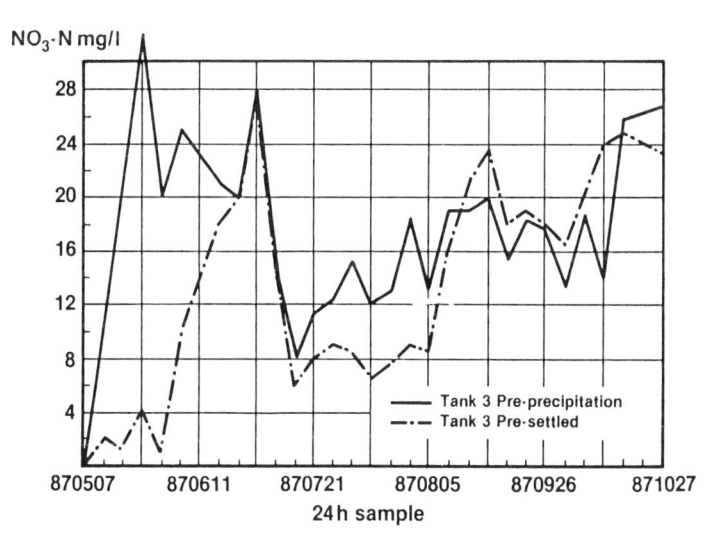

Figure 10

Another pilot test was conducted at the Örebro wastewater plant where the effluent from the plant was treated in a fixed film system with a specific surface area of 250 m² per m³. The water was very low in organic matter and suspended solids. Average values were below 10 mg/l for both parameters. The NO_3-N production rate over a period from September 1985 to February 1987 showed about 1.5 g NO_3-N/m²·d. There was no return sludge pumping. This high nitrification rate occurred even at low water temperatures. A remarkable fact was that no pH drop was shown over the biofilter, even when all ammonia nitrogen was oxidized. The activated sludge is alkaline in summer, because of nitrification, with pH values below 6 as a result.

In January 1987 the efficiency of ammonia nitrogen oxidation decreased drastically. The reason was probably a snail invasion on the biofilter. The small snails

At the same sewage plant tests on pre-precipitated water and fixed film were also done. In an aeration tank, Inka system, 25% of the volume was filled with Munter's Plasdek, $1 \text{ m}^3 = 100 \text{ m}^2$. There was no return sludge pumping. Figure 11 shows removal rates at different loadings. The highest organic load, 84 g $COD/m^2 \cdot d$, showed effluent values of 89 g COD/m^2 (28 g BOD/m^3). Figure 12 shows the effluent values at different organic loads. To achieve effluent values below COD 55, (15 g BOD), the COD load should not exceed 35 g $COD/m^2 \cdot d$. This corresponds to 3.5 kg COD/m^3 of tank volume if the entire aeration volume is filled with the blocks. About the same can be expected from an activated sludge system, but the fixed film system is without return sludge pumping. A low nitrification rate was found when the COD load was about 25 g $COD/m^2 \cdot d$. The production of $N-NO_3$ increased to about 0.6 g $N-NO_3$ at both 18 g $COD/m^2 \cdot d$ and 11 g $COD/m^2 \cdot d$.

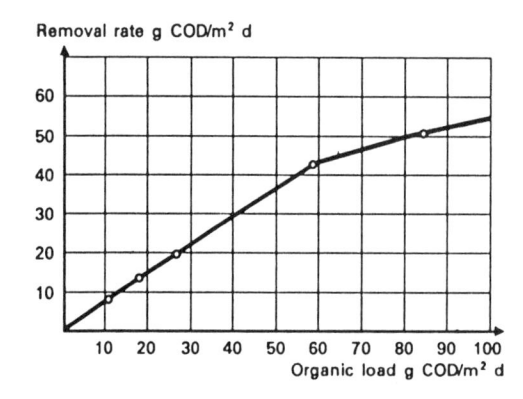

Figure 11

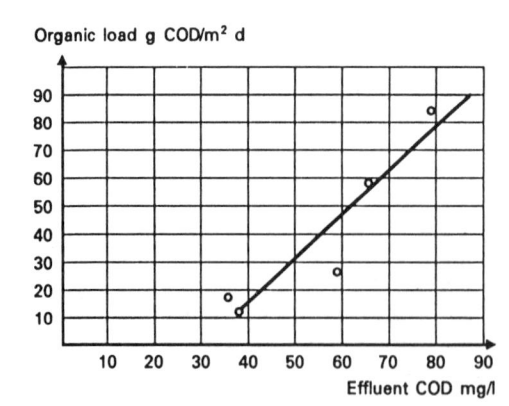

Figure 12

At the Norrköping wastewater plant pre-precipitation reduces the BOD at the activated sludge stage to 80%. This provides a system with more than 50% nitrogen removal even during winter. A pre-denitrification zone, retention time 1.5 hours, is followed by an aerated tank with 5 hours' retention time and about 10 days' sludge age. Because of pre-precipitation, the post-precipitation was shut off but is now used for denitrification, with a separate sludge system. Industrial waste products were used as a carbon source, Figure 13.

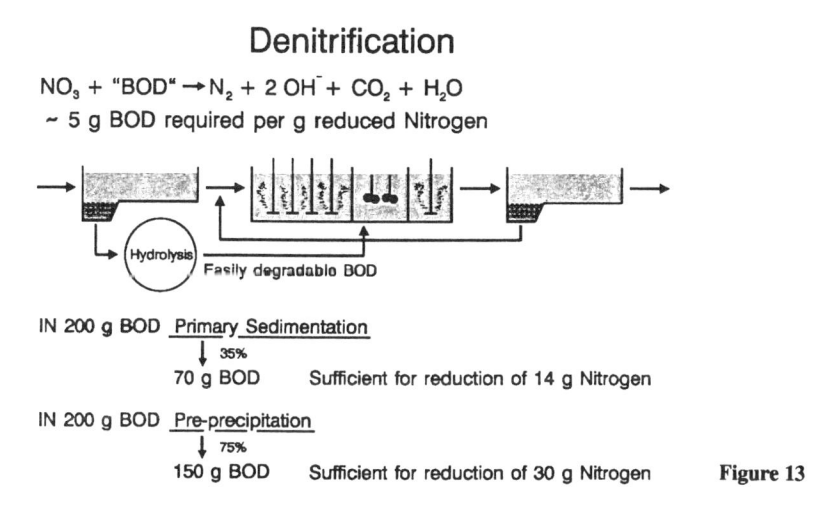

Denitrification

$NO_3 + \text{"BOD"} \rightarrow N_2 + 2 OH^- + CO_2 + H_2O$

~ 5 g BOD required per g reduced Nitrogen

IN 200 g BOD Primary Sedimentation
 ↓ 35%
 70 g BOD   Sufficient for reduction of 14 g Nitrogen

IN 200 g BOD Pre-precipitation
 ↓ 75%
 150 g BOD   Sufficient for reduction of 30 g Nitrogen

Figure 13

Denitrification

Denitrification may be carried out using either one or two main principles: pre-denitrification or post-denitrification. The latter process takes place downstream of the nitrification stage, the former upstream of it. There are microorganisms that can use both dissolved oxygen and nitrate as an electron acceptor in their oxidation process, but they prefer oxygen since that process has a higher energy yield. To utilize nitrate the environment must be anoxic, i.e. without dissolved oxygen. A degradable carbon source is also required, unless endogenous respiration is used, which requires very long retention times. Methanol is a conceivable source of carbon that provides a high rate of oxidation but it costs about 0.0017 USD per gram reduced nitrogen. Thus, the methanol cost of total nitrogen reduction in a denitrification process is about 6 USD per person per year. This high expense makes it important to use cheap carbon sources, e.g. industrial waste products or internal carbon sources. The use of primary sludge is possible, but the oxidation rate is low for unprocessed sludge and, furthermore, the quantity produced in conventional treatment is barely enough. To remove BOD upstream of a nitrification stage and then return the BOD to the denitrification stage as a source of carbon is an interesting process alternative. The quantity of organic substance precipitated consists of particles down to about 0.1 μ. The biological oxidation rate of the fraction added through pre- precipitation is higher than that of normal primary sludge, and it is also possible to treat the sludge by acid fermentation to increase the availability further. The higher the oxidation rate shown by the carbon source, the smaller the required denitrification volume. This is very important for cold waters.

If pre-precipitation or conventional biological treatment is used, approximately equal quantities of nitrogen are removed from the process with the sludge in pre-sedimentation and the biological sludge. Less is incorporated into the biological sludge for pre-precipitation since production is less, but the production of primary sludge has increased, which increases the amount of nitrogen in this sludge. Figure 14 gives an example. BOD influent = 200 g BOD, primary effluent is 130 measured as BOD and 70 g BOD per m³ primary sludge is produced. If 5 g BOD is required per gram reduced nitrogen and all primary sludge can be used, it is sufficient to reduce 14 g of nitrogen. It is very doubtful whether this will work, since a large part of this BOD is also difficult to degrade. With normal pre-precipitation 75% of the influent BOD is removed over the primary tanks, which involves a fraction of BOD that is readily degradable compared to the normal primary sludge. 150 g BOD should be enough for a reduction of 30 g of nitrogen.

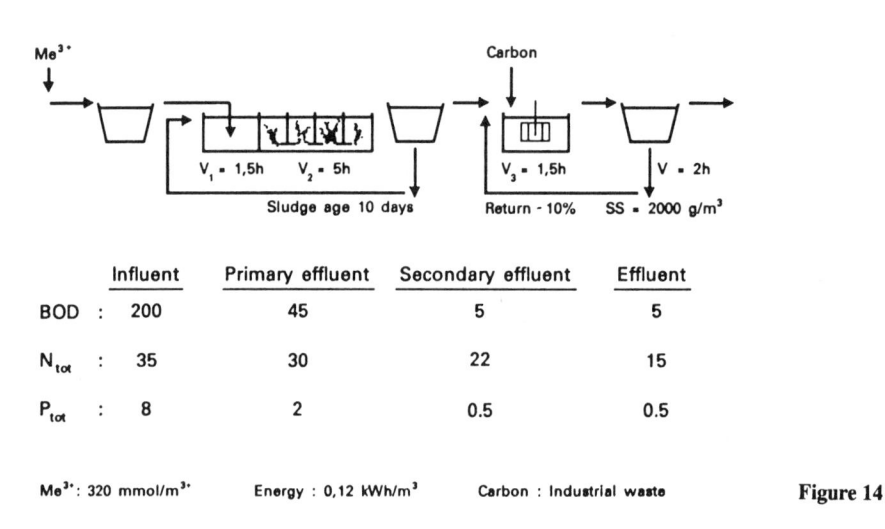

	Influent	Primary effluent	Secondary effluent	Effluent
BOD :	200	45	5	5
N_{tot} :	35	30	22	15
P_{tot} :	8	2	0.5	0.5

Me^{3+}: 320 mmol/m³   Energy : 0,12 kWh/m³   Carbon : Industrial waste   **Figure 14**

If the anoxic zone upstream of the nitrification stage is constructed, the influent wastewater may serve as a source of carbon. Nitrate is returned by the return sludge and possibly by a separate return operation. Pre-precipitation has the advantage that only the most readily degradable part of the BOD is present. The anoxic zone can be utilized in terms of volume and retention time. For the highest efficiency, utilization of the wastewater's internal carbon sources, pre- and post-denitrification should be combined.

Economic Aspects

The Swedish Water Work's Association (VAV) has made an economic study on what would be required for Swedish sewage plants to reach 75% nitrogen removal [6]. The size of the given plant is 100,000 pe and the flow 40,000 m³/day. The retention time in the activated stage has to be increased, both in the aeration and anoxic zones, to 12 hours. The total cost of building 1 m³ of aeration tank volume is estimated at 500 USD. The capital cost, 12% rate of interest, depreciation of 20 years for buildings and 10 years for machinery, gives 0.08 USD per m³ of treated water. With pre-precipitation it is possible to double the sludge age without tank expansion. With a submerged biological filter and pre-precipitation, 2 g of ammonia nitrogen can be oxidized per m²/day at a BOD load of 4 g/(m²·day) according to the results from Helsingborg's pilot plant. The retention time for the oxidation of both ammonia and nitrogen should then be about 3 hours if 60% of the volume is filled with submerged biological blocks with a specific surface area of 1 m³ = 150 m². The cost of this material is about 115 USD per m³. Three years depreciation, 12% rate of interest give 0.005 USD/m³ of treated water. For nitrification a volume saving of 3 – 4 hours' retention time or 0.025 – 0.035 USD per m³ of sewage water is possible. That is 30% of the cost of the aeration tank. For denitrification the volume saving is more difficult to predict when both pre- and post-denitrification is used. The difference in the systems is the distribution of particles between 100 μ and 1 μ which have a low oxidation rate. See Figure 6. In the pre-precipitation processs this organic matter is extracted over the primary tanks, hydrolyzed and used in the post-denitrification process. In the conventional system this organic portion is not used in the pre-denitrification stage because the retention time is too short. It is oxidized in the aerobic stage. If methanol is substituted for this organic matter, which is about 25% of the influent, then the cost of treating municipal sewage water comes to 0.017 USD. This will pay for the precipitant in pre-precipitation, which normally costs about the same.

References

[1] Karlsson, I. (1985): Chemical Phosphorus Removal in Combination with Biological Treatment. Management strategies for phosphorus in the environment. International Conference Lisbon. ISBN 0-948411-00-7. Edited by Lester, J.N. and Kirk, P.W.W.
[2] Levine, A.D., Tchobanoglous, G., Asano, T. (1985): Benefits of Particle Size Management for Biological Wastewater Treatment. University of California, Davis. Prepared for presentation at the 1985 ASCE National Conference of Environmental Engineering, Boston, Mass.
[3] Ødegaard, H. (1987): Particle Separation in Wastewater Treatment. Presented at the 7th European Sewage and Refuse Symposium, Munich, The Norwegian Institute of Technology, Norway
[4] U.S. Environmental Protection Agency (1975): Process Design Manual for Nitrogen Control
[5] Balmat, J.L. (1957): Biochemical Oxidation of Various Particulate Fractions of Sewage. Sewage and Industrial Wastes 7, Vol. 29, p. 757
[6] Holmström, H. (1987): Beräkning av kvävereduktionskostnader för kustnära avloppsreningsverk (in Swedish). VAV

I. Karlsson
Boliden Kemi AB
Water Treatment
Kungstensgatan 38
113 59 Stockholm
Sweden

Chemically Supported Oil and Grease Removal in Municipal Wastewater Treatment Plants

H. Roggatz and R. Klute

1 Introduction

The wastewaters from the metal-processing industry often contain high concentrations of grease and oil of mineral origin. Some of the grease and oil is present in the form of droplets with a particle diameter of more than 20 µm, and can be separated in the grease traps. Most of the grease and oil, however, is finely dispersed in an emulsion and cannot be separated in the grease traps without first breaking up the emulsion.

The grease and oil present in the emulsions are used as cooling agents and lubricants in metal processing. In addition, emulsions are produced when metallic surfaces are degreased using aqueous degreasing and cleaning solutions. The emulsions are oil/water emulsions which are produced from mineral oil through the addition of emulsifiers, stabilizers, preservatives, etc.

In all cases, the wastewaters containing grease and oil must be purified by pretreatment prior to introduction into the municipal sewer system to the degree that they do not exceed the limits set in the authorization regulations. According to the 40th Wastewater Regulation of 5 Sept. 1984 concerning the minimum requirements for introducing wastewater into lakes and rivers (direct pollutor) in the FRG, the limit for hydrocarbons was fixed at 10 and 5 mg/l for the metal processing industry and the vehicle and machine manufacturing industries, respectively [1]. This limit initially applied only to the direct introduction of wastewater into lakes and rivers, but due to the amendment to the Water Resources Policy Act of 23 Sept. 1986, which promotes consistency between regulations for direct and indirect introduction, this limit will also apply to the introduction of wastewater into public wastewater treatment plants in the future (indirect pollutor) [2].

As a consequence of unauthorized introduction of wastewater, interruptions in operation during the pretreatment of wastewaters in the plants, as well as a large number of small factories that, up to now, have released wastewater in an uncontrolled and polluted form into the sewer system, wastewater at municipal treatment plants contains often high concentrations of grease and oil. For example, investigations at the Schweinfurt treatment plant showed concentrations of grease and oil in the raw wastewater which varied between 27 and 2047 mg/l.

Since these substances, as mentioned previously, are present primarily as emulsions, for the most part they pass the grease trap and are then partially removed in the primary sedimentation together with the primary sludge; some of these substances enter the biological treatment area of the treatment plant. On the one hand, this leads to a significant increase of grease and oil in the primary sludge, which makes the latter unsuitable for agricultural applications. On the other, this can disrupt biological decomposition processes in activated sludge tanks or in the trickling filters.

At the Schweinfurt treatment plant, chemical precipitation/flocculation with AVR is used so that increased phosphorous elimination, as well as improved removal of COD and BOD_5 in times of high pollutant loads, can be achieved. Thus, the question was addressed, whether addition of the precipitation/flocculation chemical before the grease trap could break up the emulsified grease and oil, and thus bring about the removal of most of these substances in the grease trap of the treatment plant.

For this investigation, jar tests were used initially to study the disruption of emulsions which are typical for the metal-processing plants in the Schweinfurt area. These tests were carried out with concentrated emulsions in order to demonstrate the possibilities for pretreatment in the Schweinfurt plants themselves, as well as with samples diluted with municipal wastewater. In addition, laboratory tests were carried out to investigate the influence of different types of phase separation on the removal efficiency. In the third part of the investigation, experiments were attempted in which AVR was dosed prior to the grease trap in the Schweinfurt treatment plant.

2 Materials and Methods

The non-volatile lipophilic substances were used as an assay parameter in the investigations of emulsion disruption and for the elimination of grease and oil in the wastewater. Detection of lipophilic substances was carried out in accordance with the German Standard Methods for the Examination of Water, Wastewater and Sludge [3]. The detection method is based on the extraction of the water sample with 1,1,2-trichlorotrifluoroethane ($C_2Cl_3F_2$). In this method, the substances extracted with this solvent are separated and determined gravimetrically following evaporation of the solvent. Other organic compounds, such as waxes or certain emulsifiers, can also be determined in addition to the grease and oil.

First, jar tests were used to determine the effect of the chemical dose on the disruption of grease and oil emulsions. Mixing of the chemicals took place at a stirring speed of 200 rpm. This was followed by a slow stirring phase of 30 min. at 20 rpm. Following a standing time of 60 min. to allow for phase separation, samples were taken for determining the lipophilic substances in the supernatant.

Second, experiments were carried out which examined the separation of the demulsified grease and oil using dissolved air flotation and pressure flotation. A laboratory dissolved air flotation device (see Figure 1) was used, whereby the chemicals were mixed with the wastewater for a period of 30 sec. at a stirring speed of 200 rpm. After a slow stirring phase of 5 min. at 20 rpm, the stirring device was removed and water under pressure, which was saturated with air for 15 min. at 5 bar, was added at a speed of 1000 cm^3/min. Following the 5 min. flotation time,

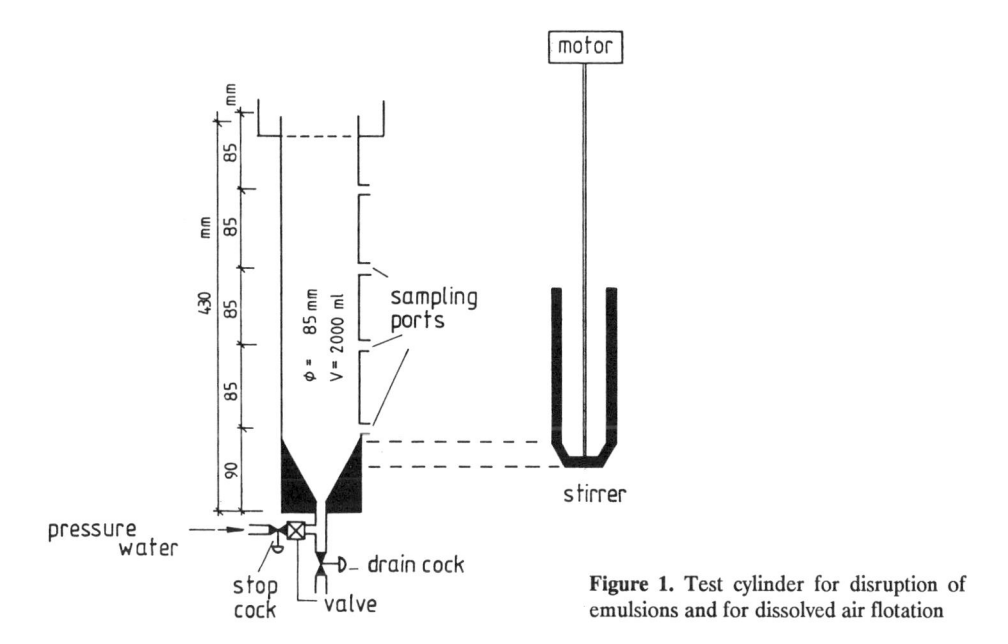

Figure 1. Test cylinder for disruption of emulsions and for dissolved air flotation

samples were taken from the middle of the supernatant. Phase separation by pressure flotation was examined using an aerated bench scale reactor (see Figure 2). This reactor was equipped with a stirrer for mixing the chemicals with the wastewater and a peripheral skirt to collect the flotated grease and oil.

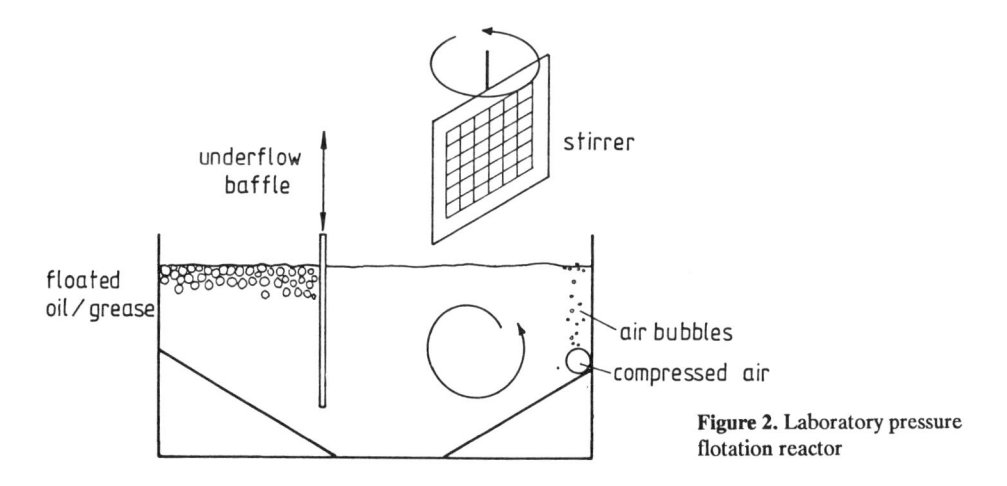

Figure 2. Laboratory pressure flotation reactor

The Schweinfurt sewage treatment plant, in which the large-scale experiments for improving the removal of grease and oil were carried out, is presently composed of a mechanical and a high-load biological step in the form of four trickling filters, which are operated in two steps. In order to improve the quality of the discharge from the treatment plant, chemical precipitation/flocculation with AVR is carried out. The chemicals are dosed in proportion to the quantity of water at the inlet to the primary settling tank, whereby the dosage was set to around 100 to 170 g/m³, depending on the pollutant load in the treatment plant. By using pre-precipitation and pre-flocculation, it is possible to stay within the limits of the legal value for treatment plant discharges of 100 mg COD/l in the 2 hr. mixed sample [4]. The treatment plant was designed to handle an additional 170,000 to 200,000 population equivalents in the future. 60% of the wastewater comes from industrial sources, mainly from plants in the metal processing industry and in the fruit and vegetable processing areas. Figure 3 shows a schematic diagram of the treatment plant.

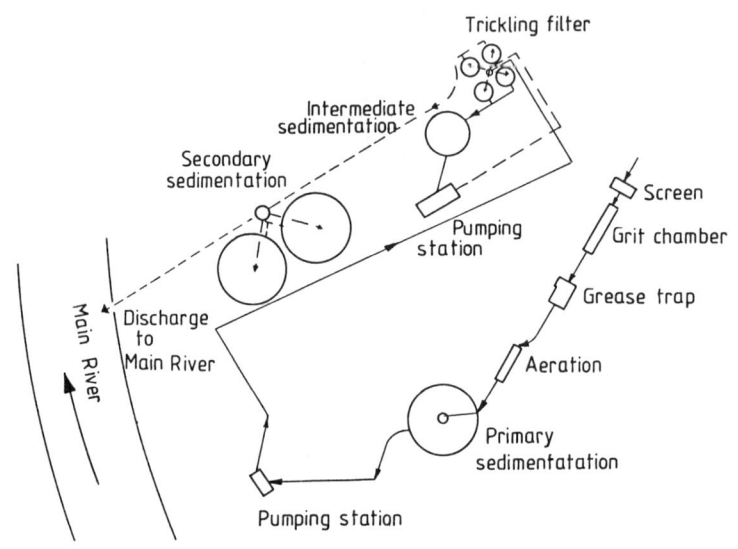

Figure 3. Schematic diagram of the Schweinfurt Municipal Sewage Treatment Works

3 Results and Discussion

Figure 4 shows the results from jar tests, in which a concentrated emulsion was disrupted with iron(III) salt. It can be seen that increasing doses of iron are effective in breaking up the emulsified grease and oil, thus making it possible to remove

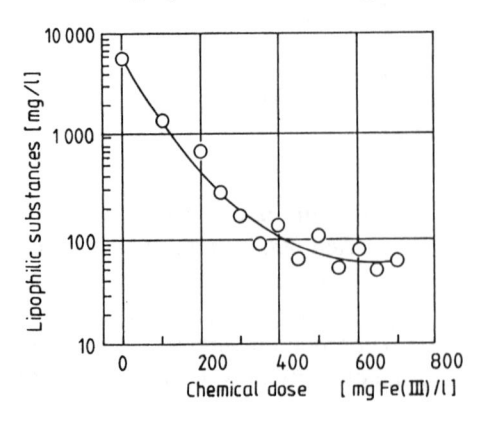

Figure 4. Breaking up a concentrated emulsion using Fe(III) salt

them. In the emulsion shown here, the lipophilic substances in the supernatant could be reduced by more than 95% with an iron dosage of about 400 mg/l. Higher dosages do not disrupt the emulsions further or improve elimination. The results show that breaking up emulsified grease and oil, and the removal of these substances from the concentrated emulsion, which usually takes place in the plant, represents an effective measure for purifying the wastewater at the source.

Systematic investigations of the lipophilic substances in Schweinfurt wastewater showed that the values of the random samples taken prior to the grease trap were between 27 and 2057 mg/l, and thus show a significant variation, as mentioned previously. In order to conduct tests for disrupting emulsions under reproducible experimental conditions, a standard wastewater was used. This standard wastewater is raw wastewater from a communal treatment plant which handles only domestic wastewater, to which used oil was added.

Figures 5 and 6 show elimination as a function of chemical dose for two different concentrations of lipophilic substances in the standard wastewater. Two different used oil emulsions were used for the experimental series, which explains the gradual change in the level of elimination. The two series of experiments show the same trend, however, namely, an improved elimination of lipophilic substances with

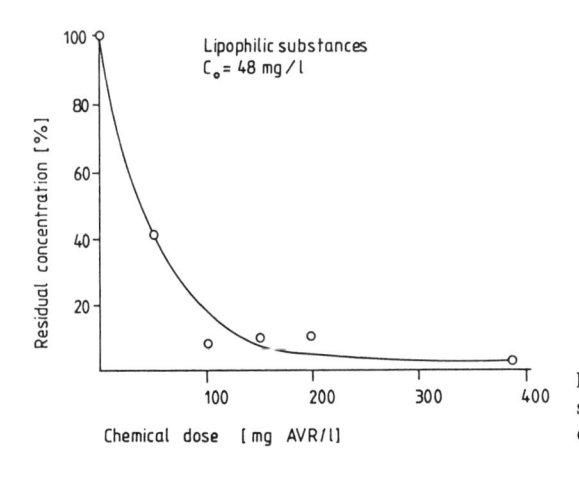

Figure 5. Elimination of lipophilic substances from the wastewater as a function of chemical dose

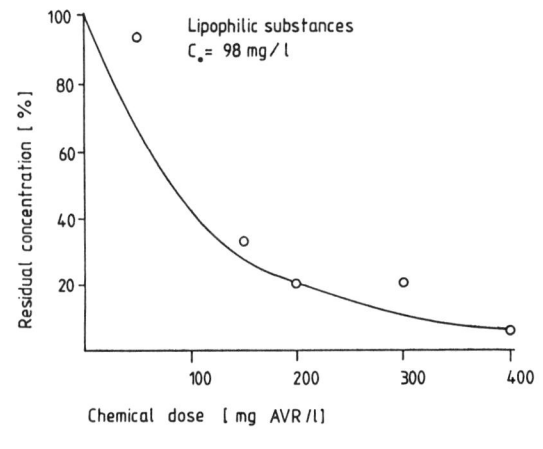

Figure 6. Elimination of lipophilic substances from the wastewater as a function of chemical dose

increasing AVR dosage. This is particularly pronounced for low AVR concentrations, for which the residual concentration of lipophilic substances lies on the order of 10 to 20% with a chemical dosage of 200 mg/l. By increasing the dose of chemicals to 400 mg/l, a further reduction of grease and oil to 95% is possible.

The results of these experiments show that the addition of chemicals leads to a disruption of grease and oil emulsions, and thus makes the removal of these substances from the wastewater possible. The dose of chemicals required for breaking up the emulsion and removing most of the grease and oil depends mostly on the type of emulsion. The emulsifiers used, as well as other chemical additives, the age of the emulsion, the stability of the emulsion, etc. play an important role in this case. These parameters were not examined in the scope of this investigation, since wastewater contains a mixture of many different emulsions. In any case, an AVR dose of about 100 to 200 mg/l can be said to disrupt emulsions to a significant degree.

In the jar tests for emulsion disruption and separation of grease and oil, the separation of solid/liquid phases was carried out over a 30 minute sedimentation period. The application of dissolved air flotation or pressure flotation can be used as an alternative method for separating the phases. Figure 7 shows the residual concentrations of lipophilic substances for different doses of chemicals. A comparison of the data shows that the removal again is a function of chemical dose and is dependant on the specific method of flotation and flotation conditions. Pressure flotation appears to result in lower residual grease and oil concentration of the treated wastewater compared with dissolved air flotation. This result cannot be generalized, however, since no optimization of the process parameters was carried out for flotation.

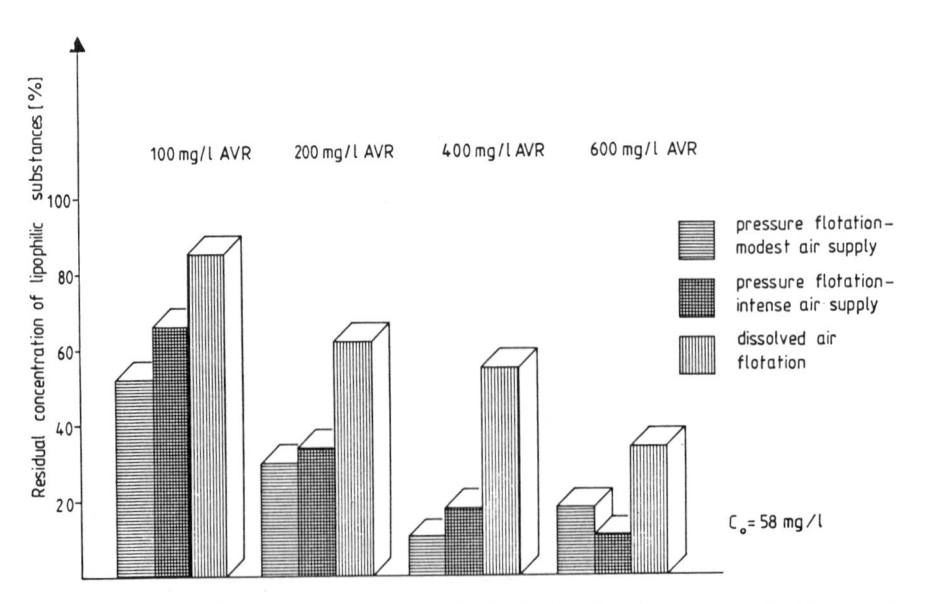

Figure 7. Comparision of pressure and dissolved air flotation for grease and oil removal from municipal wastewater

A chemical dose of 50 g AVR/m³ wastewater was chosen for the investigations for improving the removal of grease and oil in the grease trap of the Schweinfurt treatment plant. The chemicals were added just before the grease trap. Elimination could be improved from a mean value of 28% to a mean value of 74% through the use of chemicals, whereby the variation in elimination due to the significant variation of the inflow concentrations ranged from 42% to 88% (see Figure 8).

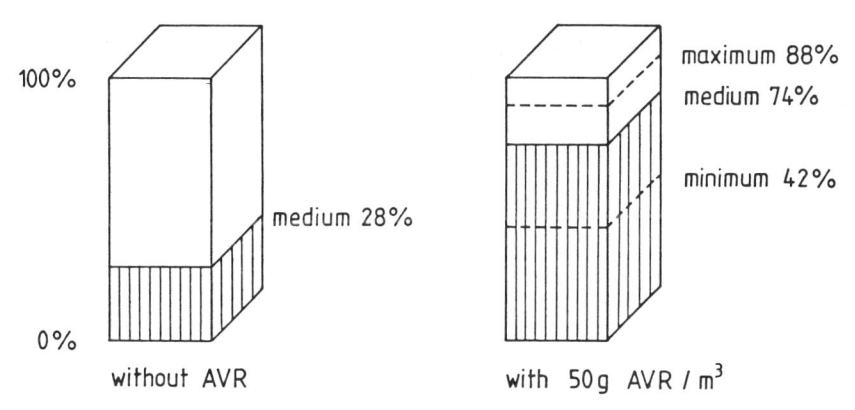

Figure 8. Improved removal of grease and oil at the Schweinfurt Treatment Plant by disrupting emulsions with AVR

The results of these investigations show that a significant part of the grease and oil in the Schweinfurt wastewater is present in the form of emulsions, and thus cannot be removed by the grease trap in the treatment plant. Since these emulsified substances have solely industrial origins, as shown by the measured peak loads, and since they are often present in the form of concentrated emulsions, pretreatment at the source is without a doubt the most appropriate measure for decreasing the pollution of the municipal wastewater and sludge. As long as such pretreatment measures are not taken, pollution can be decreased at the treatment plant by dosing 50 g AVR/m³ just before the grease trap. The resulting disruption of the emulsions allows a significantly greater portion of grease and oil to be removed in the grease trap.

4 Practical Experiences

As a result of these investigations, AVR is permanently dosed in the inlet to the grease trap at the Schweinfurt treatment plant, whereby the dose was fixed at 50 g AVR/m³ wastewater. The resulting disruption of the emulsified grease and oil led to an improved elimination of lipophilic substances in the grease trap. A consequence of this improved elimination is that the amount of lipophilic substances present in the sludge has decreased significantly. Prior to the implementation of this measure, values of 1000 to 1500 mg lipophilic substances/l wet sludge were measured; now this value is on the order of 500 mg/l.

References

[1] Vierzigste Allgemeine Verwaltungsvorschrift über Mindestanforderungen an das Einleiten von Abwasser in Gewässer (Metallbearbeitung Metallverarbeitung) - 40. AbwasserVwV - vom 5. September 1984

[2] Gesetz zur Ordnung des Wasserhaushalts (Wasserhaushaltsgesetz - WHG) vom 23. September 1986

[3] Deutsche Einheitsverfahren zur Wasser-, Abwasser- und Schlammuntersuchung: H17 Bestimmung von schwerflüchtigen, lipophilen Stoffen (Siedepunkte > 250°C)

[4] Roggatz, H. (1987): Betriebliche Erfahrungen mit der Anwendung der chemischen Fällung. In: Berichte aus Wassergütewirtschaft und Gesundheitsingenieurwesen der TU München, Bd. 76

H. Roggatz
Städtisches Baureferat
Gustav-Adolf-Straße 6
8720 Schweinfurt

R. Klute
Institut für Siedlungs-
wasserwirtschaft
Universität Fridericiana
7500 Karlsruhe

Chemical-biological Treatment
Versus Chemical Treatment - A Case Study

A. Ilmavirta

Introduction

In Finland, there are 578 municipal wastewater treatment plants (1987), out of which 61 use chemical treatment (wastewater flow 13%), Figure 1. Most of the chemical treatment plants were constructed in the 1970's.

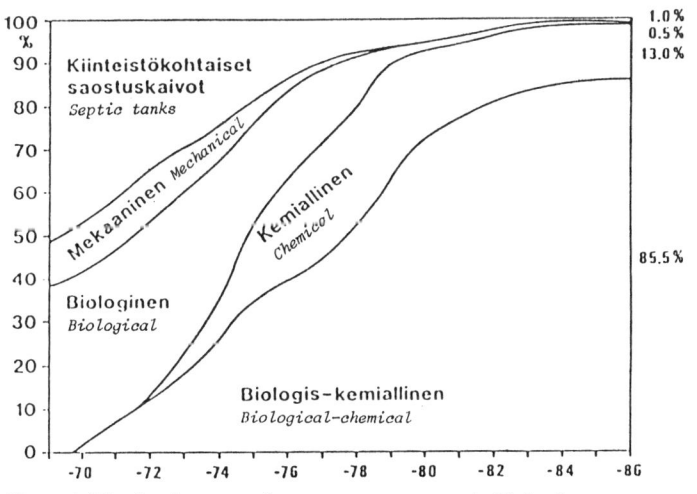

Figure 1. The development of wastewater treatment in Finland

In the 1980's the water authorities decided to change most of the chemical wastewater treatment plants to chemical-biological plants. The main reason is the BOD loading of wastewater effluent. Unfortunately the plans and constructions have not fully utilized the existing constructions.

The cost-benefit relationship has been poor in many cases. The over-dimensioned treatment plants have led to very long detention times. The phosphorus loadings have increased as a result of the re-dissolution of phosphorus.

One of the purposes of the Valkeakoski case study was to examine the possibilities of changing a chemical wastewater treatment plant to a chemical-biological plant by using only existing constructions. The test period is from December 1986 to December 1988.

The Goals of the Case Study

The goals of the Valkeakoski case study are as follows:
- to determine the possibilities of changing chemical wastewater treatment plants to chemical-biological plants by using only existing constructions
- to determine the suitability of moving aerators and round aeration tanks when upgrading the treatment process
- to make a comparison of chemical treatment versus chemical-biological treatment
- the influence of process solution to the recipient of Valkeakoski
- to assess the criteria of decision-making when upgrading chemical treatment plants
- to make a cost-benefit analysis in Valkeakoski and make a rough extension to Finnish circumstances.

Case Study Arrangements

One half of the Valkeakoski treatment plant was changed to chemical-biological treatment by replacing flocculation devices with aeration pumps, Figure 2. Sludge-removing pumps from chemical treatment are now used as return sludge pumps. The dosage of chemicals is the same for chemical and biological-chemical treatment.

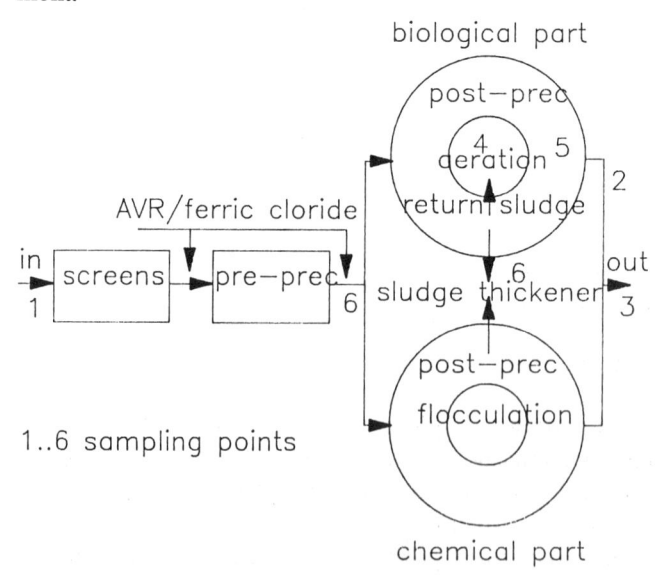

Figur 2. Valkeakoski wastewater treatment plant, a case study

The grid chambers are transformed into the pre-precipitation basin by removing pre-aeration devices and all internal walls.

The sampling points are shown in Figure 2. The chemicals used are AVR and Finnferri (ferric cloride).

An efficient flow balancing unit is installed in the pre- precipitation basins. The inflow quantity is distributed 40/60% to chemical-biological and chemical treatment, respectively.

The by-pass constructions are also changed from underflow dams to overflow dams. After this change the quantity and pollutant concentrations of overflow water decreased drastically.

The History of Valkeakoski Treatment Plant

Effluent and influent characteristics of the Valkeakoski treatment plant are shown in Figures 3 through 6. The proportion of leakage water in the influent is very high and that is the reason for the low concentration of pollutants in the influent water.

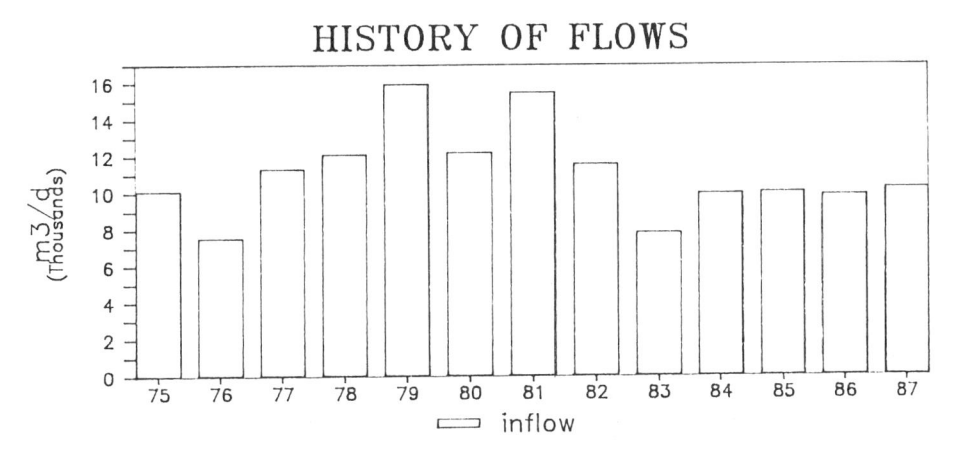

Figure 3. Mean annual inflow

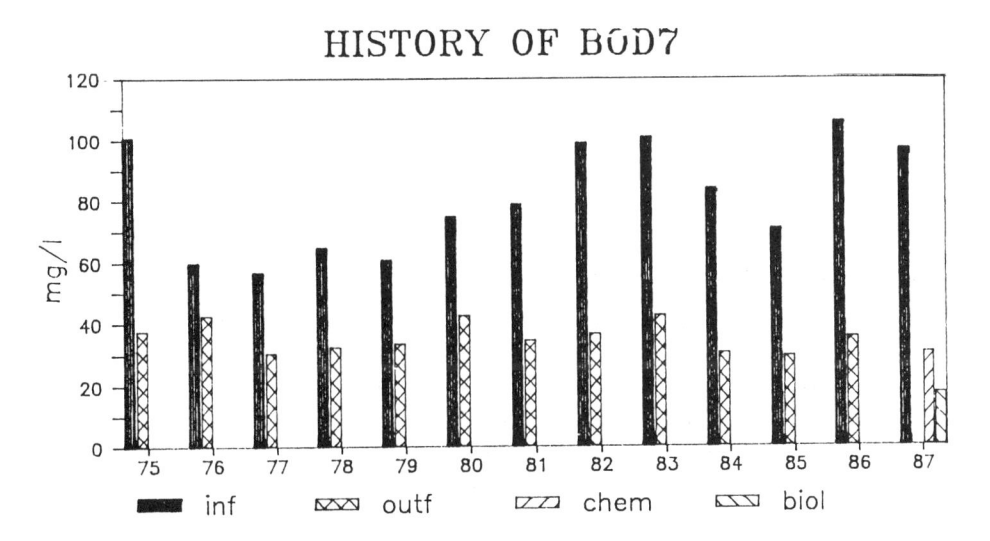

Figure 4. Mean annual BOD_7

HISTORY OF TOT-P

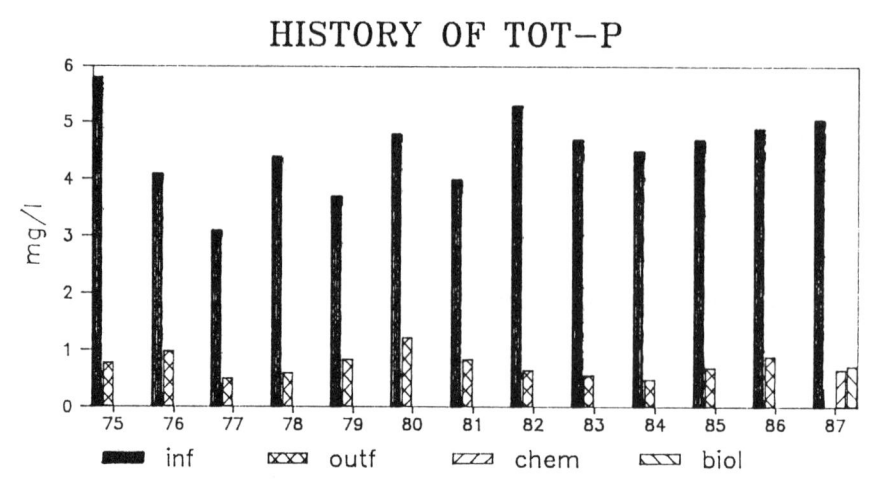

Figure 5. Mean annual tot-P

HISTORY OF SS

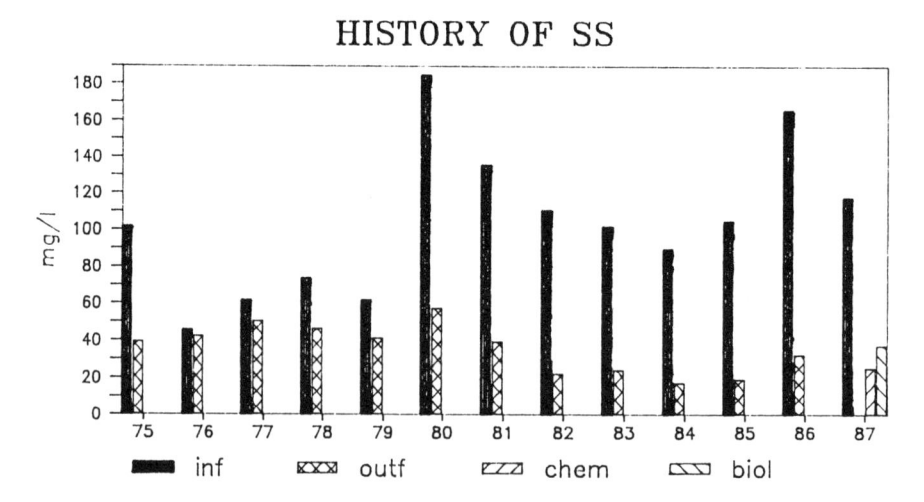

Figure 6. Mean annual SS

The Results of the Case Study

Representative results from the year 1987 are as follows:
1) At point 1 in Figure 7 the quantity of overflow is minimized with the help of the overflow weir and the flow balancing constructions. Although the inflow has been increasing all the time, the pollutant load of overflows has dropped dramatically, i.e. the process capacity has increased.
2) At point 2 in Figure 10 the pre-precipitation basins were put into use. Only after flow balancing was implemented (point 1) were the effluent concentrations sufficiently low (Figures 11 through 15). The time series are presented as seven day moving averages.

FLOW
7d moving average

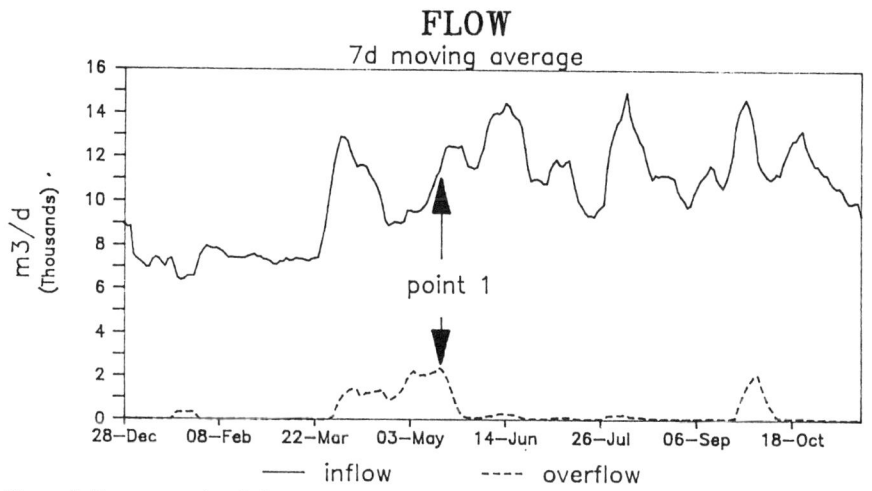

Figure 7. Treatment plant inflow and overflow series (1987)

MCRT. For acceptable values of total P and BOD, the sludge age should be greater than 2.8 days, Figure 8.

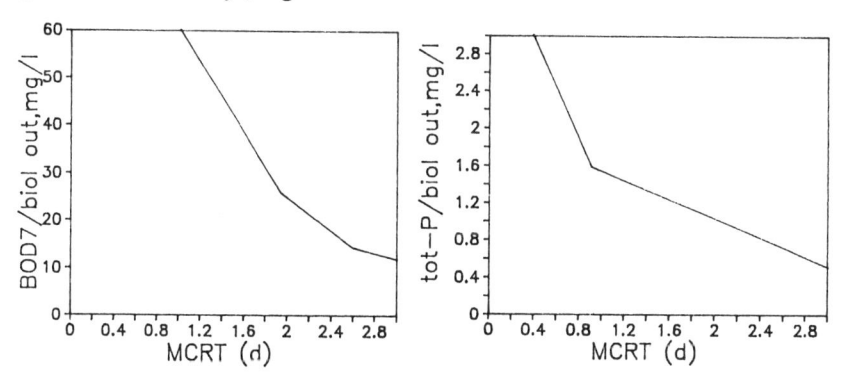

Figure 8. The effluent BOD and tot-P as a function of sludge age (chemical-biological treatment)

F/M. Sludge load has to be lower than 0.8 kg BOD/kg MLSS/d for acceptable results, Figure 9.

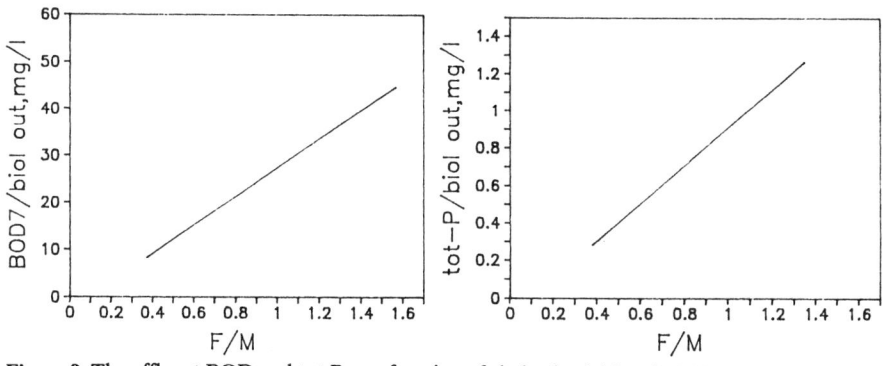

Figure 9. The effluent BOD and tot-P as a function of sludge load (chemical-biological treatment)

BOD₇. The most interesting parameter is BOD_7. If we succeed in removing a significant part of the BOD_7 effluent load, the most important goal is achieved. At the same time acceptable P (and SS) loads must be guaranteed.

The BOD effluent time series of chemical-biological treatment and of chemical treatment (the latter values are calculated from mass balances) are given in Figure 11. A significant improvement can be noticed when the results without flow balancing are compared to those with flow balancing (Figure 11).

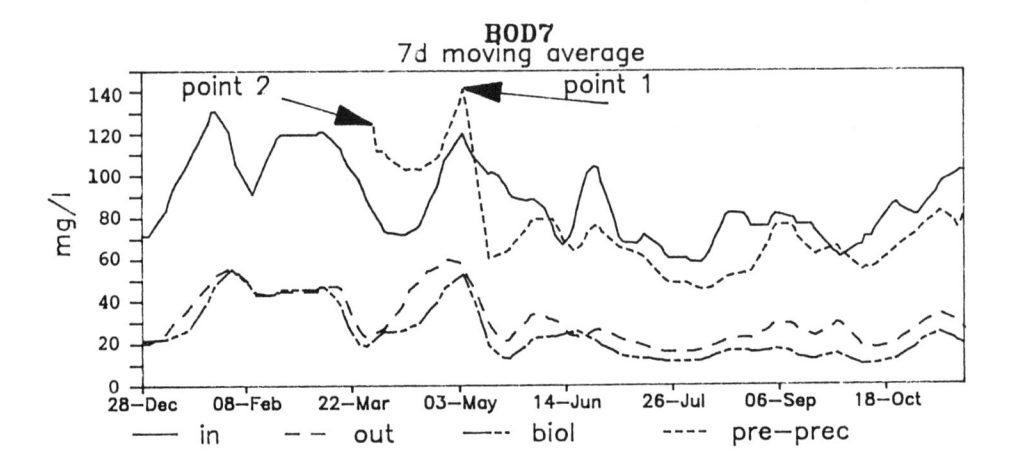

Figure 10. The BOD time series (1987)

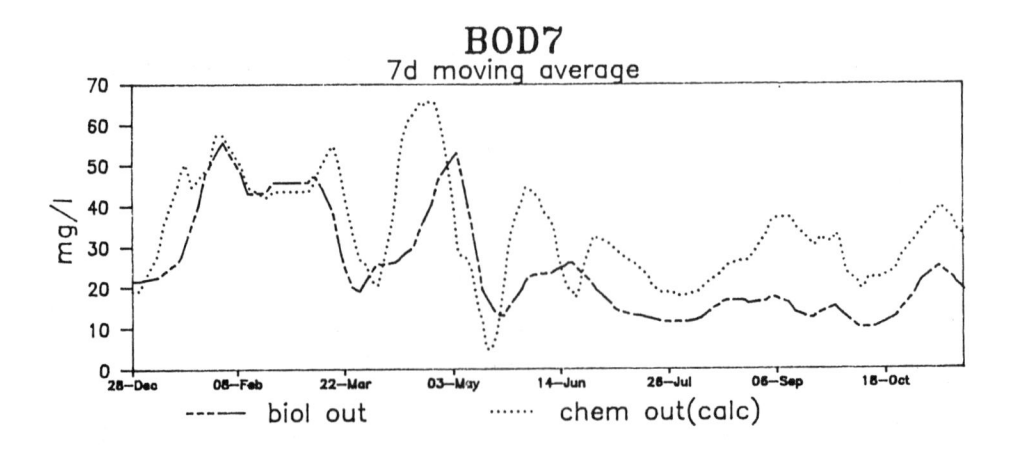

Figure 11. The effluent BOD time series (1987)

Total P. In the case of total P, there are no significant differences between chemical and chemical-biological treatment (Figures 12 and 13).

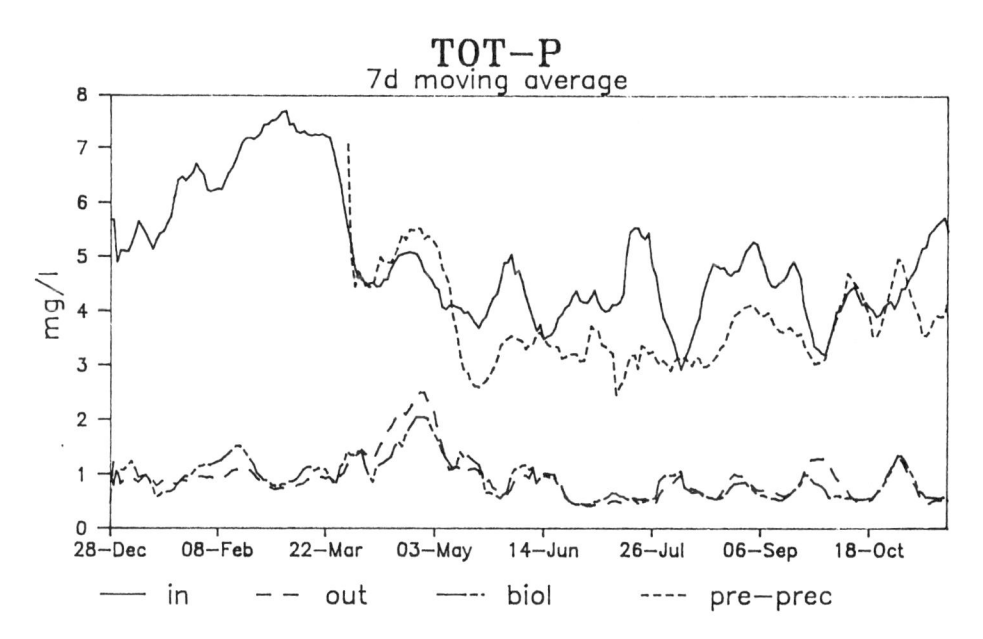

Figure 12. The tot-P time series (1987)

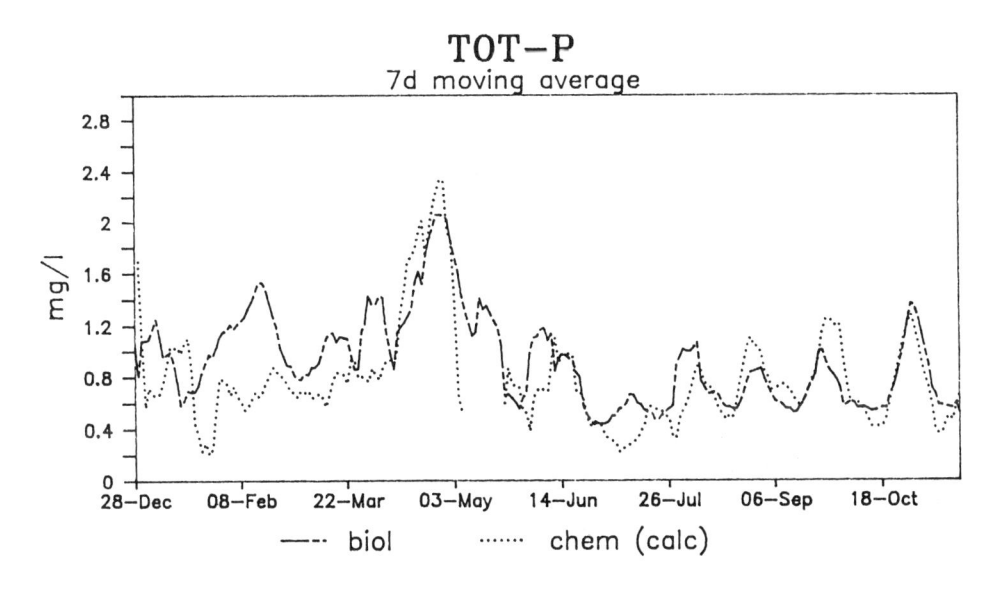

Figure 13. The effluent tot-P time series (1987)

For suspended solids the results from chemical treatment were a little better than those from chemical-biological treatment (Figures 14 and 15).

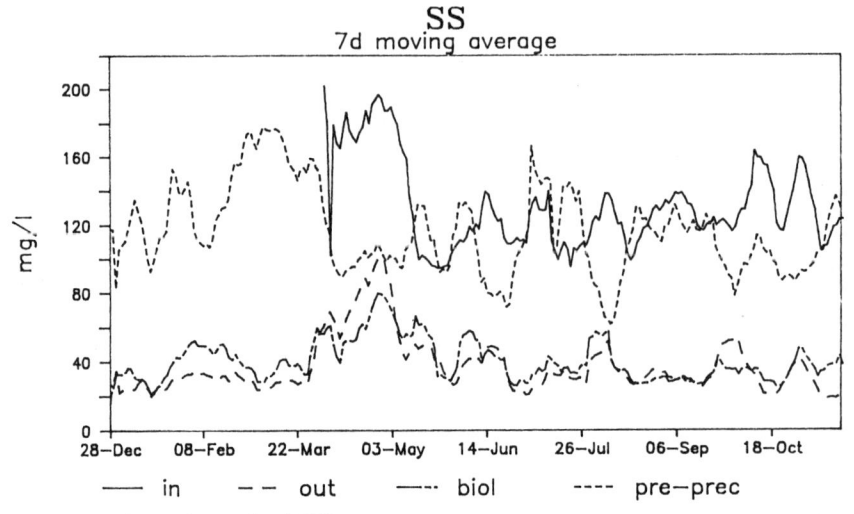

Figure 14. The SS time series (1987)

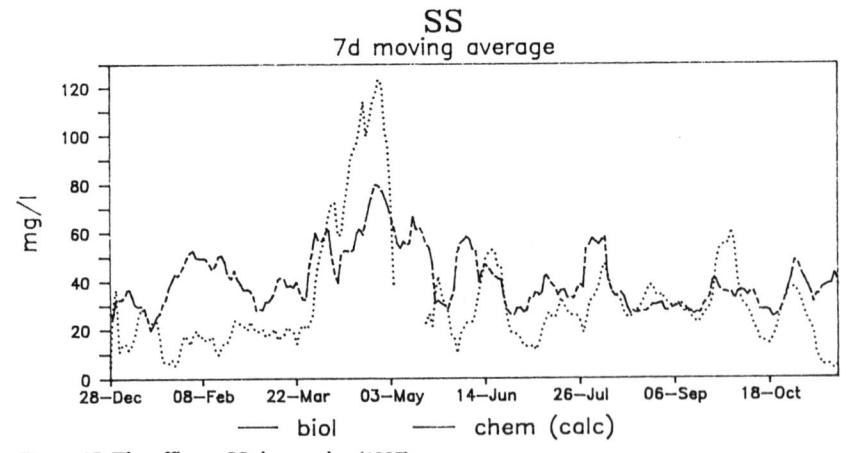

Figure 15. The effluent SS time series (1987)

Pre-precipitation. The use of pre-precipitation basins is described in Table 1.

Table 1

Area	122 m^2	
AVR	50 - 120 g/m^3	
Finferri	50 - 250 g/m^3	
Surface load	3 - 6 m/h	Day average
	4 - 8 m/h	Hour maximum
Detention time	7 - 15 min	Day average
	4 - 10 min	Hour maximum

Conclusion

The case study of the Valkeakoski wastewater treatment plant is not accomplished as a pilot plant project but as a full-scale project performed by permanent personnel. This means that all the practical difficulties are reflected in the results presented above.

The pre-precipitation results cover a period of 7 months. The final test period will cover 2 years.

The present estimates of acceptable values of the control parameters are as follows:

- MCRT > 2.8 d
- F/M   < 0.8 kg BOD/kg MLSS/d

The results of Valkeakoski are very promising. The probable savings as a result of efficient utilization of existing capacity are 15 to 20 million FIM. This approach could be adopted in other chemical treatment plants. The nationwide savings would be significant.

A. Ilmavirta
Oy Nixdorf Computer Ab
Louhelantie 10
01600 Vantaa
Finland

Reuse of Chemical Sludge for Conditioning of Biological Sludges

Y. Watanabe and A. Toyoshima

Abstract

This paper presents the effects of chemical sludge, produced by coagulation processes, on the settleability and dewaterability of biological sludge, produced by the activated sludge processes. It was demonstrated that the settleability of the mixed biological and chemical sludges was significantly better compared with that of the unblended biological sludge. The average settling velocity and density of the discrete particles of the mixed sludges were much higher than those of the biological sludge. The improvement of the mixed sludges' settleability was influenced by both mixing and ALT ratios of the chemical sludge. It was also determined that the dewaterability of the mixed sludges was improved and the specific resistance was lowered by one order of magnitude compared with the biological sludge.

This paper also presents experimental results on the adsorption of orthophosphates on chemical sludge. The experimental investigation revealed that the amount of orthophosphates adsorbed per unit mass of chemical sludge was a function of pH and the amount of Al or Fe contained in the chemical sludge. The maximum number of moles of orthophosphates adsorbed per mole of Al or Fe contained in the chemical sludge was about 0.2. The surface coordination reaction between the orthophosphates and the hydroxo-Al or Fe complexes attached to the chemical sludge was verified by acid-base titration.

Introduction

In Japan, the total amount of excess sludge (biological sludge) produced by the activated sludge processes was about 2.2×10^6 m^3/year in 1984. In many treatment plants, the settling and thickening properties of these excess sludges have been deteriorating, mainly because of the decreasing solid content. An example of the effect is illustrated in Figure 1. There are also a considerable number of water treatment plants in Japan that use coagulation processes. They have produced chemical sludge of approximately 3×10^5 tons (dry basis) per year. The biological and chemical sludges have been treated separately, but advantages were discovered when they were treated together. Watanabe et al. (1987) have pointed out the following advantages of the combined treatment of chemical and biological sludges:
(1) The mixing of chemical sludge with biological sludge increases the solid content

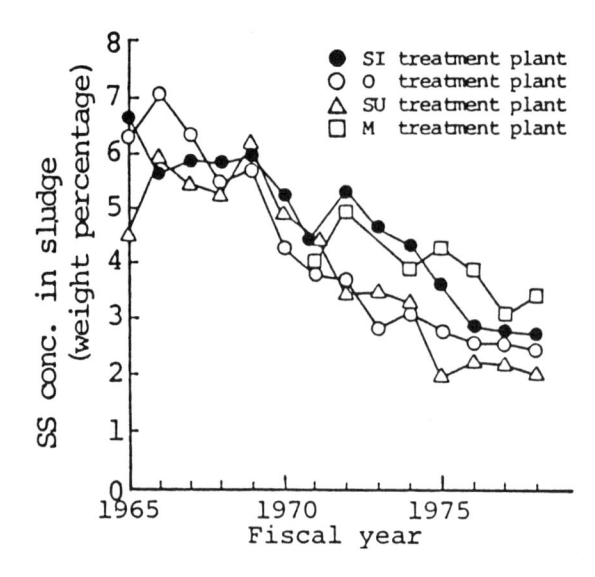

Figure 1. Trends in solid content of excess sludge

of the biological sludge, resulting in an improvement of the settleability and dewaterability of the mixed sludges as compared with that of the original biological sludge. (2) Chemical sludge contains the hydroxo-Al or Fe complexes which are capable of adsorbing orthophosphates through surface coordination reaction. Therefore, the mixing of chemical sludge with biological sludge can prevent the release of phosphate from the biological sludge during the thickening process, under anaerobic conditions.

This paper is based on the experimental investigations concerning the above-mentioned advantages of the combined treatment of chemical and biological sludges.

Materials and Methods

Biological Sludge. Biological sludge used in the experiments was the excess sludge produced by a conventional activated sludge process that treats municipal sewage. The MLSS of the excess sludge ranges from 4500 to 6900 mg/l.

Chemical Sludge. Chemical sludge used was produced by coagulating kaoline with Al(III) or Fe(III). Polymerized aluminium chloride (PAC) and ferric sulfate were used as coagulants. Tambo and Watanabe (1979) presented the floc density function of chemical flocs coagulated with Al(III) or Fe(III) (Fig. 2 and Eq. 1).

$$\rho_e = \rho_f - \rho_w = a \, d_f^{-K_\rho} \tag{1}$$

ρ_f, ρ_w the densities of the floc and water, respectively
ρ_e the buoyant or the effective density of the floc
d_f the diameter of the floc
a, K_ρ constants

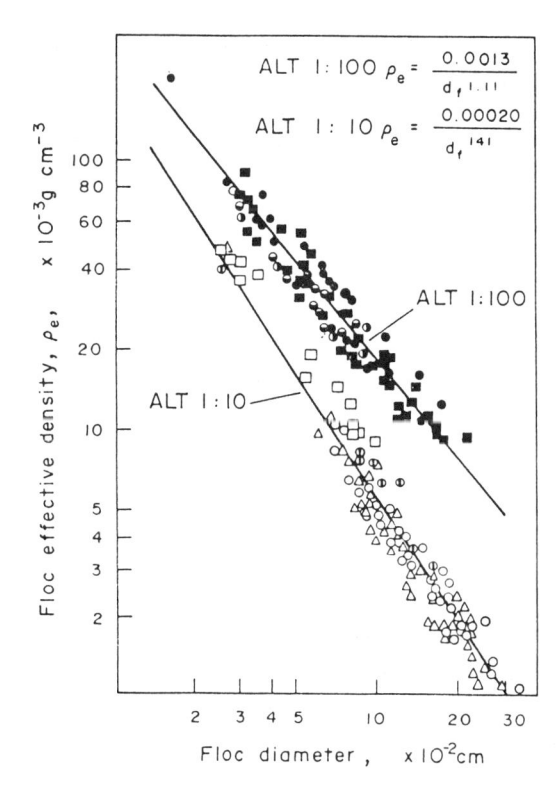

Figure 2. Floc density function

In the figure: ALT 1:100 $\rho_e = \dfrac{0.0013}{d_f^{1.11}}$; ALT 1:10 $\rho_e = \dfrac{0.00020}{d_f^{1.41}}$

Axes: Floc effective density, ρ_e, $\times 10^{-3}$ g cm^{-3}; Floc diameter, $\times 10^{-2}$ cm

Two constants in the floc density function were related to the ALT or FET ratio, which is defined as follows:

$$\text{ALT ratio} = \frac{\text{Aluminium ion concentration added}}{\text{Suspended particles concentration}}$$

$$\text{FET ratio} = \frac{\text{Ferric ion concentration added}}{\text{Suspended particles concentration}}$$

Chemical sludge coagulated at various ALT or FET ratios was used to investigate its effects on the settleability and dewaterability of the biological sludge. The adsorption of orthophosphate on the chemical sludge coagulated at various ALT or FET ratios was also examined.

Experiments. Settleability of the mixed sludges with various mixing ratios of chemical sludge at varying ALT or FET ratios was determined by measuring the subsidence velocity of the solid-liquid interface in the batch settling test. The mixing ratio is defined as the volume ratio of the chemical sludge in the mixed sludges before settling. A one liter graduated cylinder and a transparent column with diameter and effective height of 10 and 200 cm, respectively, were used. A Buchner tunnel apparatus was used to measure the dewaterability of the sludges and then the specific resistance of the sludge cakes was evaluated. The batch experiments concerning orthophosphate adsorption on the chemical sludge was carried out using jar tests. Water temperature was kept at 30°C. The acid-base titration was conducted by the method proposed by Sigg and Stumm (1980) to estimate the number of moles of OH⁻ displaced during the surface coordination reaction between orthophosphate and Al or Fe contained in the chemical sludge. The zeta potential of the sludge particles was measured by a Lazer Zee Meter, Model 501 (Penken Inc.).

Results and Discussion

Effects of Chemical Sludge on the Physical Properties of Biological Sludge

Batch Settling Curves of Mixed Sludges. Figure 3 shows the effect of chemical sludge on the subsidence velocity of the solid-liquid interface during the batch settling test of the sludges. In the experiment, excess sludge concentration was kept at 4715 mg/l, and the chemical sludge, coagulated at the ALT ratios of 1/100 and 1/10, was added at concentrations of 1500 and 900 mg/l. Figure 3 also demonstrates the effects of chemical sludge on the settling and thickening properties of the excess sludge and can be summarized as follows: (1) The settleability was improved after mixing the chemical and biological sludges. The improvement in the settleability of the mixed sludges was influenced by the ALT ratio and the amount of the chemical sludge. (2) The volume of the mixed sludges was lower than that of the excess sludge.

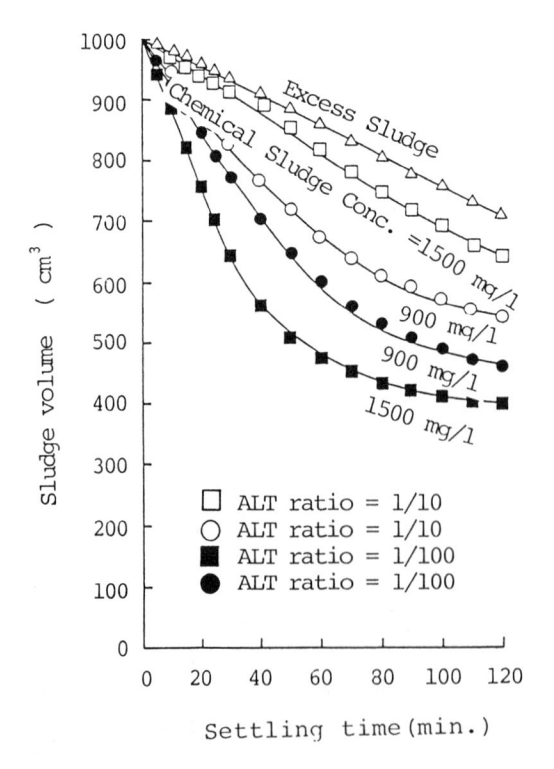

Figure 3. Batch settling curve

In order to investigate the effect of the content and the ALT ratio of the chemical sludge on the settleability of the mixed sludges, the batch settling tests with various mixing ratios of chemical sludge coagulated at various ALT ratios were carried out. Figure 4 shows the relationship between the mixing ratio and the chemical sludge content by weight in the mixed sludges.

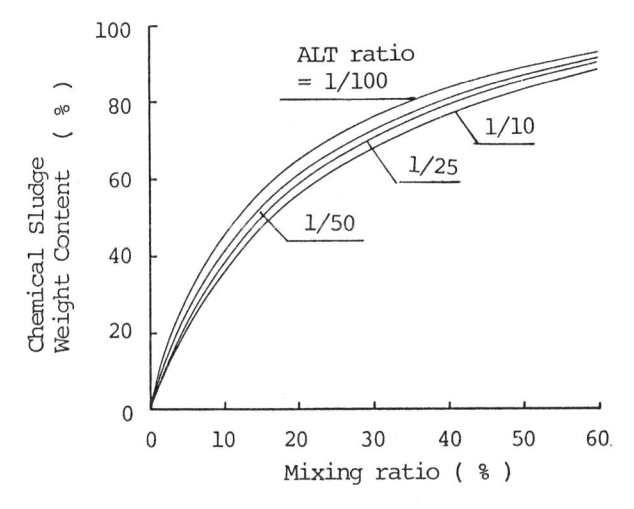

Figure 4. Chemical sludge content

Figures 5 to 8 present the batch settling curves at various ALT ratios, obtained from a 1 liter graduated cylinder. Figures 9 and 10 show the batch settling curves of the mixed sludges with the chemical sludge coagulated with Al(III) and Fe(III), respectively. The results were taken from a settling column 200 cm high.

Figure 5. Batch settling curve (ALT ratio = 1/10)

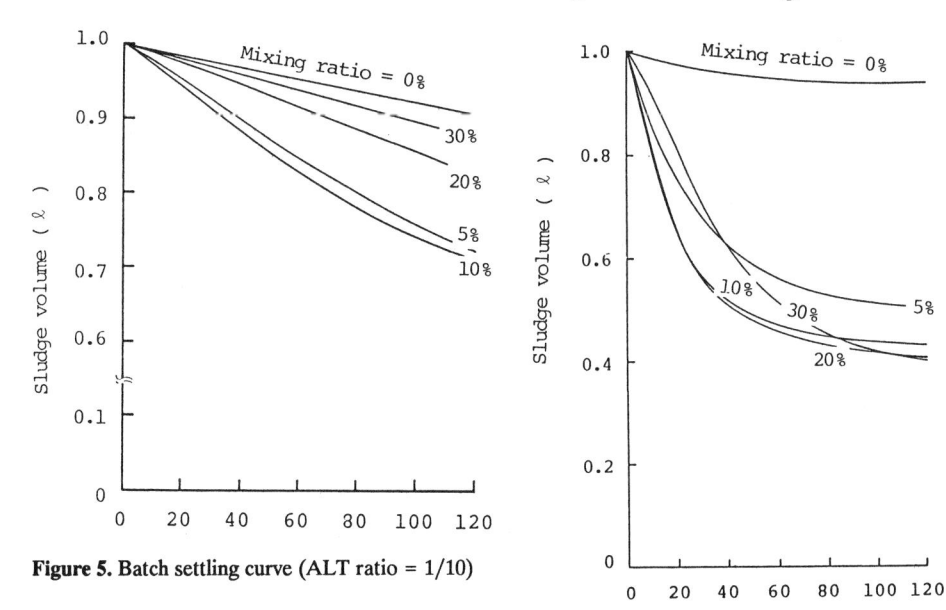

Figure 6. Batch settling time (ALT ratio = 1/25)

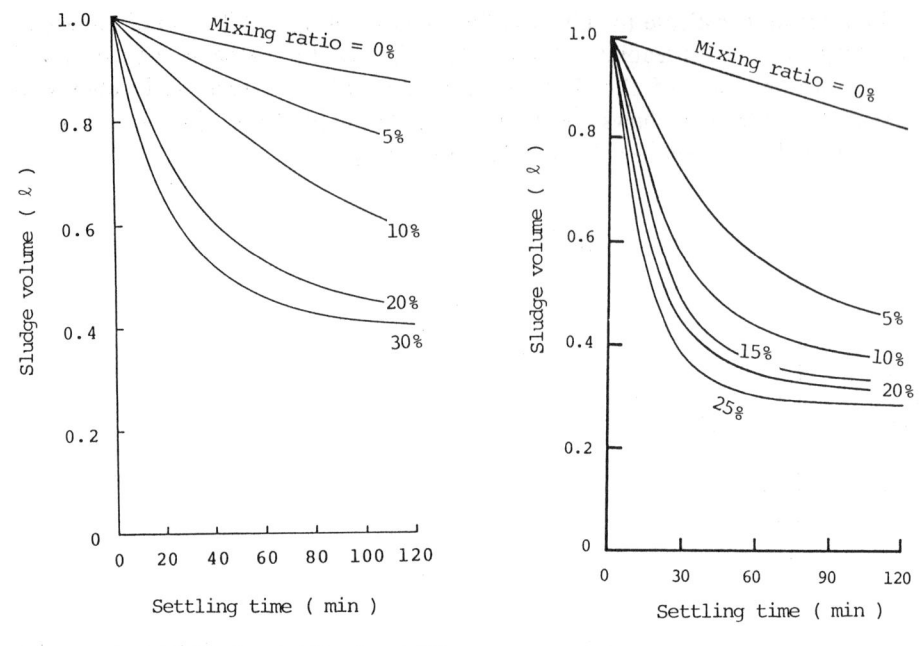

Figure 7. Batch settling time (ALT ratio = 1/50)

Figure 8. Batch settling time (ALT ratio = 1/100)

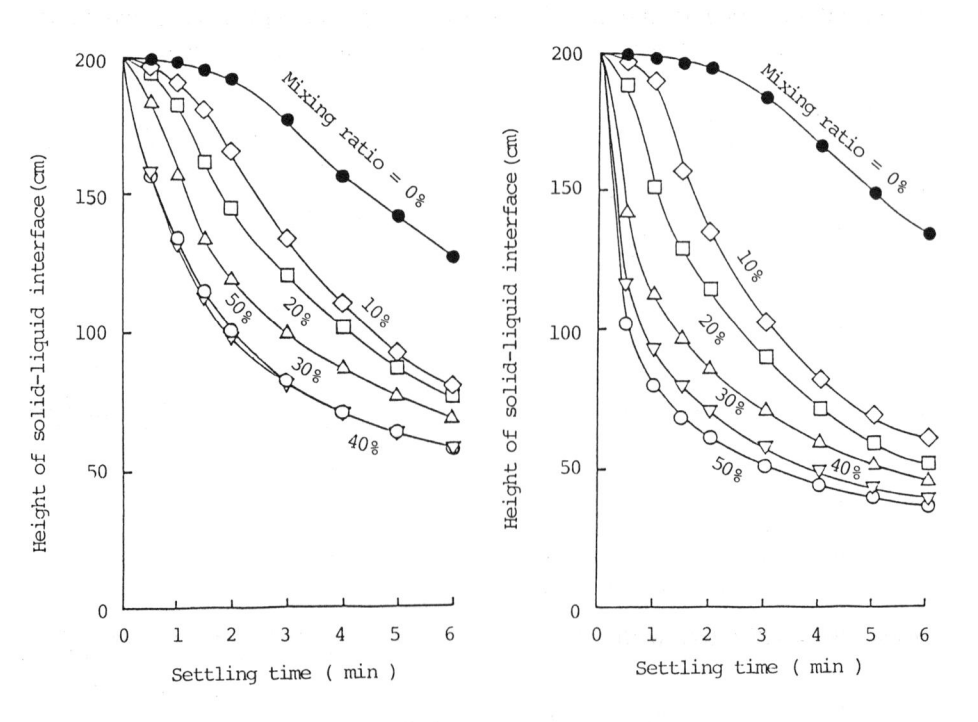

Figure 9. Batch settling curve (ALT ratio = 1/100)

Figure 10. Batch settling curve (FET ratio = 1/25)

Hindered Settling Velocity and Average Settling Velocity of the Sludge Particles.
The linear portion in the batch settling curve corresponds to sludge subsidence at the initial concentration at uniform velocity, i.e., hindered settling zone. The hindered settling velocity at each experimental condition was determined using the settling curves shown in Figures 5 to 8. Figure 11 gives the computed hindered settling velocity as a function of both mixing and ALT ratios of the chemical sludge. Knowing the hindered settling velocity at a given porosity ratio, i.e., volume of liquid per unit volume of the suspension, the settling velocity of the discrete particles comprising the suspension can be estimated by Equation 2.

$$(W_c/W)^n = \epsilon \qquad (2)$$

W_c the velocity of hindered settling at a given porosity ratio ϵ
W the settling velocity of the discrete particles
n   a constant with magnitude close to 0.2

Using Equation 2, the average settling velocity of the discrete particles comprising the sludges at a fixed mixing ratio of 25% was estimated. The hindered settling velocity at various porosities and at a given ALT ratio was measured. Figure 12 shows the double log plots of W_c and ϵ. The porosity ratio was estimated by the method proposed by Tambo and Abe (1969). The average settling velocity of the discrete particles can then be estimated as the W_c where ϵ equals one. Table 1 summarizes the calculated average settling velocities of the discrete particles. At a mixing ratio of 25% chemical sludge and at ALT ratios of 1/100 and 1/50, the hindered settling velocity of the mixed sludges was almost 20 times higher than that of the excess sludge (see Fig. 11). However, the average settling velocity of the discrete particles was only 3 times higher than that of the excess sludge. This is mainly due to the decrease in porosity ratio of the mixed sludges caused by the reduction of sludge volume. The reduction in the porosity ratio of the mixed sludges is estimated to be about 40% in the above cases from Equation 2.

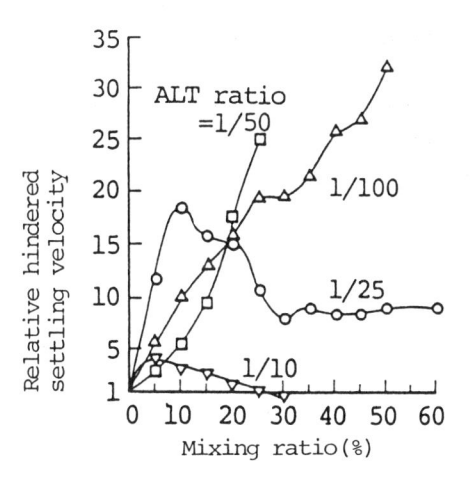

Figure 11. Relationship among ALT ratio of chemical sludge, mixing ratio and relative hindered settling velocity

Table 1. Settling velocity of discrete particles comprising mixed sludge (mixing ratio = 25%)

ALT ratio	1/100	1/50	1/25	1/10	Raw sludge
Settling velocity W [10^{-2} cm/s]	19	21	9.2	2.2	6.3

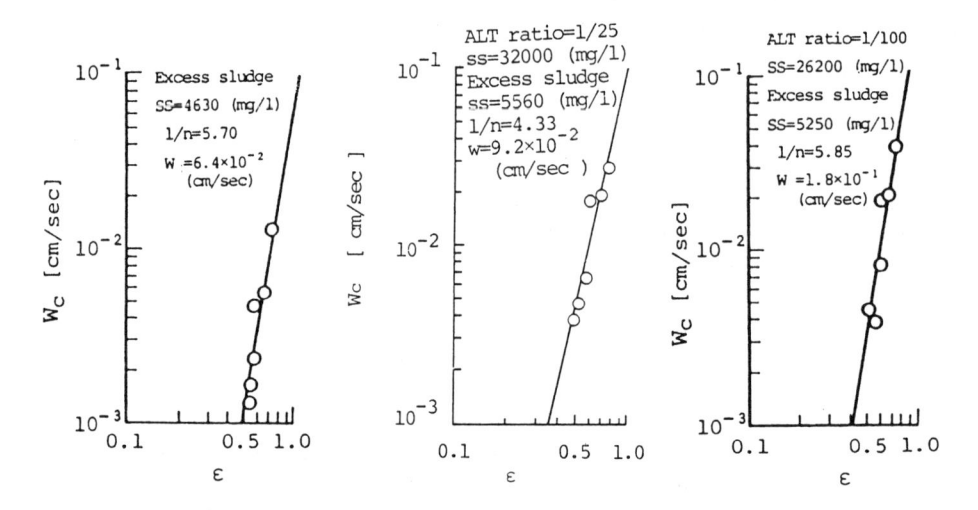

Figure 12. Hindered settling velocity as a function of porosity ratio

Sludge Volume Index (SVI) of the Mixed Sludges and Density of the Mixed Sludge Particles. Figure 13 shows the relationship between the relative SVI and the mixing ratio taken at various ALT ratios. The relative SVI is defined as the SVI ratio of the mixed sludges and the excess sludge. The density of the sludge particles was also measured. The settling velocity and diameter of a discrete particle were measured using photographs taken by a camera equipped with auto-bellows and a multi-strobe that flashes at a given time interval. Two images of a particle were photographed, and its diameter and settling velocity were computed on the negative. The effective density of the particle ρ_e was then calculated by substituting the values determined for W and d_f in the Stoke's Settling Equation (Eq. 3).

$$W = (g/18\mu)(\rho_f - \rho_w) d_f^2 \qquad (3)$$

Figure 14 shows the relationship between the effective density and the diameter of the sludge particles.

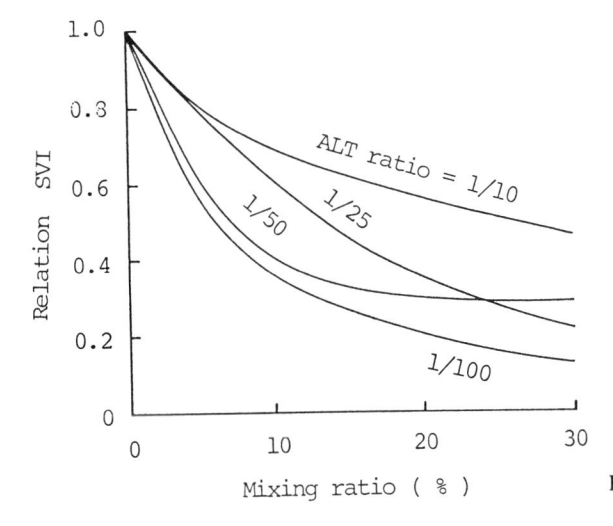

Figure 13. Decreasing SVI

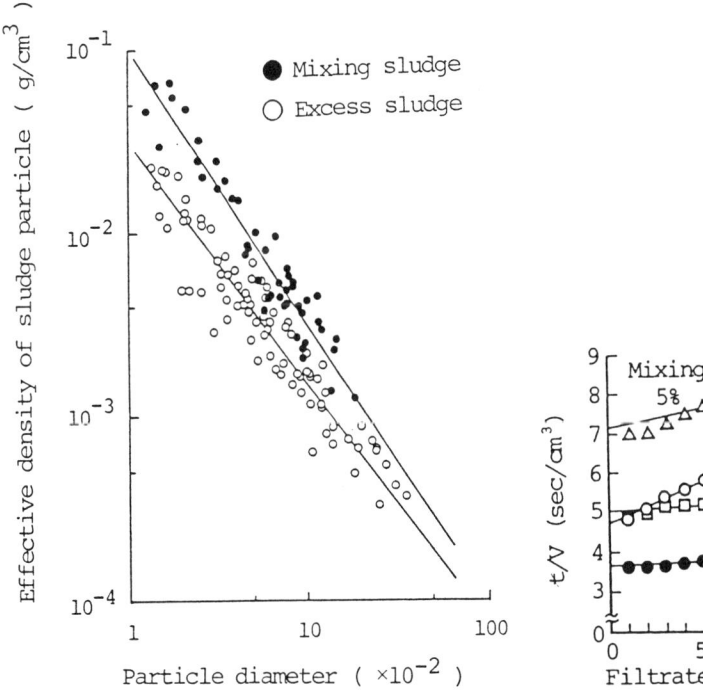

Figure 14. Floc density function of sludge particle

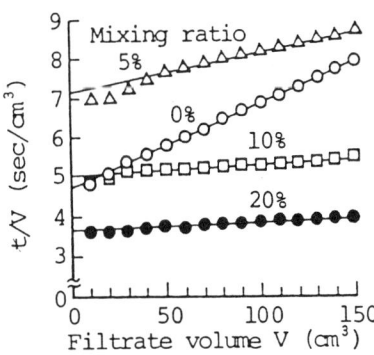

Figure 15. t/V vs. V in Eq. 4

Dewaterability of the Mixed Sludges

The settled sludges were dewatered in the Buchner funnel apparatus. The volume of the filtrate was measured with increasing filtration time, and these data were plotted and are shown in Figure 15. The specific resistance of the mixed sludges was then determined by the Carman-Ruth Equation (Eq. 4).

$$t/V = (\mu/2)(\omega r V/PA^2) + (\mu R_m/PA) \tag{4}$$

- V the volume of filtrate
- t filtration time
- P pressure
- A area
- μ the viscosity of the filtrate
- r the specific resistance of the sludge cake
- R_m resistance of the filter media
- ω the weight of solid per unit volume of filtrate

Figure 16 exhibits the relationships among the specific resistance, the mixing and ALT ratios of the chemical sludge. The dewaterability of the mixed sludges was significantly improved compared with that of the excess sludge.

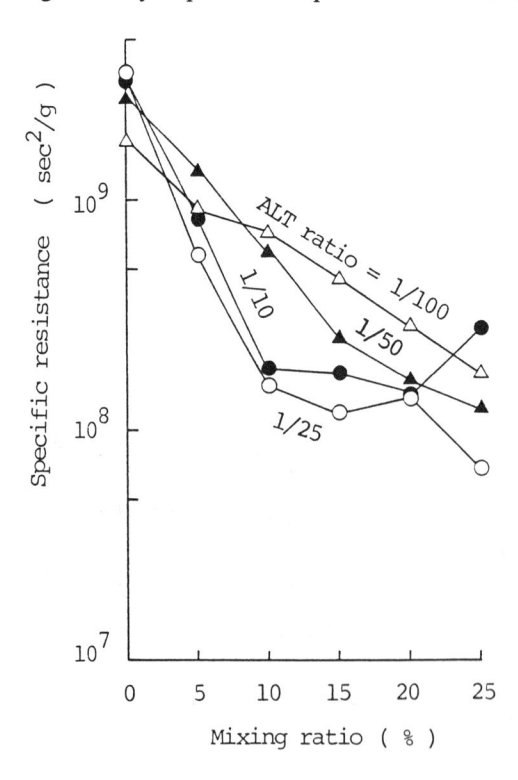

Figure 16. Decreasing specific resistance

Adsorption of Phosphate on the Chemical Sludge

Adsorption Capacity. Figure 17 gives the batch curves of orthophosphate adsorption onto the chemical sludge. The initial orthophosphate concentration was 50 mg/l as

PO_4^{-3} and the ALT ratio of the chemical sludge was 1/10. At equilibrium, the effect of the orthophosphate concentration on the amount of orthophosphate adsorbed on the chemical sludge can be expressed by Langmuir's adsorption isotherm, as shown in Figure 18. The maximum number of moles of orthophosphate adsorbed per unit mole of Al or Fe contained in the chemical sludge, as a function of the initial pH, is reported in Figure 19. After adsorption, the pH increased approximately by 1.5 units in the chemical sludge. For the chemical sludge coagulated by Al(III), the adsorption capacity of orthophosphate remained essentially the same at the initial pH ranges of 3.0 to 6.5 (correspond to equilibrium pH ranges of 4.5 to 8.4). This is also confirmed by Figure 20. It shows that no desorption of orhtophosphate occured at equilibrium pH ranges of 3.0 to 8.0. As illustrated theoretically later, 1 mole orthophosphate is adsorbed by 1 mole Al contained in the chemical sludge. However, Figure 19 demonstrates that about 20% of the Al attached to the chemical sludge is effective in the surface coordination reaction with orthophosphates. In Figure 21, the relationship between the maximum amount of orthophosphate adsorbed per unit mass of the chemical sludge and the ALT ratio of chemical sludge at an initial pH of 5.0 is presented. The figure shows that the orthophosphate adsorption capacity of chemical sludge is proportional to its Al content.

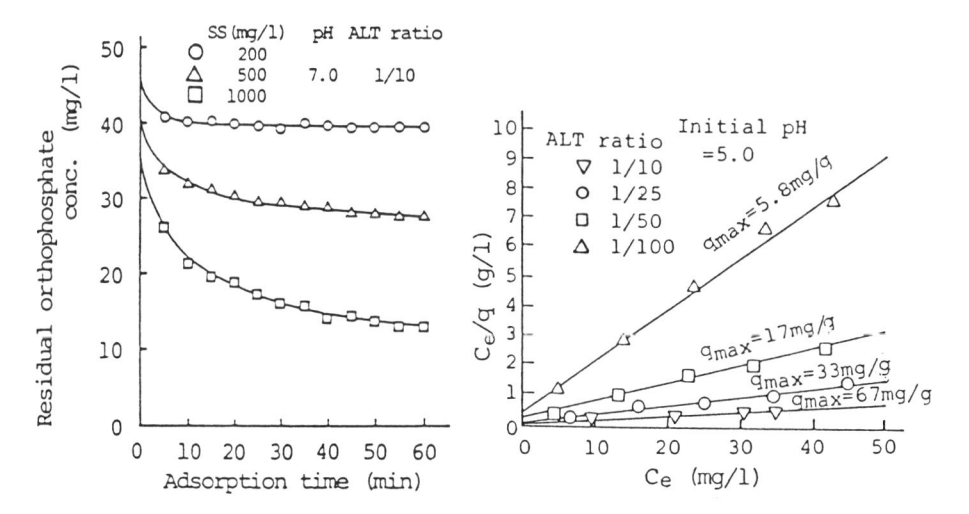

Figure 17. Adsorption of phosphate onto chemical sludge

Figur 18. Langmuir isotherm of phosphate adsorption

Adsorption Mechanism. Solid particles containing oxides or hydroxides of Si, Al and Fe react with cations and anions through surface coordination in natural water systems. The reaction that occurs between orthophosphate and hydroxo Al complexes in a neutral or weak acid pH range may be illustrated as follows:

$$\equiv Al - OH + H_2PO_4^- \rightleftharpoons AlH_2PO_4 + OH^- \qquad (5)$$

From Equation 5, it is evident that adsorption of 1 mole orthophosphate displaces or releases 1 mole of OH^-. The validity of the equation was investigated by an

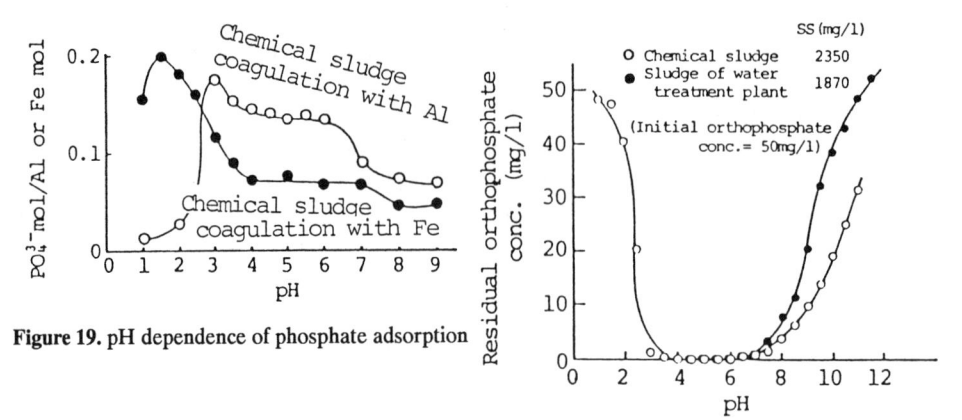

Figure 19. pH dependence of phosphate adsorption

Figure 20. Desorption of phosphate with changing pH

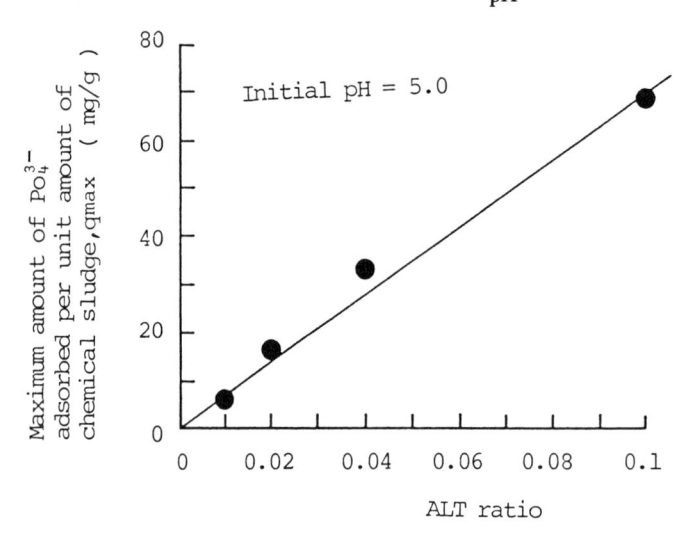

Figure 21. Effect of ALT ratio on maximum amount of phosphate adsorbed by chemical sludge

acid-base titration experiment. In the titration of the mixed solution of chemical sludge and orthophosphates, the displacement of OH⁻ during the hydrolysis of Al and orthophosphate as well as the surface coordination reaction must be taken into account (Eq. 5). From this concept the following equation was developed:

$$[OH]_{total} = [OH]_1 + [OH]_2 + [OH]_3 \tag{6}$$

$[OH]_{total}$ total molar concentration of OH⁻ displaced by the adsorption of orthophosphate on the chemical sludge
$[OH]_1$ molar concentration of OH⁻ displaced during the surface coordination reaction
$[OH]_2$ molar concentration of OH⁻ displaced during the hydrolysis of orthophosphate
$[OH]_3$ molar concentration of OH⁻ displaced during the hydrolysis of Al

$[OH]_{total}$, $[OH]_2$ and $[OH]_3$ were determined by acid-base titration. The $[OH]_1$ was calculated by substituting the values determined for $[OH]_{total}$, $[OH]_2$ and $[OH]_3$ in Equation 6. Figure 22 shows the acid-base titration curves obtained for the solvent, orthophosphate solution, chemical sludge and mixed solution of orthophosphate and chemical sludge coagulated with Al(III). At a fixed pH, $[OH]_{total}$, $[OH]_2$ and $[OH]_3$ were estimated as the difference along the abscissa between the solvent and mixed solution of orthophosphate and chemical sludge, the solvent and the chemical sludge, and the solvent and orthophosphate solution, respectively.

Figure 23 gives the acid-base titration curves for the chemical sludge coagulated with Fe(III). Table 2 presents the computed values of $[OH]_{total}$, $[OH]_2$ and $[OH]_3$ as well as the calculated values of $[OH]_1$. The number of moles of orthophosphate adsorbed is in good agreement with the calculated moles of $[OH]_1$.

- ● Residual orthophosphate conc.
- ☐ Chemical sludge coagulated by Al or Fe + orthophosphate
- ○ Chemical sludge coagulated by Al or Fe
- △ Orthophosphate
- — Solvent

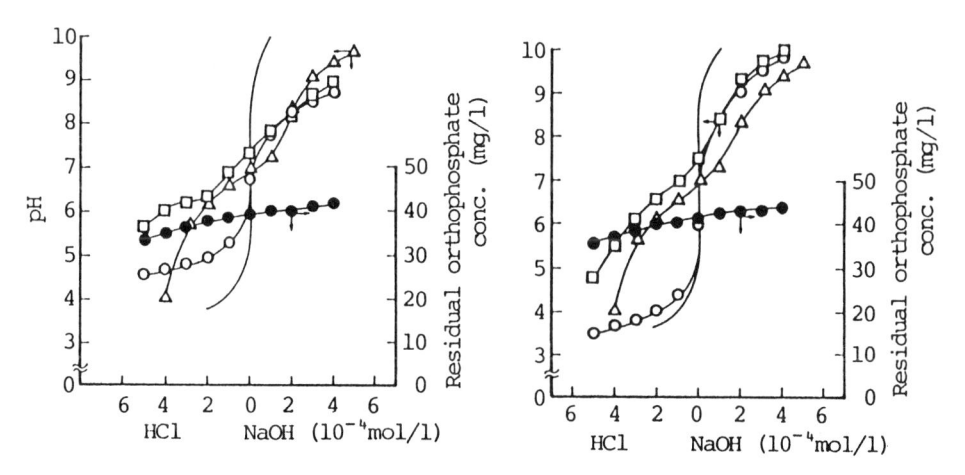

Figure 22. Acid-base titration curve of chemical sludge coagulated with Al(III)

Figure 23. Acid-base titration curve of chemical sludge coagulated with Fe(III)

Table 2. Molarity of exchanged hydroxyl ion and of adsorbed orthophosphate ion

Coagulant	pH	Adsorbed orthophosphate conc. m mol/l	$(OH)_{total}$ m mol/l	$(OH)_2$ m mol/l	$(OH)_3$ m mol/l	$(OH)_1$ m mol/l
Al	5	0.211	0.710	0.175	0.355	0.180
	6	0.156	0.405	0.010	0.225	0.180
	7	0.119	0.095	0	0	0.095
Fe	5	0.148	0.460	0	0.332	0.128
	6	0.121	0.330	0	0.230	0.1
	7	0.094	0.095	0	0	0.095

Adsorption of anions on the chemical sludge particles decreases the net charge of the particle surface. Thus, the zeta potential of the chemical sludge particles with and without the orthophosphate adsorption was measured. Figure 24 shows the relationship between the zeta potential and the number of moles orthophosphate adsorbed per unit mole of Al contained in the chemical sludge. In Figure 25, the pH dependence of the zeta potential with and without orthophosphate adsorption is illustrated.

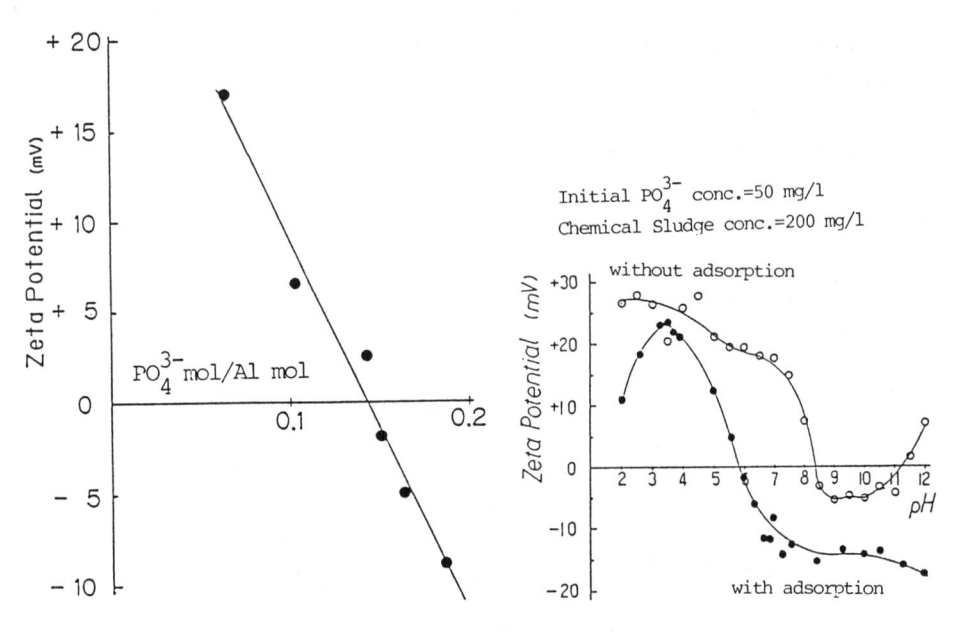

Figure 24. Change in zeta potential with adsorbing phosphate

Figure 25. Change in zeta potential with changing pH

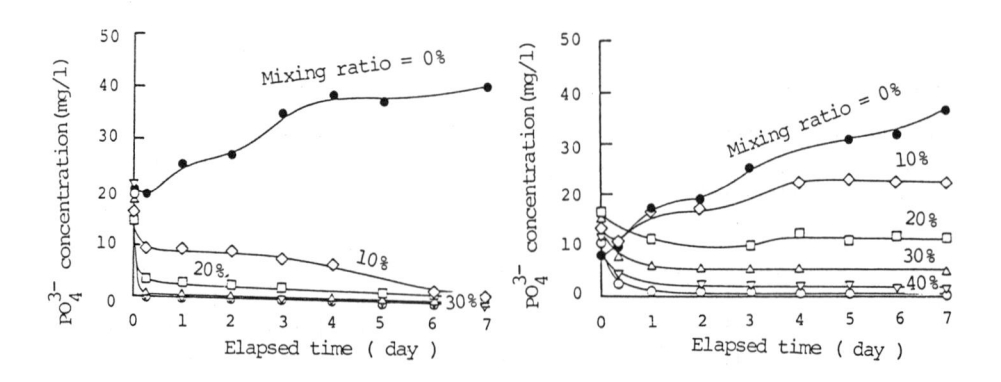

Figure 26. Effect of chemical sludge on phosphate release (ALT ratio = 1/100)

Figure 27. Effect of chemical sludge on phosphate release (FET ratio = 1/10)

Control of Phosphate Release from Biological Sludge under Anaerobic Conditions.
Since the chemical sludge adsorbs the orthophosphate, under anaerobic conditions phosphate release from the biological sludge during the thickening process is controlled by the mixing ratio of the chemical sludge and the biological sludge. Figures 26 and 27 exhibit the control of phosphate release at various mixing ratios of the chemical sludge coagulated with Al(III) and Fe(III). The figures show that chemical sludge coagulated with Al(III) could control the release of phosphate better than the chemical sludge coagulated with Fe(III).

Summary and Conclusions

The experimental investigation confirms the following advantages of the combined treatment of chemical and biological sludges:

1) Mixing of chemical sludge with biological sludge substantially improved the settleability of the biological sludge. The settleability of the mixed sludges was influenced by both the mixing and ALT ratios of the chemical sludge. Chemical sludge with lower ALT ratios produced a better settleability. The average settling velocity of the discrete particles in the mixed sludges was about 3 times higher than that of the biological sludge at a mixing ratio of 25% and ALT ratios of 1/100 and 1/50. The density of the mixed sludge particles was much higher than that of the biological sludge.
2) Mixing of chemical sludge with biological sludge improved the dewaterability of the biological sludge. The specific resistance of the mixed sludge decreased by one order of magnitude with that of the biological sludge.
3) Hydroxo-Al or Fe complexes contained in the chemical sludge adsorbed the orthophosphates by means of a surface coordination reaction. The adsorption capacity of the chemical sludge was proportional to the amount of Al or Fe contained in the chemical sludge. Adsorption remained constant at equilibrium pH ranges of 4.5 to 8.5. The maximum number of moles of orthophosphate adsorbed per unit mole of Al or Fe was about 0.2. The surface coordination reaction between orthophosphates and Al or Fe contained in the chemical sludge was investigated by acid-base titration. The experiment verified that the displacement of 1 mole of OH means the adsorption of 1 mole of orthophosphate. Under anaerobic conditions, phosphate release from biological sludge during the thickening process was controlled by mixing the two sludges.

References

[1] Sigg, L., Stumm, W. (1980): The Interaction of Anions and Weak Acids with the Hydrous Goethite (α-FeOOH) Surface. Colloid and Surface 2, 101-117
[2] Stumm, W., Morgan, J.J. (1981): Aquatic Chemistry. 2nd edn., Wiley-Interscience, pp. 625-640
[3] Tambo, N., Abe, S. (1969): Behaviour of Floc Blanket in the Up-Flow Type Sedimentation Basin. J. of Japan Water Works Ass. *415*, 7-18 (in Japanese)

[4] Tambo, N., Watanabe, Y. (1979): Physical Characteristics of Flocs - The Floc Density Function and Aluminum Floc. Water Research, Vol. 13, 409-419
[5] Watanabe, Y. et al. (1987): Improvement of Physical Properties of Biological Sludge and Chemical Adsorption of Orthophosphate by Chemical Sludge. Proc. of Environ. & Sani. Eng. Research, Vol. 23, 149-156 (in Japanese)

Y. Watanabe
Department of Civil Engineering
Miyazaki University
Miyazaki 889-21
Japan

A. Toyoshima
Department of Civil Engineering
Kinki University
Higashi Osaka 577

Influence of Sludge from Chemical Biological Wastewater Treatment on Nitrification and Digestion

C.F. Seyfried, H.-D. Kruse and F. Schmitt

1 Introduction

In the future all major sewage treatment plants in the Federal Republic of Germany should be equipped with both nitrification and P-elimination. Lots of Literature about and practical experience with nitrogen oxidation as well as with phosphate elimination exists. However there are only few publications about the influence of precipitation on nitrification and sludge treatment. In the following report the present state of knowledge will be presented.

2 Influence on Nitrification and Denitrification

2.1 Pre-precipitation

A pre-precipitation can only negatively affect nitrogen elimination increased amounts of undissolved carbonaceous compounds precipitate at the same time. For this reason denitrification, which is necessary for the stabilisation of the nitrogen oxidation process (for instance in the case of insufficient acid capacity), may not be adequately supplied with carbonaceous compounds. The often encountered problem of a primary treatment removing so much substrate that an extensive denitrification in a preceeding stage is no longer possible is worsened by a pre-precipitation. On the other hand, the addition of lime as precipitant can compensate for the insufficient acid capacity.

2.2 Simultaneous Precipitation

2.2.1 Influence of Precipitation Sludge on Sludge Age

In calculating sludge age of nitrification, the increase in solids caused by the precipitant should also be taken into account. Figure 1 shows the increase in solids in simultaneous precipitation as a function of the amount of precipitant - relative to the excess sludge. The relatively cheap simultaneous precipitation can, if both nitrification and denitrification are required, lead to such an increase in the necessary volume of the aeration tank, due to the increase in sludge production, that it is no longer cheaper than other precipitation processes. This effect is reduced when the solid concentration of the activated sludge can be raised from 3.3 g/l to 4 g/l by improved settling properties (Figure 1). The calculations also show that the influence of simultaneous precipitation on the enlargement of the aeration tank volume is greater when the BOD_5 concentration is lower; two stage plants are thus not suitable for simultaneous precipitation in the second stage. In Figure 2 the same calculations are presented for simultaneous sludge stabilisation. Here the influence of a simultaneous precipitation on the size of the aeration tank is considerably lower.

2.2.2 Influence on the Acid Capacity

In cities and communities with soft drinking water the acid capacity of the wastewater is often not large enough to guarantee nitrification without a sinking of the pH value which would then lead to a breakdown of the biological stage. The acid capacity should therefore generally be calculated along with the nitrification and denitrification process. If a phosphate precipitation is planned then it should be included in the acid capacity calculation.

– Iron or aluminium ions form hexa-aquo-complex compounds having the general formula of $Me(H_2O)_6^{3+}$, as soon as they get into a hydrous solution. This complex compound reacts like an acid by protolysing further with the water,

$$Me(H_2O)_6^{3+} \dashrightarrow 3H^+ + Me(OH)_3 + 3H_2O,$$

depending upon the pH value in the solution.

– Since the hydroxides as complex compounds of low solubility precipitate they do not contribute to the acid capacity; there are thus, during the formation of hydroxides, 3 acid equivalents to be taken into consideration per mole of metal salt added.

– In the presence of phosphate ions, precipitation products of the general composition of $Me_n(OH)_{3m}(PO_4)_{n-m}$ are formed, where m lies between 0 ($MePO_4$) and n ($Me(OH)_3$). The chemical precipitation is usually not carried out in a mole ratio of M:P = 1:1 but rather with an excess of metal ions in order to attain acceptable residual P-concentrations. The ratio of M:P is referred to as β and depends upon the attainable efficiency.

It can be assumed for simplification purposes that the composition of the precipitation products is given by the formula $Me_\beta(PO_4)_f(OH)_{3(\beta-f)}$.

$(PO_4)_f$ is the share of precipitated phosphate in the concentration of total phosphate. At high precipitaion efficiencies the residual phosphate concentration can be neglected so that the formula is reduced to $Me_\beta(PO_4)(OH)_{3(\beta-1)}$.

Thus the following are to be used in the calculation:

Precipitation as metal   hydroxide: 3 acid equivalents
Precipitation as metal   phosphate: 2 acid equivalents

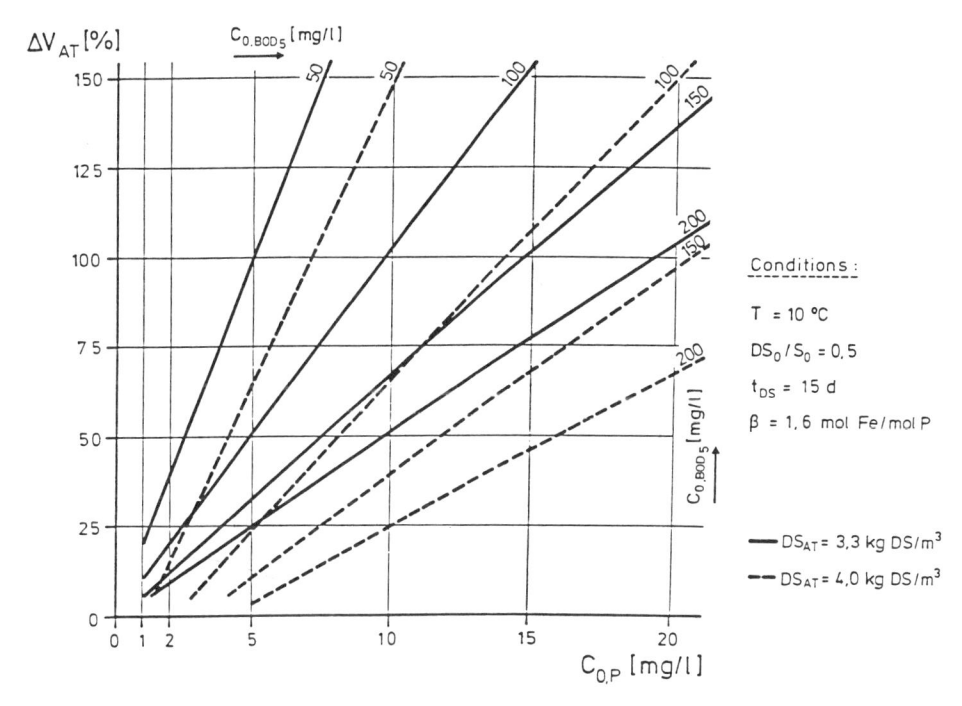

Influence of Sludge from Chemical Biological Wastewater Treatment on Nitrification and Digestion 309

Figure 1. Increase of the required aeration tank volume for nitrification by simultaneous P-precipitation versus influent-concentration of BOD_5 and P_{TOT} (sludge loading = 0.1 kg $BOD_5/m^3 d$)

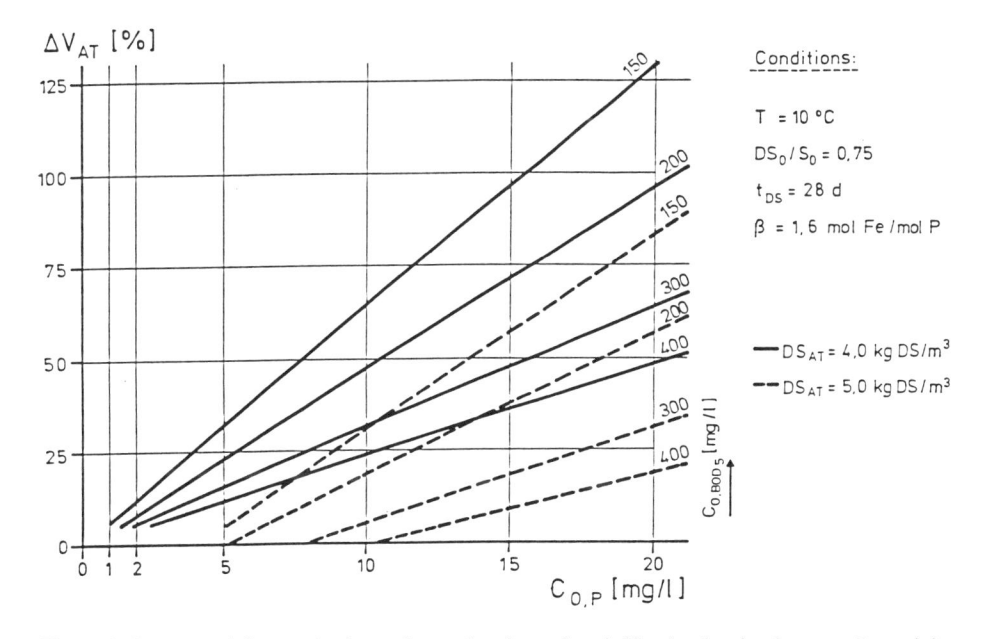

Figure 2. Increase of the required aeration tank volume for nitrification by simultaneous P-precipitation versus influent-concentration of BOD_5 and P_{TOT} (Sludge loading = 0.05 kg $BOD_5/m^3 D$)

The amount of acid capacity consumed during precipitation can be calculated, using a simplified formula, to be

$$\text{acid capacity} = 3(\beta - 1)\text{Me} + 2\text{Me} \ [\text{mmol/l}]$$

or simplified

$$\text{acid capacity} = (3\beta - 1)\text{Me} \ [\text{mmol/l}]$$

Simultaneous Precipitation with Ferrous Salts. During the oxidation of ferrous to ferric salts as it occurs in simultaneous precipitation, the following reaction takes place

$$Fe^{2+} + \tfrac{1}{4}O_2 + 2OH^- + \tfrac{1}{2}H_2O \longrightarrow Fe(OH)_3.$$

Thus, to form $Fe(OH)_3$ out of ferrous salts, 2 equivalents of acid capacity are needed instead of 3 equivalents as in the reaction with ferric salts. The calculation of the acid capacity results thus in:

$$\text{acid capacity} = 2(\text{Me}) \ [\text{mmol/l}]$$

2.2.3 Influence of the Precipitant on the Nitrifying Bacteria

In experiments on a semi-technical scale Gujer and Boller [2] have determined an activation of 15% for $FeCl_3$ while for $FeSO_4$ they found an inhibition of the nitrifying bacteria of up to 35%. Kayser obtained similar results using $FeSO_4$ in both the nitrification stage and in the highly loaded first biological stage, in which case the nitrification in the second stage was negatively affected. Our research at the sewage treatment plant Hannover-Gümmerwald produced the results shown in Figure 3: nitrification was slowed down by 20% by adding $FeSO_4$ while when $FeCl_3$ was used a considerable activation of the nitrosomonas was detected [8, 16].

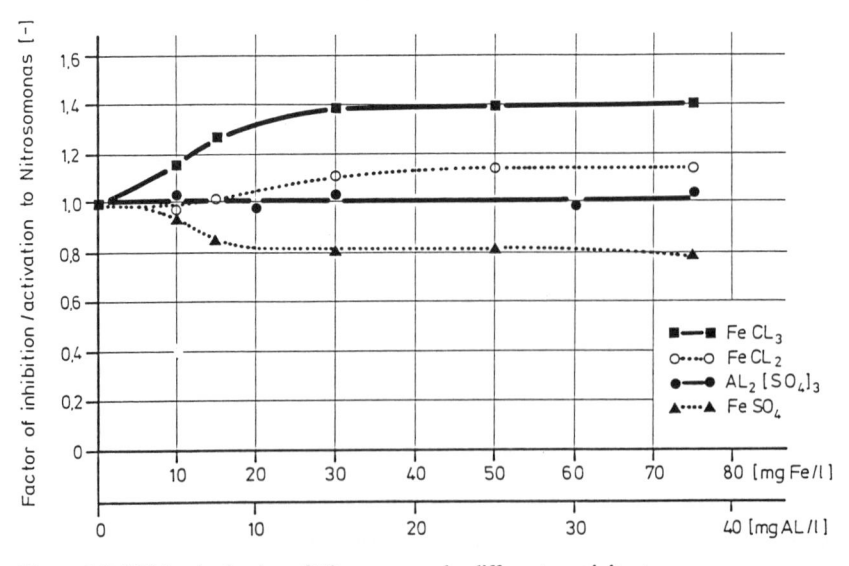

Figure 3. Inhibition/activation of Nitrosomonas by different precipitants

When the inexpensive $FeSO_4$ is used its inhibiting influence must thus be compensated for, as much as possible, with a relatively higher sludge age. A scientific explanation for these effects does not yet exist.

3 Influence of the Sludge Digestion

3.1 Introduction

Basically the effects on sludge digestion are due to 2 factors:
- the change in the chemical composition of the sludge due to the precipitant
- the increase in the volume of generated sludge and in the amount of solids.

The precipitant and the precipitation sludge can have the following effects on the anaerobic digestion process and the further sludge treatment:
- Change in the pH value and the buffer capacity: alkaline precipitants can raise the share of NH_3 in the total amount of ammonia nitrogen and inhibit the methanogenic bacteria. Acidic precipitants can, on the other hand, disturb the neutral environment of the anaerobic bacteria and, at the same time, raise the portion of molecular (toxic) hydrogen sulphide.
- The use of precipitants containing sulphate raises the hydrogen sulphide production. In this manner the anaerobic process can be inhibited; furthermore, it negatively affects the use of digester gas (e.g. gas engines).
- The gas yield decreases.
- Increased re-dissolution of phosphates and thus possibly a recycling of the phosphates in the sewage treatment plant.

The increase in the volume of sludge can lead to a shortening of the digestion time and as a result to an anaerobic stabilization of lower quality.

3.2 Results from Experiments and Studies with Precipitation Sludge

O'Shaugnessy et al. [9], Mosebach [7], Hierse [6], Peschen [10] and Fayoux [1] have published studies about the anaerobic stabilization of precipitation/flocculation sludges. With the exception of Peschen [10] who examined the influence of lime precipitation sludges all the others only used sludge from the precipitation with trivalent metal salts.

Table 1 presents a short survey of the experimental results from these authors. O'Shaugnessy et al. [9] published a particularly thorough study. The study's contents (pH, alkalinity, organic acids, gas production, decrease in organic solids, H_2S-formation, re-dissolution of aluminum, dewaterability) were very extensive as was also the length of the study, 19 months on a technical scale. Neither of the 2 precipitants containing aluminium negatively influenced the anaerobic stabilization. A re-dissolution of phosphate could not be detected either.

Mosebach [7] even detected an increase in the fixation of phosphate in the digesters in his continuos experiments with iron-bearing sludges from simultaneous

Table 1. Survey of the test conditions and influences of the precipitation sludges on the anaerobic stabilization

	O'Shaughnessy et al. [9]	Mosebach [7]	Author Hierse [6]	Peschen [10]	Fayoux [1]
Scale	Full-scale	Full-scale batch tests	Batch tests	Full-scale	Semitechnical scale
Precipitant	$Na_2Al_2O_4$ $Al_2(SO_4)_3$	Fe^{3+} $\beta = 2$ to 9	$FePO_4$ added	$Ca(OH)_2$	$FeCl_3$ $(Fe(OH)_3)$
Type of precipitation/flocculation	Simultaneous	Simultaneous	--	Pre-precipitation	Pre-precipitation
Digestion time [d]	19	--	Max. 36	--	20
Phosphate re-dissolution	No	No [1]	Scarely [2]	--	--
Sulphate influence	No	--	Yes	--	--
Decrease in gas production	No	No	No [3]	--	No [4]
Precipitant re-dissolution	No	--	--	--	--
Comments	[5]	--	--	[6]	--

[1] However low P re-dissolution in batch tests
[2] At $\beta \geq 1.5$ negligible
[3] At $FePO_4$ proportion < 10% of the total solid content of the dosed mixed sludge
[4] At Fe proportion < 10% of the solid content of the dosed sludge
[5] Test period 19 months altogether
[6] Only data about the increase in solids, or rather, the improvement in dewaterability

precipitation, with a mole ratio in the sludge from $\beta = 2$ to 9. An interference with the digestion processes (gas yield, pH) was not observed. Hierse [6] determined from digester-batch tests that sludges containing iron phosphate, as are produced in simultaneous precipitation, do not negatively influence sludge digesting.

An inhibition of the digestion process first occurs at shares of $FePO_4$ in the total residue of the supplied sludge that are larger than 10%. At 20% $FePO_4$ an inhibition of the gas production could be observed which lasted almost 2 weeks.

During normal digestion times of more than 20 days only a small amount of re-dissolution was observed. This re-dissolution can, however, be neglected at mole ratios greater than $\beta = 1.5$. The phosphate concentration in the liquid phase of the digester content always stayed below the concentration that would have been theoretically possible due to the reduction of the iron in the ferric phosphate. According to the author this is due to the formation of released phosphates on the still available metal ions like, for instance, Fe^{2+}, Al^{3+}, Mg^{2+}, etc. The re-dissolution of the phosphorus during the anaerobic process is presumably mainly caused by the presence of organic acids. When the digestion process ist not working sufficiently a higher concentration of organic acids is observed. In Figure 4 the phosphate concentration in the liquid phase of the digester content is shown according to Hierse [6], as a function of the digestion time, i.e. as a function of the residual concentration of organic acids. One can see that, at too short digestion times of below 20 days,

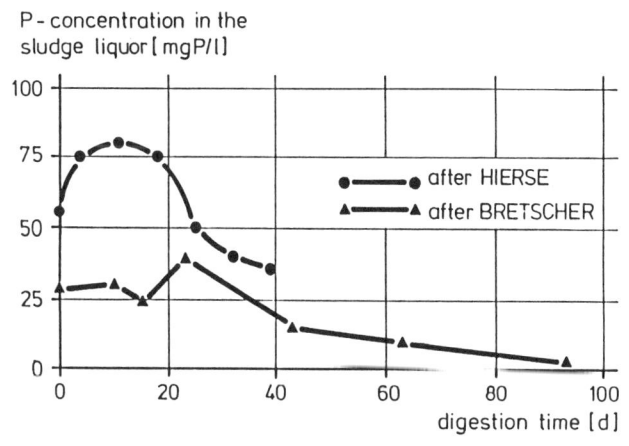

Figure 4. P-concentration in the digest sludge liquor versus digestion time

considerable concentrations of re-dissolved phosphorus are to be found. In such cases, a separate phosphate precipitation in the sludge liquor must be planned as long as metal salts are not used as a flocculant in mechanical sludge dewatering to achieve a precipitation of the re-dissolved phosphorus in the sludge liquor. According to Hierse [6], the re-dissolution of the phosphorus in the digester depends upon the concentrations of calcium ions and sulphite. An increase in the calcium concentration reduces the re-dissolution rate presumably because of the production of apatite. The calcium carbonate is slowly dissolved in the anaerobic tank due to the biologically generated CO_2 and is therefore available for the formation of apatite. This could be the reason why the proportion of bound phosphorus even increases during longer digestion times in Hierse's study (see Figure 5).

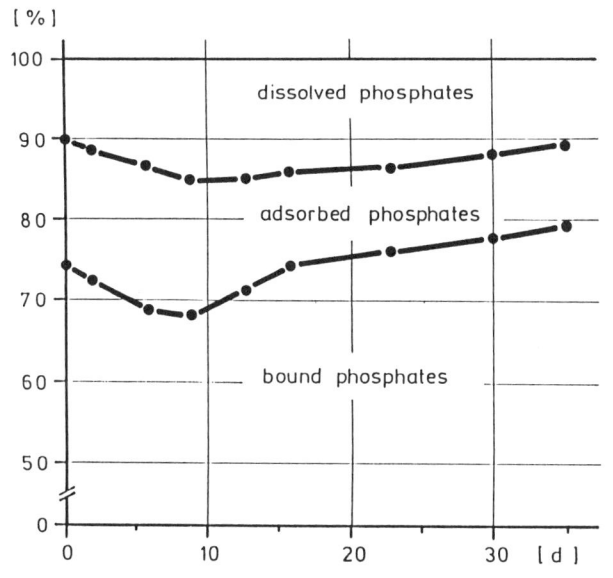

Figure 5. Change of the phosphorus components in digested sludge during the digesting process in weigth-% [6]

A higher concentration of sulphur can promote re-dissolution by means of the formation of sulphite. Both Hierse [6] and Schüßler [14] rule out an inhibition of the anaerobic process by phosphate. On the other hand, the precipitation of phosphate increases the dewaterability of the digested sludge. In any case the sludge liquor from the post-thickener and the sludge dewatering should be collected and the phosphates present in it should be precipitated out with lime or another precipitant in order to avoid an undesirable recycling of the phosphorus.

In contrast to the full-scale tests performed by O'Shaughnessy, Hierse found in his batch tests an obvious inhibition of gas production due to the increased sulphate concentration, or rather, due to the sulphide which arose out of the sulphate. Those results are not astonishing since in batch tests the anaerobic bacteria do not have enough time to adapt. In our study the anaerobic bacteria showed a good ability to adapt to H_2S.

Fayoux [1] wrote the most recent publication about iron containing sludges in anaerobic stabilization. He examined pre-precipitation sludge which was precipitated using $FeCl_3$ and mixed sludge which was provided with iron hydroxides. No influence on the anaerobic process was detectable with either sludge when the Fe-portion was less than 10% (relative to the solid content of the supplied sludge) and digestion time was below 20 days. The Fe-concentration was even 18% in the pre-precipitation sludge, where inhibition occurred. Typical Fe-concentration lies between 5 and 6% – relative to solid content – in anaerobic tanks at sewage treatment plants with simultaneous precipitation. The amount of dissolved iron in the digester lies between 50 and 80 mg/l in pre-precipitation sludge while in iron hydroxide mixed sludge it falls between 200 and 300 mg/l. The author suspects that this higher amount of dissolved iron is the cause for an inhibiting effect on the anaerobic process.

Peschen [10] reports about his good experiences with sludge from a pre-precipitation with lime; especially the dewaterability of the digested sludge is improved.

3.3 Practical Results

Mertsch et al. [19] performed a comprehensive investigation of sludge treatment in 148 municipal sewage treatment plants with precipitation stages. More than half of the plants treated the precipitation sludge together with the other sludges in one digester. In none of the plants negative effects from the precipitation sludge on the digestion process were reported. The specific gas yield lay at an average of 17.1 l/(population equivalent · d) and thus within the range of the gas yield of other municipal sewage treatment plants. In our own study in a municipal sewage treatment plant gas production did not change either when ferric chloride or ferrous sulphate were used as precipitants [17].

3.4 Excess Sludge from Biological P-elimination

Concrete studies on the properties of excess sludge from plants with biological P-elimination do not exist yet. Sekoulov's [15] opinion is that all phosphate in a digester gets dissolved again. Rensink et al. [12] assumes that approx. 1/3 of the phosphates in the sludge remain and 2/3 are re-dissolved. Our studies of the P-balance of a sewage treatment plant with digester and biological P-elimination are not finished yet. Experiments with lime precipitation of dissolved phosphate in the digested sludge or in the sludge liquor showed that the lime consumption was not only higher in the precipitation of digested sludge than of the liquid phase of the digester content but also that blockages in the dewatering centrifuges were to be found.

Precipitation in sludge liquor produced the results shown in Figure 6; just the pH value is relevant and there, only a value of over 11 pH leads to a safe and complete P-precipitation. Only the part of lime that is soluble in water is important for the alkalinization. The proportion of water soluble CaO can be calculated approximately by using the following formula: (100 – stated CaO content) x 2. This product, when subtracted from 100 gives the approximate water soluble proportion.

Example: (100 - 88) x 2 = 24; 100 - 24 = 76% CaO soluble in water.

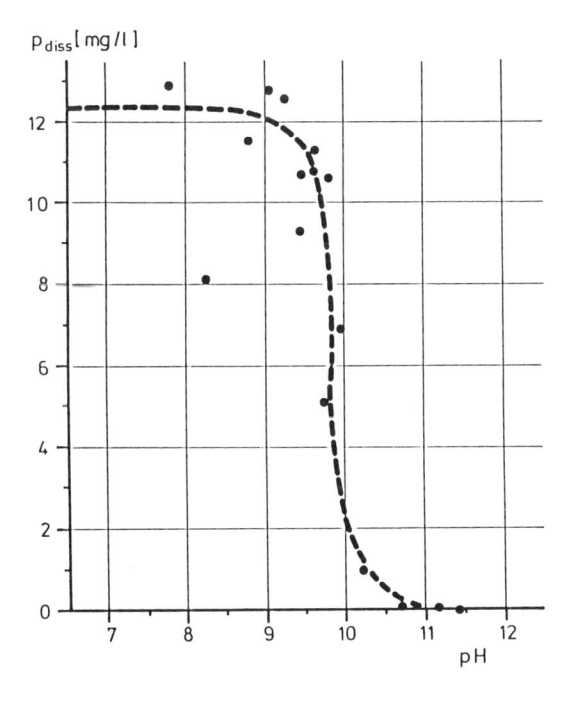

Figure 6. P-precipitation tests with digested sludge liquor by lime

4 Summary

The chemical precipitation of phosphates in municipal wastewaters is a process proven world-wide; biological P-elimination is also being employed more and more so that meanwhile here too results from practical experience with large plants are available. The effects of the P-elimination on nitrification and denitrification are often taken too little into account. The effects of the P-precipitation on the sludge age, the acid capacity and the "activity" of the nitrifying bacteria are presented. The biological P-elimination with a supporting simultaneous precipitation seems to be the most appropriate solution, when nitrification and denitrification are both necessary at the same time.

To summarize, the following statements about the properties of the sludges from P-precipitation and biological elimination can be made:

— · An additional phosphate re-dissolution due to the use of chemical precipitation cannot be confirmed at normal dosages. However, in anaerobic treatment plants a phosphate re-dissolution of more than 60% must be expected out of the sludge from biological P-elimination.
— An inhibition of the anaerobic digestion process, together with a lowered gas yield, could not be found at normal dosages of precipitant. Inhibition begins only at Fe-concentrations greater than 10% of the solid content of the supplied mixed sludge.
— The use of a precipitant containing sulphate for more than a couple of months did not affect the anaerobic digestion process in tests on a technical scale. This result was confirmed in practice. The sulphur content in the digester gas may however slightly increase.
— The digestion process with the addition of sludges from the precipitation is as stable as the anaerobic digestion process with solely municipal sludge.
— Apparently, according to a statistical analysis (Mertsch et al. [19]), the additional sludge in the digester also does not exert as strong an influence as generally thought.

References

[1] Fayoux, CH. (1984): Study of the Anaerobic Digestion of Sludge Containing Iron Hydroxides. 6. Europäisches Abwasser- und Abfallsymposium München 1984, S. 445-463
[2] Gujer, W., Boller, M. (1979): Der Einfluß der chemischen Flockung und Fällung auf das Belebtschlammverfahren. Schriftenreihe d. Inst. f. SWW d. Universität Karlsruhe 20
[3] Hahn, H.H. (1983): Schlämme aus der Abwasserfällung/-flockung. Schriftenreihe d. Inst. f. SWW d. Universität Karlsruhe 32
[4] Hahn, H.H. (1987): Wassertechnologie — Fällung, Flockung, Separation. Springer, Heidelberg
[5] Hahn, H.H. (1988): Schlamm aus der Fällungs- und Flockungsstufe. Schriftenreihe des Fachgebietes Siedlungswasserwirtschaft 2, Universität Kassel
[6] Hierse, W. (1982): Untersuchungen über das Verhalten phosphathaltiger Schlämme unter anaeroben Bedingungen. Inst. f. Wasserversorgung, Abwasserbeseitigung und Raumplanung d. TH Darmstadt, WAR 11
[7] Mosebach, K.H. (1975): Phosphatrücklösung bei der Ausfaulung von Simultanschlämmen. Wasser und Abwasser in Forschung und Praxis 11
[8] Nyhuis, G. (1985): Beitrag zu den Möglichkeiten der Abwasserbehandlung bei Abwässern mit erhöhter Stickstoffkonzentration. Veröffentl. d. Inst. f. Siedlungswasserwirtschaft u. Abfalltechnik d. Universität Hannover 61
[9] O'Shaughnessy, J., Nesbitt, J.,Long, D., Kountz, R. (1974): Digestion and dewatering of phosphorus enriched sludges. Journal Water Pollution Control Fed. 76, p. 1.915

[10] Peschen, N. (1983): Auswirkung der Kalkfällung auf die Entwässerung und Verwertung der Schlämme. Schriftenreihe d. Inst. f. SWW d. Universität Karlsruhe 32

[11] Peter, A., Sarfert, F. (1983): Praktische Betriebserfahrungen bei der Behandlung von Schlämmen aus der Simultanfällung/-flockung. Schriftenreihe d. Inst. f. SWW d. Universität Karlsruhe 32

[12] Rensink, J.H., Donker, J.J. (1983): Biologische Phosphorelimination aus Abwasser. GWF-Wasser/Abwasser 125, Heft 5

[13] Sampson, G. (1983): Auswirkungen der Fällung und Flockung auf den Schlammanfall und die Kosten der Schlammbehandlung. Berichte aus Wassergütewirtschaft und Gesundheitsingenieurwesen d. TU München 41

[14] Schüßler, H. (1982): Phosphatelimination in kommunalen Kläranlagen — Technik und Kosten. Veröffentlichungen d. Inst. f. Siedlungswasserwirtschaft u. Abfalltechnik d. Universität Hannover 49

[15] Sekoulov, J. (1987): Verfahren zur biologisch-chemischen Phosphatentfernung in Kombination mit Nitrifikation und Denitrifikation. Hoechst-Symposium, Berichtsband Knapsack

[16] Seyfried, C.F., Nyhuis, G. (1985): Untersuchungen zu Nitrifikation/Denitrifikation auf der Kläranlage Gümmerwald/Hannover. Forschungsbericht d. Inst. f. Siedlungswasserwirtschaft u. Abfalltechnik d. Universität Hannover

[17] Seyfried, C.F., Schüßler, H. (1980): Anfall und Behandlung von Schlämmen aus der Fällungsreinigung. Forschungsbericht d. Inst. f. Siedlungswasserwirtschaft u. Abfalltechnik d. Universität Hannover

[18] Seyfried, C.F. (1988): P-Fällung — Regeln der Technik und Kosten. Schriftenreihe des Fachgebietes Siedlungswasserwirtschaft 2, Universität Kassel

[19] Mertsch, V., Klute, R., Hahn, H.H. (1984): Schlamm aus der Abwasserfällung/-flockung. KA 31, Heft 11, S. 920-926

C.F. Seyfried, H.-D. Kruse,
F. Schmitt
Institut für Siedlungswasserwirtschaft
und Abfalltechnik
Universität Hannover
Welfengarten 1
3000 Hannover
Fed. Rep. of Germany

Pretreatment of Sludge Liquors in Sewage Treatment Plants

B. Paulsrud, B. Rusten and R. Storhaug

Abstract

Recycling of liquors from sludge processing facilities can sometimes create serious problems in sewage treatment plants. A literature review has shown that separate treatment of sludge liquors is feasible and can be a solution to the problems that occur. Norwegian data on pretreatment of sludge liquor from primary-chemical sewage treatment plants receiving septage are presented. Both the activated sludge process and the rotating biological contactor have been tested, and these processes can produce an effluent quality which causes no harm when returned to the plant inlet.

Introduction

Performance problems at sewage treatment plants are often attributed to the recycling of sludge liquors (sidestreams) generated in the solids treatment facilities (i.e. supernatant from digesters, thickeners and holding tanks or filtrate/centrate from dewatering equipment). Although the flows of these sidestreams generally are small compared to plant influent flows, sludge liquors can contain significant concentrations of organic and inorganic materials, either in solution or in suspended solids form (Holmstrøm [1]). Recycling of these sidestreams within the treatment plant without regard to their effects can cause process disturbances, violation of discharge limits, release of odours, and increased operation and maintenance costs.

Several general approachs to prevent or solve problems that may result from sludge processing liquors can be identified:

– Modify solids treatment and disposal systems to eliminate particular sludge liquors.
– Modify previous solids processing steps to improve liquor quality from a particular solids treatment process.
– Change the timing, return rate, or return point for reintroducing sludge liquors into the wastewater process.
– Modify wastewater treatment facilities to accommodate sludge liquor loadings.
– Provide separate sludge liquor treatment prior to recycling.

Separate treatment of anaerobic digester supernatant and thermal conditioning liquor have been practiced for many years (EPA [2]). Keefer and Kratz [3], Rudolfs and Fontenelli [4] and Howe [5] studied chemical treatment of digester supernatant using ferric chloride, lime, caustic soda, sulfuric acid, chlorine, bentonite clay and zeolite. Malina and Di Filippo [6] proposed a treatment scheme for digester supernatant including CO_2-stripping (by air), lime/ferric chloride coagulation plus settling followed by ammonia stripping.

Straight aeration of digester supernatant at plant scale has also been attempted [7, 8, 9]. Even where the supernatant after aeration was not settled prior to return, it was found that wastewater treatment operation improved, probably as a result of better settling in the primary clarifiers. Aerobic biological filters appear to be feasible methods of biologically treating digester supernatant. The Greater London Council studied aerobic biofilter treatment of supernatant liquor using coke as the filter medium [10]. At a 1:1 dilution with clarified plant effluent, 85 to 90 percent BOD_5 removal and 60 percent ammonia removal were obtained.

The Metropolitan Sanitary District of Greater Chicago has conducted several investigations involving nitrification and nitrogen removal from digested sludge lagoon supernatant, using both attached and suspended growth biological processes [11, 12].

In Sweden, Tendaj-Xavier [13] has performed nitrification studies of digester supernatant using an activated sludge plant. About 60 percent ammonia removal was obtained. The Swedish Water and Wastewater Works Association (VAV) has presented a control manual for sludge treatment processes, and this publication also includes several suggestions for separate treatment of digester supernatant [14].

Thermal conditioning liquors are even more troublesome to handle than the digester supernatant due to a significant fraction of nonbiodegradable COD. Erickson and Knopp [15] used the activated sludge process for heat treatment liquor. They reported a COD reduction of 83% (2,000 mg/l of COD still remaining in the effluent) and a BOD_5 reduction of 98% with an aeration time of 41 hours. Anaerobic biological filtration of heat treatment liquor has been tested for use at the City of Los Angeles Hyperion wastewater treatment plant [16]. Results of a two-month test at a hydraulic detention time of 2 days, showed average BOD_5 and COD removals of 85 and 76%, respectively. Also chlorine oxidation [2] and aerobic, thermophilic digestion [17] has been studied for the treatment of thermal conditioning liquor.

In Norway several primary-chemical sewage treatment plants are receiving septage (septic tank pumpings) in their sludge handling systems. This practice increases the sludge liquor flows and reduces their quality, thus creating an adverse impact on the coagulation/flocculation process when recycling the liquor to the plant inlet.

A research program was conducted during 1983–85 to cope with these problems, and most of the work was devoted to separate pretreatment of the sludge liquor. Both the rotating biological contactors and the activated sludge process were studied in pilot plants, and a full scale demonstration project was run for 6 months with a batch operated activated sludge process.

Treatment in Rotating Biological Contactors

The treatment of sludge liquor from septage dewatering was studied by Rusten [18] in two pilot-scale RBCs, one operating at 5.4°C and the other at 14.5°C. Both RBCs consisted of 4 stages and tests were run for approx. 2 months.

In addition the effects of recycling both untreated and RBC-treated septage liquor into a coagulation/flocculation process were studied in jar tests using alum as the coagulant. The septage liquor came from a filter-belt press and was less concentrated than typical septage liquor due to dilution by wash water from the press. The characteristics of the three batches of septage liquor used during the test period are summarized in Table 1.

Table 1. Characteristics of the three batches of septage liquor used for feeding the pilot scale RBCs (mean values) [18]

Parameter	Batch 1	Batch 2	Batch 3
Suspended solids [mg/l]	775	230	877
COD total [mg/l]	1941	729	2402
COD filtered [mg/l]	452	324	506
BOD$_7$ total [mg/l]	-	29	652
Total P [mg/l]	22.2	11.8	24.7
Total N [mg/l]	104	57	132
NH$_4$-N [mg/l]	50	27	62
Alkalinity [meq/l]	-	4.0	6.2
pH	7.1	7.4	7.2

Figure 1 shows total organic removal rates versus total organic loads for the two RBCs. Most of the reduction of organic matter in the RBC systems took place in the first stage. Removal rates as high as 80 – 90 g COD$_T$/m²d were observed at 14.5°C. A maximum removal efficiency of about 85% total COD was found at a temperature of 14.5°C and an organic load of about 80 g COD$_T$/m²d.

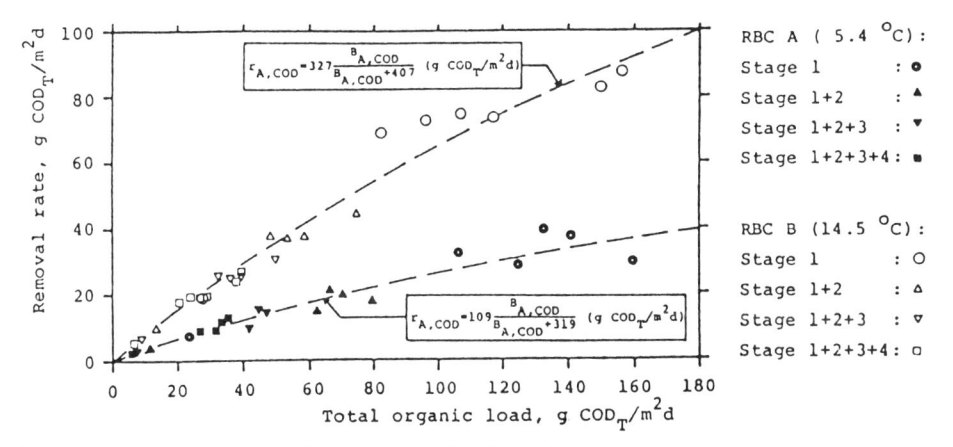

Figure 1. Separate treatment of septage liquor in pilot-scale RBCs. Shows total organic removal rates versus total organic loads, based on data from all 4 stages. $r_{A,COD}$ = organic removal rate, $B_{A,COD}$ = organic load [18]

Based on the removal rates of total COD at 5.4°C and 14.5°C, respectively, a temperature coefficient (θ) of 1.10 – 1.11 was found.

The results indicated simultaneous nitrification and denitrification in the RBC unit operated at 14.5°C. The low temperature RBC unit showed no sign of nitrification.

Jar-tests showed that nitrification of the septage liquor was required to reduce the consumption of alum for phosphorus removal.

Low supernatant phosphorus and suspended solids concentrations could be obtained in the jar-tests with both treated and untreated septage liquor mixed with raw sewage, provided sufficient alum was added.

Previous RBC treatment was, however, essential for attaining a good quality supernatant with respect to COD. High rate treatment in a RBC unit (<30% removal of COD_T) was sufficient to have a very positive effect on the residual concentration of total and filtrable COD after mixing with municipal wastewater and precipitation with alum.

Treatment with the Activated Sludge Process

Pilot Scale Tests

Six continuously operated activated sludge pilot plants were run in parallel to treat septage liquor (Storhaug [19]). Three plants were operated at 8°C and the other three at 16°C. For each temperature three different organic loadings were used. The septage liquor was taken from centrifuge dewatering of septage at a fullscale sewage treatment plant. This liquor was more concentrated than the liquor used in the RBC experiments (see Table 2). In this project as well, the effects of the activated sludge treatment on the coagulation/flocculation process were studied in jar tests.

Table 2. Characteristics of septage liquour used in pilot scale activated sludge plants [19]

Parameter	No. of samples	Mean	Standard deviation
Suspended solids [mg/l]	5	2334	547
COD total [mg/l]	6	5417	928
COD filtered [mg/l]	6	1172	485
TOC [mg/l]	3	594	129
TOC filtered [mg/l]	6	137	42
BOD_7 total [mg/l]	6	1875	435
Total P [mg/l]	6	58.7	8.2
PO_4-P [mg/l]	6	39.2	10.9
Alkalinity [meq/l]	6	17.8	4.8
pH	6	8.09	0.14

Data from all six pilot plants are summarized in Table 3.

Table 3. Organic loadings, concentrations and removal rates for the treatment of septage liquor in six parallel activated sludge pilot plants (mean values) [19]

	Pilot plant					
	1	2	3	4	5	6
Temperature [°C]	8	8	8	16	16	16
F/M-ratio [kg BOD_7/kg MLVSS d]	0.41	0.27	0.13	0.47	0.28	0.15
Sludge age [days]	1.8	2.8	7.2	2.0	2.9	12.3
BOD_7						
In [mg/l]	1875	1875	1875	1875	1875	1875
Out [mg/l]	421	101	65	115	112	51
Removal [%]	78	95	97	94	92	97
COD						
In [mg/l]	5417	5417	5417	5417	5417	5417
Out [mg/l]	1510	596	475	715	710	476
Removal [%]	72	89	91	87	97	91
Alkalinity						
In [meq/l]	17.8	17.8	17.8	17.8	17.8	17.8
Out [meq/l]	14.5	9.8	4.7	11.8	8.3	1.3
Removal [%]	19	45	74	34	53	93

The pilot plant operated at 8°C and with a F/M-ratio of 0.41 kg BOD_7/kg MLVSS·d was obviously overloaded. Otherwise the pilot plants performed excellently at both temperature levels. Plant no. 6 experienced stable nitrification throughout the whole test period, and this resulted in very low effluent alkalinity. Plant no. 3 also attained nitrification during the first part of the experiments.

The jar tests with both treated and untreated septage liquor, mixed with municipal wastewater, documented that the alkalinity of the liquor determined the required alum dosage for satisfactory phosphorus removal. This nitrification of the septage liquor is favourable in order to reduce the consumption of coagulants, when such liquor is recirculated in a primary- chemical sewage treatment plant.

When mixing nitrified septage liquor with plant influent, the necessary alum dosage will be determined by the phosphorus concentration of the sludge liquor.

Similar to RBC treatment, pretreatment of septage liquor in the activated sludge process significantly reduced the residual concentrations or organic matter after mixing with municipal wastewater and precipitation with alum.

Full-Scale Demonstration Project

The demonstration project was carried out with an activated sludge unit operated as a sequencing batch reactor and located at Sørumsand sewage treatment plant (Paulsrud & Nedland [20]). This is a secondary precipitation plant (using alum) designed for 6,000 pe., but with an actual load of about 2,900 pe.

Septage is discharged to the main sewer a short distance upstream from the treatment plant. The volume of septage discharged on a daily basis is equivalent to 5 – 10% of the incoming sewage.

All the sludge produced at the plant is dewatered in a centrifuge. During the 6 months' test period, the sludge liquor (centrate) was collected in a split box and a certain amount of sludge liquor was pumped to the batch activated sludge plant.

Figure 2 shows the schematic diagram for this process. The activated sludge unit had a total volume of 30 m^3. Every morning a maximum of 17.5 m^3 of treated sludge liquor was pumped out of the tank, and replaced with an equivalent volume of untreated liquor. Air was supplied through coarse bubble aerators at the bottom of the tank.

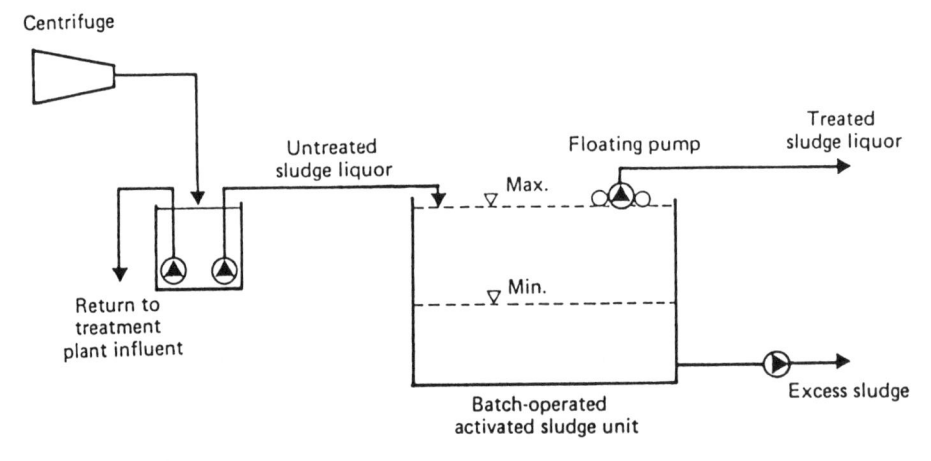

Figure 2. Schematic diagram of a full-scale, batch-operated, activated sludge system for separate treatment of sludge liquor [20]

Sludge dewatering units at Norwegian treatment plants are normally in operation only during regular working hours, from Monday to Friday. Thus, we had an ideal situation for batch operation of the activated sludge unit. A regular cycle was like this:

— Aeration was stopped early in the morning, and the sludge was allowed to settle for 1 – 1.5 hours.
— Treated sludge liquor was pumped to the treatment plant influent.
— Aeration was resumed and the tank was filled with untreated sludge liquor.
— Aeration was continued until the following morning.

On Saturdays and Sundays the activated sludge unit was aerated all day. The process cycle was controlled by two simple timers.

Results from a one month period with intensive monitoring are shown in Table 4. During this period the aeration tank had a pH of 6.8 – 7.2 and a temperature of 12 – 13°C. The organic loading (F/M-ratio) was between 0.049 and 0.137 kg BOD$_7$/kg MLVSS·d, with a mean value of 0.094 kg BOD$_7$/kg MLVSS·d. A COD/BOD$_7$-ratio of 1.8 was found for untreated sludge liquor. Treated sludge liquor was not analysed for BOD.

Table 4. Performance data for activated sludge treatment (batch operation) of sludge liquor [20]

Parameter	Influent		Effluent		Removal efficiency [%]	
	Range	Mean	Range	Mean	Range	Mean
COD [mg/l]	990 - 3050	1539	62 - 760	244	63 - 96	86
SS [mg/l]	688 - 1403	1407	110 - 640	319	46 - 88	70
Total N [mg/l]	68 - 237	110	20 - 111	60	1 - 73	45
Total P [mg/l]	6.0 - 22.9	16.5	0.1 - 10.9	3.4	17 - 99	76
PO_4-P [mg/l]	1.1 - 2.4	1.8	0.1 - 1.0	0.4	29 - 95	78
Alkalinity [meq/l]	2.8 - 17.2	4.7	1.0 - 8.0	3.1	---	33

In general the removal of organic matter was very good. A fairly good reduction of suspended solids was also obtained. At the low F/M-ratios used in this study nitrification took place. The data are not consistent, but it is assumed that some of the nitrogen was used for cell synthesis, and that some was removed by nitrification/denitrification due to the batch operation of the activated sludge process.

Previous pilot-scale experiments have shown that the alkalinity and the phosphorus concentration determine the chemical addition required for phosphorus removal. Table 4 shows that the alkalinity was reduced, but only by 33% on an average basis. However, the removal of phosphorus was excellent. The removal of particulate phosphorus can be attributed to the reduction in suspended solids. The extremely low concentration of soluble phosphorus may have been caused by precipitation, due to the fraction of chemical sludge in the sludge mixture.

References

[1] Holmstrom, H. (1985): Sludge Handling in Sewage Treatment Plants. National Swedish Environment Protection Board, PM 1955 (in Swedish)
[2] U.S. Environmental Protection Agency (1979): Process Design Manual for Sludge Treatment and Disposal. EPA 625/1-79-011
[3] Keefer, C.E., Kratz, H. Jr. (1949): Treatment of Supernatant Sludge Liquor by Coagulation and Sedimentation. Sewage Works Journal *12*, p. 738
[4] Rudolfs, W., Fontenelli, L.S. (1945): Supernatant Liquor Treatment with Chemicals. Sewage Works Journal *17*, p. 538
[5] Howe, R.H. (1959): What to do with Supernatant. Wastes Engineering *30*, p. 12
[6] Malina, J.F. and DiFilippo, J. (1971): Treatment of Supernatant and Liquids Associated with Sludge Treatment. Water Sewage Works, Reference Number, p. R-30
[7] Kappe, S.E. (1958): Digester Supernatant: Problems, Characteristics, and Treatment. Sewage and Industrial Wastes *30*, p. 937
[8] Erickson, C.V. (1945): Treatment and Disposal of Digestion Tank Supernatant Liquor. Sewage Works Journal *17*, p. 889
[9] Anonymous (1943): The PFT Supernatant Liquor Treater. Sewage Works Journal *15*, p. 1018
[10] Brown, B.R., Wood, L.B., Finch, H.J. (1972): Experiments on the Dewatering of Digested and Activated Sludge. Water Pollution Control *71*, p. 61

[11] Lue-Hing, C., Obayashi, A.W., Zenz, D.R., Washington, B., Sawyer, B.M. (1974): Nitrification of a High Ammonia Content Sludge Supernatant by Use of Rotating Discs. Presented at the 29th Annual Purdue Industrial Waste Conference, West Lafayette, Indiana, p. 245

[12] Prakasam T.B.S., Robinson, W.E., Lue-Hing, C. (1974): Nitrogen Removal from Digested Sludge Supernatant Liquor Using Attached and Suspended Growth Systems. Presented at the 32nd Annual Purdue Industrial Waste Conference, West Lafayette, Indiana, p. 745

[13] Tendaj-Xavier, M. (1985): Biological Treatment of Anaerobic Digested Sludge Liquor. Avd. for Vattenvårdsteknik, KTH (in Swedish)

[14] Swedish Water and Wastewater Works Association (1986): Control of Sludge Treatment Processes. Publication VAV P 61 (in Swedish)

[15] Erickson, A.H., Knopp, P.V. (1970): Biological Treatment of Thermally Conditioned Sludge Liquors. Proceedings of the 5th International Water Pollution Control Research Conference, San Francisco, Vol II, p. 30

[16] Haug, R.T., Raksit, S.K., Wong, G.G. (1977): Anaerobic Filter Treats Waste Activated Sludge. Water & Sewage Works, p. 40

[17] Loll, U. (1977): Treatment of Thermally Conditioned Sludge Liquors. Water Research *11*, pp. 869-872

[18] Rusten, B. (1983): Separate Treatment of Septage Liquor Using Rotating Biological Contactor. Report 9/84, NTNF's Program for VAR-teknikk, Trondheim (in Norwegian)

[19] Storhaug, R. (1984): Separate Treatment of Septage Liquor with the Activated Sludge Process. Report 14/84, NTNF's Program for VAR-teknikk, Trondheim (in Norwegian)

[20] Paulsrud, B., Nedland, K.T. (1985): Separate Treatment of Sludge Liquor in a Batch Operated Activated Sludge Plant. Report 35/85, NTNF's Program for VAR-teknikk, Trondheim (in Norwegian)

B. Paulsrud, B. Rusten,
R. Storhaug
Aquateam - Norwegian Water
Technology Centre A/S
P. O. Box 6593 Rodelokka
0501 Oslo 5
Norway

Heavy Metal Removal from Sewage Sludge: Practical Experiences with Acid Treatment

M. Ried

Introduction

Sewage sludge often contains a certain amount of heavy metals, which reduce its value as a soil-conditioner or as a fertilizer. According to several laws, sewage sludge must not be used on agricultural soils any more, but has to be landfilled or incinerated if a certain heavy metal level is exceeded. Considering the amount of sewage sludge being produced yearly, this may cause economical and ecological problems. Since it is not possible to reduce the load of heavy metals before the influent arrives at the sewage treatment plant, their removal from the sludge has to be discussed.

Development

There have been several approaches concerning methods of heavy metal removal from sewage sludge in the past. Most widely discussed are acid treatment methods [1 – 6], and in the past few years, biological techniques have also been investigated [7, 8]. Since heavy metals in sewage sludge are associated with the solid sludge phase and are usually found as precipitates or adsorbed onto sludge particles, they first have to be transferred from the solid into the liquid phase and subsequently the liquid has to be separated from the solid phase. This is the reason why ion-exchange or electrolytical techniques are not directly applicable in this case. The principle of the acid treatment method was explained by Müller [9, 10] during the 2nd Gothenburg Symposium 1986; he proposed an acidification of dredged materials and sludge with hydrochloric acid at pH values around 0.5. The acid mixture is then separated in centrifuges and the centrifuge supernatant is precipitated with lime at pH = 10.

Experimental

Utilizing this principle, pilot scale experiments were carried out using various heavy metal-containing sludges. The experiments were adjusted especially for treatment of sewage sludge, and a flow chart of the process is given in Figure 1.

```
FROM DIGESTER
    |
    |     SEWAGE SLUDGE,
    |     HEAVY METAL POLLUTED
    |
[MIXING UNIT I ] <------ ACID
    |
    |
[MIXING UNIT II] <------ CONDITIONING
    |                    CHEMICALS
    |
[DEWATERING UNIT] ------> DEWATERED SLUDGE CAKE
    |                     (HEAVY METAL REDUCED)
    |     LIQUID PHASE,
    |     HEAVY METAL CONTAINING
    |
[HEAVY METAL    ] <------ LIME
[PRECIPITATION  ]
    |
    |
[DEWATERING UNIT] ------> HEAVY METAL
    |                     CONCENTRATE
    |     LIQUID PHASE,
    |     HEAVY METAL FREE
    |
TO SEWAGE TREATMENT PLANT
```

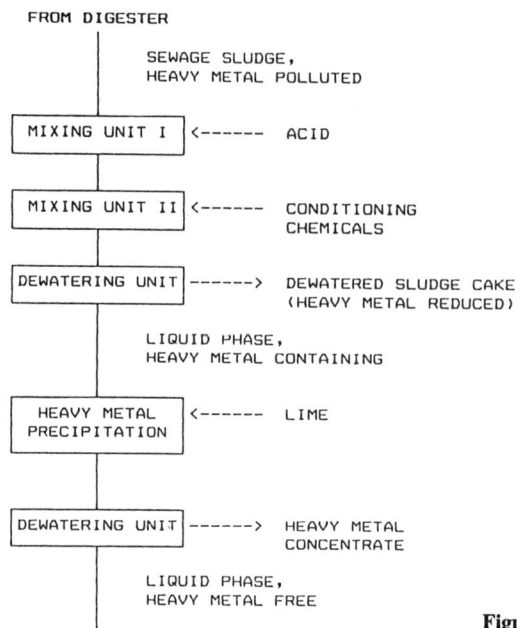

Figure 1. Flow chart of the acid treatment process for the removal of heavy metals

The idea was to collect data about the performance and the behaviour of such a process which could be transferred to fullscale operation, and to obtain information about chemical consumption and cost.

Acid Dosage

Choice of Acids

Laboratory scale experiments showed that mineral acids such as hydrochloric acid, sulfuric acid or nitric acid are suitable for removing heavy metals from sludge solids. It proved not to be possible to replace mineral acids by organic acids, because the strength of the organic acids is not sufficient to solubilize heavy metal compounds. There is no difference in the performance of the above-mentioned acids as seen in Figure 2. This figure exemplifies the change in Zn concentration in the liquid phase of the sludge after lowering the pH with either HCl, HNO_3 or H_2SO_4.

pH Values

As shown in Figures 2 and 3, the solubilization of heavy metals depends strongly on the pH, and it usually does not begin before pH = 4. Significant concentrations of heavy metals in the liquid phase require pH values of about 1. Adding acid until this pH is attained, about 80 – 95% of the nickel, cadmium and zinc will be transferred into the liquid sludge phase. The transfer of chromium and lead proved to be

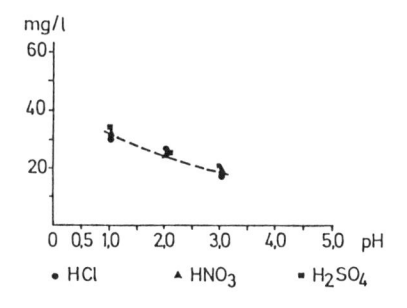

Figure 2. Effect of mineral acids on heavy metal release

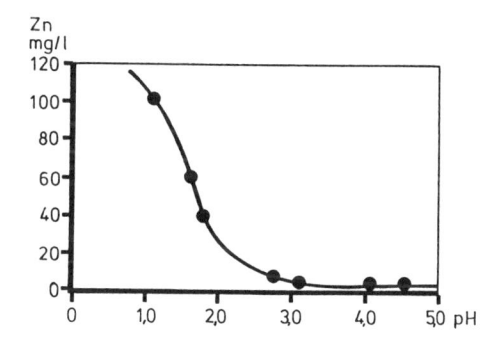

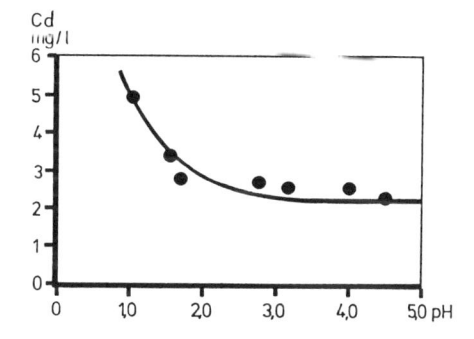

Figure 3. Zn and Cd solubilisation as a function of pH

poorer, but this might be influenced by the relatively low concentration of these elements in the sludges we investigated. There is always a certain background concentration of heavy metals in sewage sludge, which is strongly bound to the sludge solids and is thus called "residual phase" [11]. This residual phase requires HNO_3/HCl-digestion methods for liquification since it might be bound to the mineral structure of the sludge.

Acid Consumption

Lowering the pH to 1.0 requires a considerable amount of acid. Acid consumption depends on the solids content of the sludge to be treated and sludges with low water contents need less acid per unit dry matter to attain the same pH as more dilute sludges. Figure 4 shows the variation in the HCl requirements for acidifying

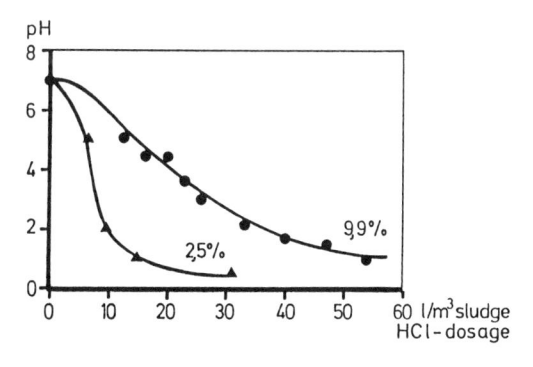

Figure 4. HCl requirements for sludges composed of 2.5 and 9.9% dry matter

anaerobically treated sewage sludges. The acid consumption during our experiments was between 0.3 and 1.2 l HCl per kg dry sludge material to adjust the pH to between 1.8 and 0.9. These figures are valid for sludges with solids contents of between 3 and 8%. Expressed as acid consumption per cubic meter liquid sludge, this comes to 20 to 55 l of concentrated HCl. Using H_2SO_4, the acid requirements are, of course, smaller and come up to approximately one half of the HCl amount.

Dewatering

The most important step after solubilizing the heavy metals is separating them together with the liquid phase from the sludge solids. Mechanical dewatering of sewage sludge is able to produce sludge cakes having a 20 – 50% solids content, which means that one half to 80% of the sludge cake still remains as liquid. This fact is of particular importance if this liquid phase contains heavy metals. For this reason, the most effective dewatering methods have to be applied, and therefore, we chose a filter press as a dewatering machine for our experiments.

Dewatering Characteristics

Digested sewage sludge which has been acidified to a pH of approx. 1 shows rather poor dewatering behaviour. In terms of CST values (capillary suction time), dewaterability behaves as in Figure 5, which shows that CST values depend on pH after dosing polymer.

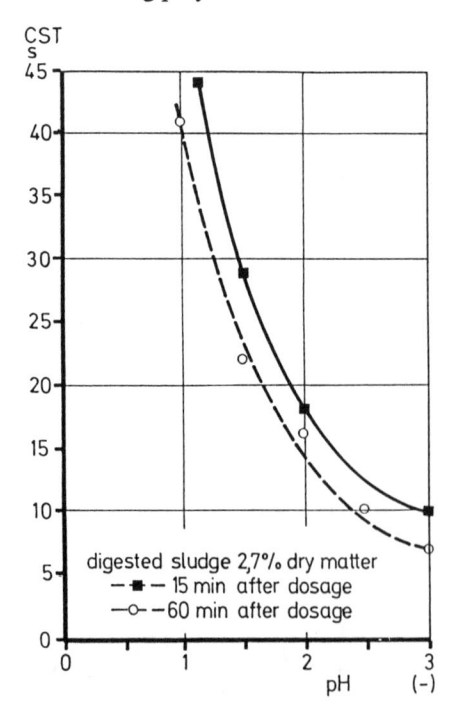

Figure 5. Capillary suction time of acidified sludge

In our experiments it was necessary to operate the filter press with sludges having CST values of less than 15 to achieve sufficient dewatering results. Since it was not possible to obtain such values in sludges having pH values of 1 or 2 after the addition of polymer, and since polymer-conditioned sludges are rather difficult to treat in a filter press, we had to look for other more suitable methods of dewatering the acidified sludge.

Experiments using conditioning materials such as sawdust or incinerator ashes did not show positive behaviour either.

Because of the fact that it was necessary to lower the pH to approx. 1 to release the heavy metals sufficiently, it was not possible to use the conventional lime and ferric salts dosage, since that causes a rise in pH with subsequent precipitation of the heavy metals. But since the heavy metals, once solubilized, do not precipitate before the pH rises above pH = 3, we succeeded in operating the filter press with a special combination of ferric salts and lime. Thus the solid content of the filter cake was between 35 and 50%.

During conditioning it was important not to raise the pH too much. Since ferric salts show an acid reaction in liquid solutions, it was possible to avoid an excessive rise in pH by alternately dosing lime and ferric salts.

With respect to filter press dewatering, the specific filtration resistance is a more suitable means of evaluating the dewaterability of sewage sludge. This parameter of chemical-treated sludge is shown as a function of pH in Figure 6. Successful filter press dewatering requires specific filtration resistances of about 20×10^{12} m/kg or less.

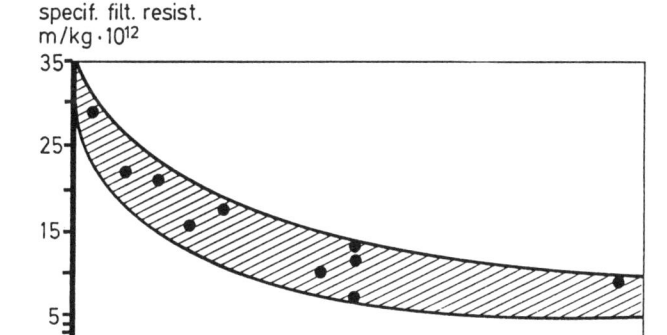

Figure 6. Specific filtration resistance of acidified sludge

To dewater acidified sludge, more chemicals are necessary than when the sludge has a normal pH. In our experiments, about 4 to 7 times the ferric chloride sulfate dosage was required to obtain sufficient dewaterability. The duration of the filtration process was also extended. While it takes about 50 minutes to dewater non-acid-treated sludge, 1 to 4 hours were the usual times required to dewater acidified sludge.

Release of Heavy Metals

Figure 7 shows metal concentrations in the effluent of the filter press as a function of pH. This shows that it is indeed possible to separate the heavy metals with the liquid phase as long as the pH is lower than 3 to 3.5.

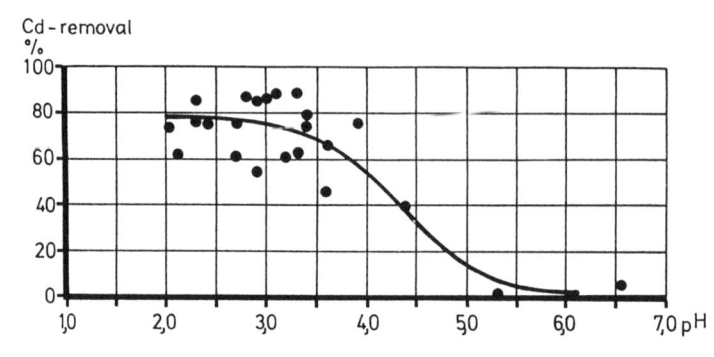

Figure 7. Heavy metal removal ratio as a function of pH

Table 1 shows the efficiency of heavy metal removal from the sludge. The process is especially effective fo Cd, Zn, and Ni, but the removal of Cu is weaker. As mentioned before, we did not investigate sludges with high levels of Pb and Cr in our pilot plant, but laboratory experiments showed that at least 50% of these elements could be removed. There is one exception in the performance of the process, and that concerns mercury. Especially in anaerobically digested sludges, mercury is bound as a sulfide which is an almost insoluble compound. Therefore, the process does not work with mercury-polluted sewage sludges.

Table 1. Efficiency of removing heavy metals [12]

Element	Concentration			Removal ratio %
	untreat. sludge mg/kg dry mat	dewatered cake mg/kg dry mat	filtrate mg/l	
Zinc	6916	646	250.0	90.7
Nickel	116	42	2.4	63.6
Cadmium	110	13.4	4.1	87.8
Copper	1299	652	42.9	49.8

Separation of Heavy Metals

The next step in the process is removing the heavy metals from the solution again. This was very easy to perform since lime precipitation turned out to work quite satisfactorily. For precipitation, the pH of the acid solution had to be adjusted to 7 to 11, depending on the heavy metals in the solution. The separation of the precipitate was also carried out in a filter press, which yielded a heavy metal concentrate having a solids content of more than 20%. The heavy metal content of the concentration and the supernatants is shown in Table 2.

Table 2. Heavy metal precipitation

Element	Concentration		
	acid filt mg/l	precipitate mg/kg dry mat	supernatant mg/l
Nickel	1.81	398.8	0.44
Cadmium	0.78	148.3	0.07
Zinc	71.9	19446	0.49
Chromium	0.67	120	0.3
Lead	5.18	51.9	1.28
Copper	12.5	47.1	0.22

Overall Effectivity

Using the acid treatment method, is is possible to remove heavy metals from sewage sludge to a certain extent. The sludge cake produced during the process is not completely heavy-metal-free, but contains a liquid and a solid acid-insoluble heavy metal residue. Before reusing the dewatered and heavy-metal-reduced sludge cake, it is advisable to neutralize it with lime or to replace the acid residue by rewashing the cake with water. In that case, the sludge cake may be recycled in agricultural soil without any problems. The supernatant from heavy metal precipitation should be returned to the sewage treatment plant influent, and the heavy metal concentrate can be dried and treated as special refuse. It will not be possible to reuse this concentrate in any way, but since the amount of concentrate produced during the treatment is less than 5% of the sewage sludge input, it will be easier to deal with. In any case, most of the sludge can be recycled and the amount of sludge refuse is reduced significantly.

Cost

Because of the high acid requirement and the conditioning chemicals, the acid heavy metal removal process is rather expensive. The most cost-effective part is acidification, so that up to now it was cheaper to landfill or to incinerate sewage sludge. To reduce cost, some approaches have been suggested to recycle the acid phase.

Conclusion

Adding hydrochloric acid to anaerobically digested sewage sludge causes at least 50 to 95% of the heavy metals to be transferred from the solid into the liquid phase. The separation of the sludge liquids by means of filter presses requires the addition of a specific combination of ferric salts and lime for conditioning. After dewatering, the sludge cake has a solids content of 35 to 50%. Reusing heavy-metal-released sludge cake on agricultural soil will be possible after neutralizing it. The process is rather expensive, but the amount of refuse to be deposited is reduced to about 5%.

Acknowledgements

This investigation was supported by the OSWALD-SCHULZE-STIFTUNG, D-4390 Gladbeck, West Germany.

References

[1] Wozniak, D.J., Huang, J.Y.C. (1982) J. Water Poll. Contr. Fed. *54*, 1574
[2] Jenkins, R.L., Scheybeler, B.J., Smith, M.L., Baird, R., Lo, M.P., Haug, R.T. (1981) J. Water Poll. Contr. Fed. *53*, 25
[3] Salotto, B.V. (1982): Engineering Assessment of Hot Acid Treatment of Municipal Sludge for Heavy Metals Removal. EPA Project Summary 600/S2-82-014
[4] Scott, D.S., Horlings, H. (1975) Env. Sci. Technol. *9*, 849
[5] Legret, M. Divet, L., Marchandise, P. (1987) Wat. Res. *21*, 541
[6] Scott, D.S. (1980): Removal and Recovery of Metals and Phosphates from Municipal Sewage Sludge. EPA Project Summary 600/2-80-037
[7] Schönborn, W., Hartmann, H. (1979) Gas und Wasserfach *120*, 329
[8] Schönborn, W., Hartmann, H. (1978) Europ. J. Appl. Microbiol. Biotechnol. *5*, 305
[9] Müller, G., Riethmayer, S. (1982) Chem. Zeit. *106*, 289
[10] Müller, G. (1986) in: Hahn, H.H., Klute, R., Balmér, P. (eds.): Recycling in Chemical Water and Wastewater Treatment. Schriftenreihe des ISWW Karlsruhe *50*, Karlsruhe
[11] Förstner, U., Calmano, W. (1982) JB vom Wasser *59*, 83
[12] Ried, M., Leonhard, K. (1987) in: Thomé-Kozmiensky, K.J. (ed.): Recycling von Klärschlamm. ef-Verlag, Berlin

M. Ried
Lehrstuhl für Wassergütewirtschaft
TU München
Am Coloumbwall
8046 Garching
Fed. Rep. of Germany

Treatment of Filter Effluents from Dewatering of Sludges by a New High Performance Flocculation Reactor

U. Wiesmann, K. Oldenstein and L. Fechter

Abstract

For the removal of turbidity and phosphorous, a flocculation aid must be added to turbid water. With current techniques, the formation of macroflocs requires between 10 and 60 min. Only a small area was available for the treatment of a filter effluent from dewatering of sludges produced by de-sludging of a lake. Therefore a cylindrical stirred tank was used that was characterized by the same narrow distribution of local velocity gradients and residence times as a new flocculation reactor. In order to achieve the water quality required for discharge into the lake (c_{ak}=2 FTU, $c(P_T)$=0.06 mg/l), a total mean residence time of only 14 min is sufficient (0.5 min for precipitation and coagulation in a tube reactor, 2 min for flocculation in the cylindrical stirred reactor and 11.5 min in a lamella separator). The filter effluent of about 60 m³/h was treated for nearly two years without any trouble.

Introduction

Around the turn of the century a new waterwork went into operation in the southern part of Berlin. The Schlachtensee, which together with the Krummen Lanke, Grunewaldsee and Hundekehlensee forms the Grunewald lake district, is situated within the region of the fall. In order to counteract a drop in the water level, water from the Havel River has been pumped into the Schlachtensee since 1920. During the last few decades the Havel became a eutrophic river with a phosphorous concentration around 1 mg/l. Because of the great importance of the Grunewald lake district as a recreation area for the inhabitants of Berlin (West), the water pumped into the Schlachtensee was treated by precipitation, flocculation, sedimentation and filtration since 1980 in order to remove phosphorous [1]. This procedure has also had a favourable effect on the water quality of the Grunewaldsee (Fig. 1).

However, the Grunewaldsee contains a layer of digested sludge several meters thick in places as a result of eutrophication. Therefore, a redissolution of phosphorous has to be exercised. Using a desludging process as a second part of the regeneration program of the Grunewald lake district, a sludge layer of 1.5 m is to be

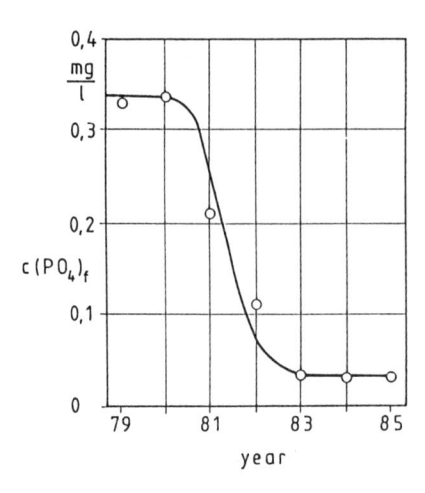

Figure 1. Annual mean ortho-phosphate concentration of the Grunewaldsee in Berlin (West)

removed in order to enlarge the water space of the lake with a maximal depth of only 3 m. In addition fish will have a better oxygen supply when a closed layer of ice covers the lake in winter.

Unfortunately, there are no sewers and no places for polders close to the lake. Therefore the sludge has to be dewatered by means of filter presses and the filtrate has to be treated to attain the water quality of the lake. Because only a small area of 30 m², was available for this purpose, a process with high efficiency was required for the removal of phosphorous and turbidity from the filtrate.

Bench Scale Investigation of a Cylindrical Stirred Tank for the Formation of Macroflocs by Flocculation

Purpose

The process for the removal of phosphorous and turbidity consists of three steps: the formation of microflocs by precipitation and/or coagulation, the formation of macroflocs by flocculation and the separation of the flocs from the water. In conventional tanks for flocculation mean residence times between 10 and 60 min are required, resulting in tank volumes up to 60 m³ for the treatment of 60 m³/h, for example.

How do the shape of the tank and the operating mode have to be changed in order to reduce the necessary tank volume?

First we have to consider that the flocculation rate increases with an increasing mean velocity gradient

$$G = \sqrt{\frac{P}{\eta V}} \tag{1}$$

- P is the power consumption caused by frictional forces of the flow
- η viscosity
- V volume of the tank

goes through a maximum and finally decreases as a result of floc break-up.

Now the first requirement can be formulated:

1) It must be possible to adjust the optimal mean velocity gradient, and the flocculation tank has to be designed in such a way that the local $\overline{G}$ value does not differ considerably from the mean value. This condition is not met in conventional flocculation units with turbine, gate impeller and paddle agitators. Moreover, single stage flocculation tanks show a wide residence time distribution. Small flocs leave the reactor after a very short residence time and large flocs stay there for too long a time without growth. Therefore, a second requirement has to be considered:

2) The flocculation tank has to show a narrow residence time distribution for all flocs.

These requirements prompted the development and testing of the cylindrical stirred tank as a flocculation reactor, because of its narrow local velocity gradient and residence time distributions [2, 3]. The stirred reactors employed consist of two coaxial cylinders, a stationary outer and a rotating inner one. Two secondary vortices already form at very low stirring speeds, and disintegrate into several smaller vortices, referred to as Taylor's vortices when the speed is increased. At increased stirring speeds, other flow patterns can be distinguished (Fig. 2a–f). For a continuous axial flow, a helical flow pattern appears which resembles a flow through a hose twisted around a cylinder.

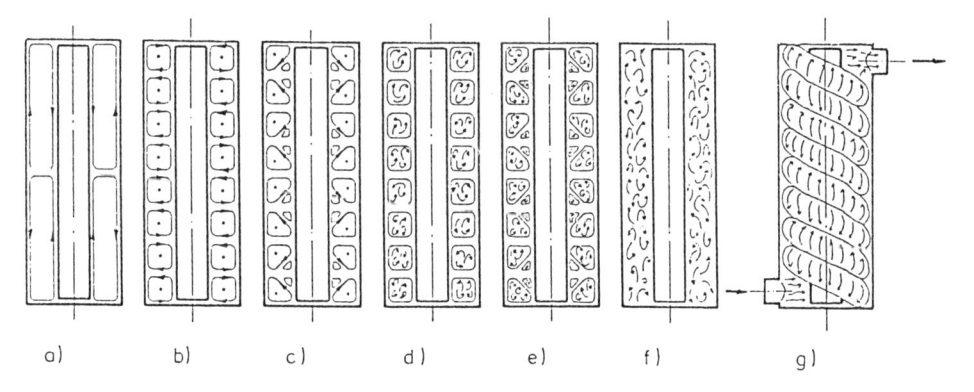

Figure 2. Flow patterns in the annular gap of a cylindrical stirred reactor; a) laminar flow, b) laminar-cellular flow without secondary vortices, c) laminar-cellular flow with secondary vortices, d) turbulent-cellular flow without secondary vortices, e) turbulent-cellular flow with secondary vortices, f) turbulent flow, g) flow pattern with vortices in a continuous flow

Experimental Plant

Figure 3 shows the laboratory set-up. Sewage plant effluent or tap water for producing a model colloid flows from a storage tank through a static mixer. Upstream of this mixer the coagulant or precipitator $FeCl_3$ is added, so that microflocs can be formed during the flow through a tubular reactor, which consists of a plastic hose with a diameter of 19 mm resulting in a mean velocity gradient of $\overline{G} = 500$ s^{-1}. After the addition of the flocculation aid Sedipur TF 2 TR, an anionic

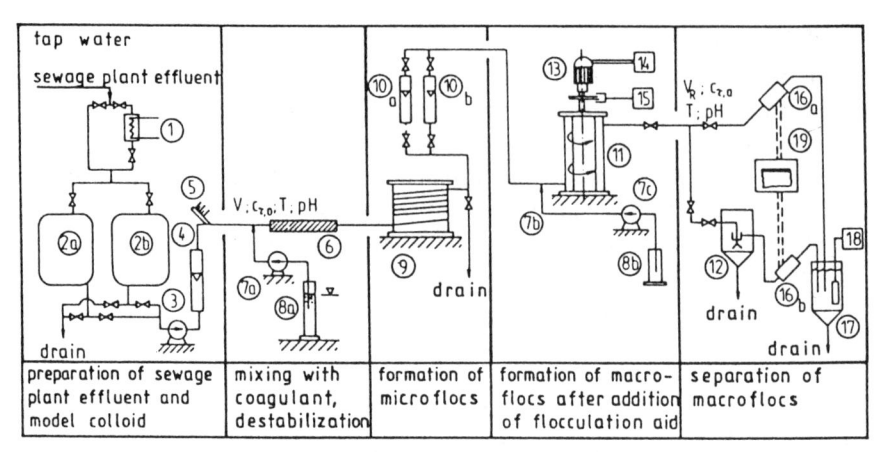

Figure 3. Laboratory-scale experimental set-up

polyacrylamide, the water flows through the cylindrical stirred reactor, where macroflocs are formed. Downstream, turbidity is measured discontinuously by two light scattering photometers in a closed cuvette after sedimentation and continuously after sedimentation in a vertical flow tank. Turbidity values, decreasing during sedimentation in the closed cuvette, are recorded. After 20 min, a practical constant value c_a is obtained, which is regarded as a measure of residual colloidal turbidity. The continuously measured turbidity c_{ak} is higher than c_a because of the presence of smaller flocs which cannot be separated.

Skimmed milk was used as a model colloid at a constant concentration of 10^{-3} l milk per l water showing turbidity of 26.2 FTU ± 10% [4]. It could be shown in previous experiments that turbid water from the Spree River and diluted skimmed milk have similar properties with regard to coagulation and flocculation.

Results

Figure 4 shows the relationship between the dimensionless energy input or Camp number as a function of the mechanical energy needed for flocculation

$$\text{Ca} = \bar{G} t_v \qquad (2)$$

and the residual turbidity c_{ak} for four different residence times between $t_v = 15$ and 120 s in the cylindrical stirred reactor. The concentrations of the coagulant and flocculation aid were $c(FeCl_3) = 116.5$ mg/l and $c(s) = 1$ mg/l, respectively. The measurements were carried out at 15°C and a gap width of $r_a - r_i = 2$ cm, which gives a dimensionless gap width

$$s = \frac{r_a^2 - r_i^2}{r_a^2} = 0.49 \qquad (3)$$

r_i is the inner and
r_a the outer radius of the cylinder gap

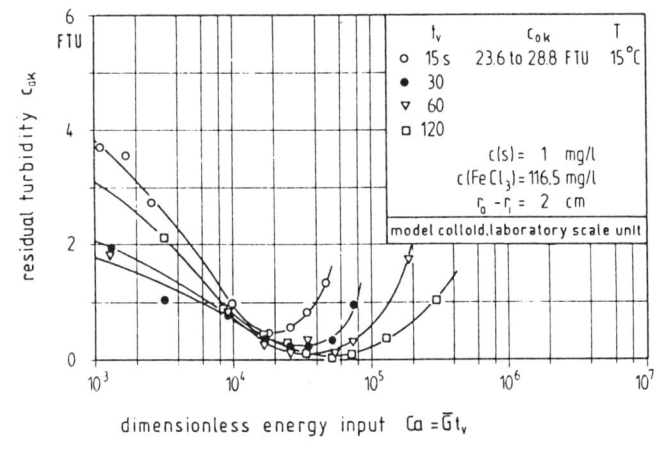

Figure 4. Residual turbidity versus dimensionless energy input for various residence times t_v; diluted skimmed milk, laboratory scale unit

Initially, the c_{ak} values decrease with increasing Ca number, reach a minimum at $10^4 <$ Ca $< 10^5$ and then increase, since the rate of floc break-up is now faster than that of floc formation. With short residence times and corresponding energy inputs, floc break-up begins at lower Ca numbers on account of higher velocity gradients or shear stresses. A definite residual turbidity can be achieved at different Ca numbers or combinations of t_v and $\bar{G}$. Optimization is based on economic considerations. For the shortest residence time of 15 s a residual turbidity of $c_{ak} < 0.5$ FTU could not be obtained.

Figure 5 shows the residual turbidity as a function of Ca number and of gap width $r_a - r_i$ for a mean residence time of 30 s. The flocculation efficiency decreases with increasing gap width on account of wider local velocity gradient and residence time distributions [2, 3, 4]. There is also an optimum gap width for a given residual turbidity. The most important result of these experiments is that a short mean residence time of only 30 to 120 s is needed to form settleable flocs. Is it possible to achieve similar results in a full scale process?

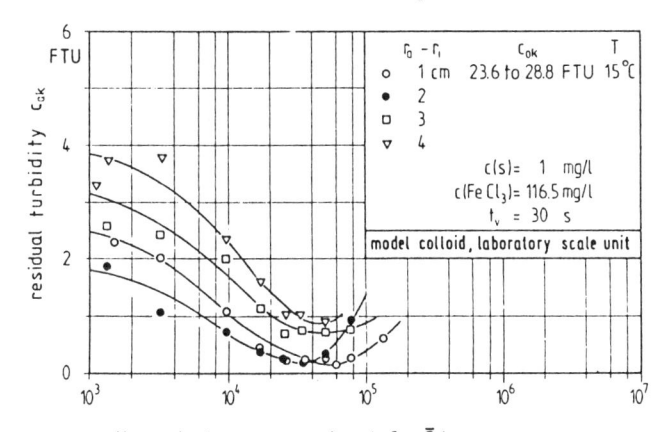

Figure 5. Residual turbidity versus dimensionless energy input for various gap widths $r_a - r_i$; diluted skimmed milk, laboratory scale unit

The Process of Desludging and Sludge Dewatering

The sludge is extracted by a floating suction dredger and pumped through a pipe to an area divided off from a parking lot close to the lake. After underground storage it passes through a screen which separates solid particles into two tube reactors for chemical sludge conditioning and two belt filter presses operating in parallel. The dewatered sludge is carried by lorries to a dump and used for covering. The filtrate is pumped to the water treatment plant via underground storage (Fig. 6).

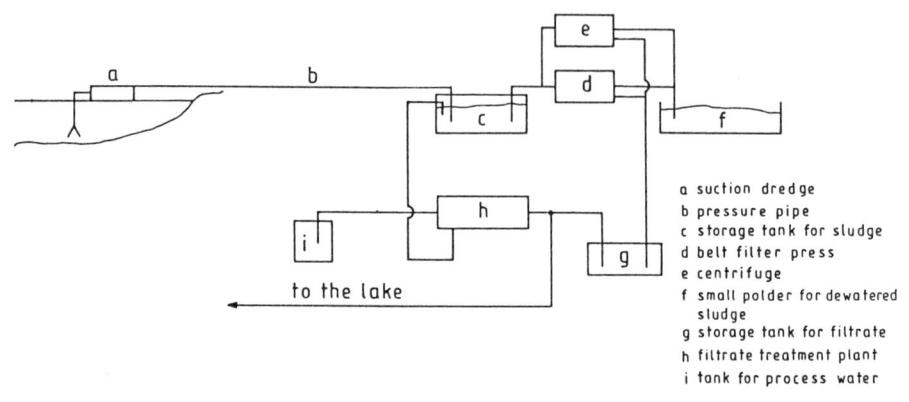

a suction dredge
b pressure pipe
c storage tank for sludge
d belt filter press
e centrifuge
f small polder for dewatered sludge
g storage tank for filtrate
h filtrate treatment plant
i tank for process water

Figure 6. Schematic diagram of the treatment plant for desludging of the Grunewaldsee, for sludge dewatering and filtrate purification

The Water Treatment Process, Results and Control

Quality of Lake Water and Filtrate

The quality of the Grunewaldsee water fluctuates, being governed by different vegetation periods. For example in November and December 1985 the following mean values were measured:

- total phosphorous concentration $c(P_T) = 0.06$ mg/l
- turbidity $c = 2$ TE/F

From August to October 1986 the $c(P_T)$ values for filtrate water quality varied daily and most of them were higher than those of the lake by a factor of ten.

Therefore treatment was necessary before the water could be discharged into the lake.

The Water Treatment Process

The plant consists of
- a facility for dosing chemicals for precipitation ($FeClSO_4$) and flocculation (Röhm SF 380 as a polyelectrolyte),

- a tube reactor for precipitation and the formation of microflocs,
- a cylindrical stirred tank reactor for flocculation,
- a countercurrent lamella separator for floc separation,
- and a station with various measuring instruments (Fig. 7).

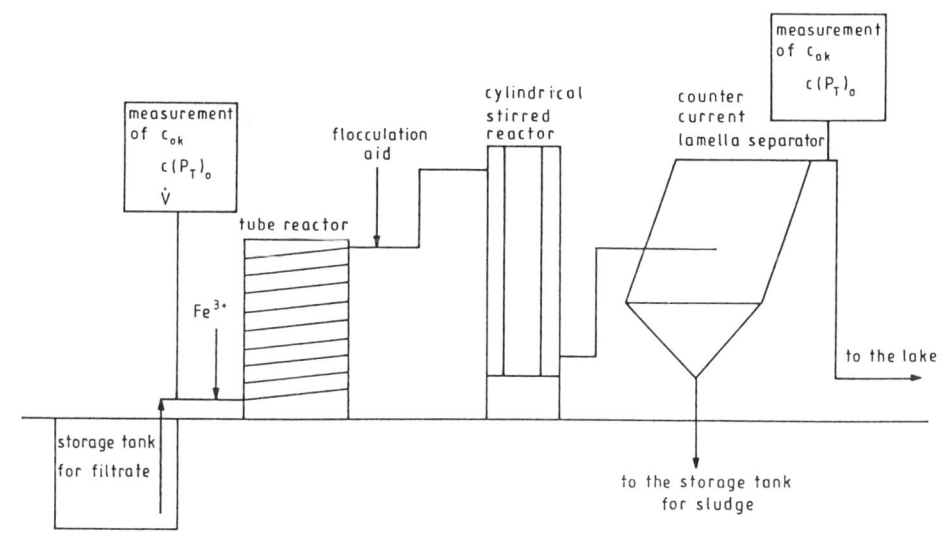

Figure 7. Full-scale set up of the filtrate treatment plant

Two plants were designed, erected and tested: a pilot plant for a flow rate of 10 m³/h (Dec. 1985), and a full scale plant for a flow rate of 60 m³/h (Sept.–Oct. 1986) [5]. For the cylindrical stirred tanks the geometric symmetry of the bench scale apparatus was maintained. The best laboratory results were obtained for a gap width of 2 cm. Therefore width $s = 0.49$ (see Eq. (3)) was maintained approximately, resulting in a gap width of 8.7 cm (pilot plant) and 29 cm (full scale plant). Both tanks were operated in the range of $Ca = 2$ to 4×10^4 corresponding to that of the bench scale reactor.

Effect of Coagulant Concentration

Figure 8 shows inlet turbidity c_{ok} and residual turbidity at inlet one c_{ak}/c_{ok} plotted against the concentration of the coagulant $c(Fe_3^+)$ with a mean residence time of 92 s (pilot plant) and 120 s (full scale plant). The upper curve represents the results from the pilot plant, the lower curve those from the full scale plant.

The higher residual turbidities of the pilot plant could be attributed to the higher inlet turbidities during Dec. 1985. Essentially, the results agree, showing residual turbidities lower than 2 FTU already at almost the same low mean residence times of 1 to 2 min and a concentration of $c(Fe_3^+) = 10$ mg/l. It could be demonstrated that the concentration of flocculation aid could be reduced from 2 to 1 mg/l Sedipur with no increase in turbidity.

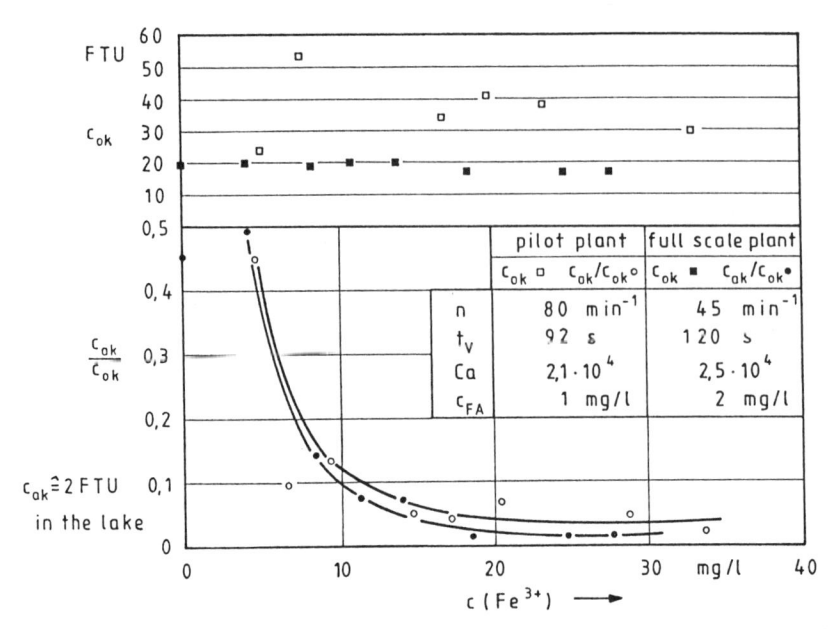

Figure 8. Inlet turbidity c_{ok} and residual turbidity at inlet one versus the concentration of the coagulant

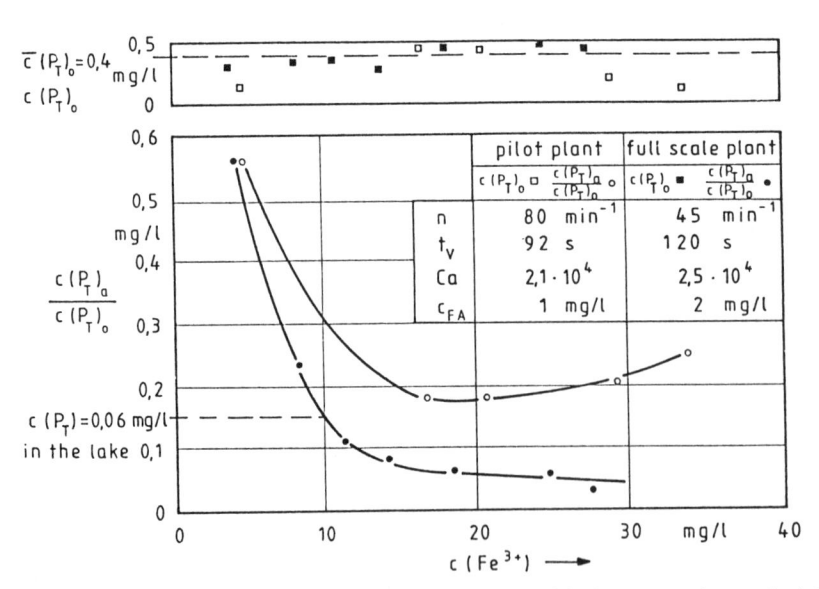

Figure 9. Total phosphorous concentration $c(P_T)_o$ and residual concentration at the inlet one versus the concentration of the coagulant

Figure 9 shows the relationship between total phosphorous concentration $c(P_T)_a/c(P_T)_0$ and the concentration of the coagulant $c(Fe^{3+})$. Total phosphorous concentration is composed of

$$c(P_T) = \underbrace{c(PO_4^{3-}\text{-}P)_f + c(PO_x\text{-}P)_f}_{\text{dissolved}} + \underbrace{c(P)_p}_{\text{colloidal}}$$

and can be removed so that $c(P_T)_a/c(P_T)_0 = 0.15$ and $c(P_T)_a = 0.06$ mg/l, which is the mean value of the Grunewaldsee, with the same concentration of $c(Fe^{3+}) = 10$ mg/l, which was enough to reduce turbidity to the required value of 2 FTU. The higher values in the pilot experiments were possibly caused by a lower temperature and a higher scattering of values at the inlet, resulting from operating conditions under which the filter press was changed frequently at that time.

A total mean residence time of only 14 min is sufficient (0.5 min for coagulation, 2 min for flocculation, 11.5 min for floc separation).

Process Control when Concentrations of Chemicals are Varied or Kept Constant

For a feedback process control, a continuous measurement of water quality is needed. So far no suitable process for the continuous measurement of the total phosphorous concentration has been developed. Therefore turbidity should be tested as an alternative. During the first run, constant concentrations of coagulant and flocculation aid of

$$c(Fe^{3+}) = 15 \text{ mg/l}$$
and $$c(FA) = 1.3 \text{ mg/l}$$

were added and an energy input of $Ca = 4 \times 10^4$ corresponding to a stirrer speed of $n = 65$ min^{-1} was set. Figure 10 shows the results for four consecutive days in

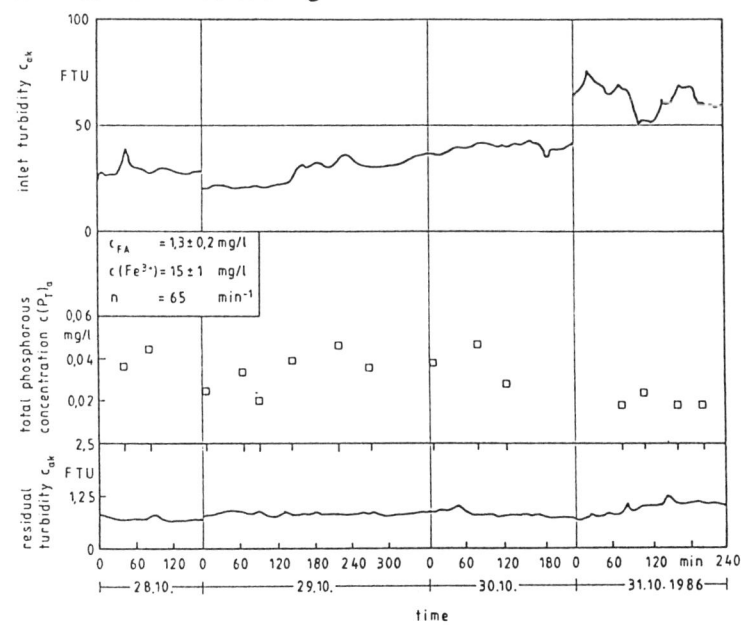

Figure 10. Inlet and residual turbidity as well as residual total phosphorous concentration versus time for four days — constant concentration of coagulant and flocculation aid

October 1986. Although inlet turbidity varied between 20 and 75 FTU, the residual values of approximately 1 FTU were nearly constant and all discharge concentrations of total phosphorous were lower than the required value of 0.06 mg/l.

The purpose of the next experiment was to show if chemicals could be saved with a feedback control by changing the concentrations of coagulant and flocculation aid. During this process the residual turbidity had to be between $0.75 < c_{ak} < 1.75$ FTU. This could be achieved without any problems for the period of five hours which was available for this test, as seen in Figure 11.

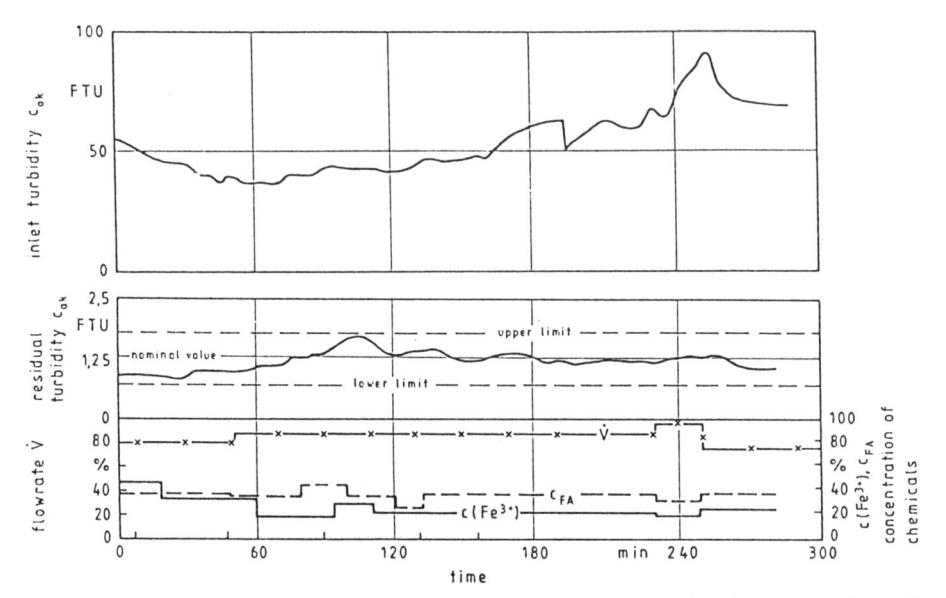

Figure 11. Inlet and residual turbidity versus time — back control by varying the amount of coagulant and flocculation aid. $100\% = 30$ mg/l Fe^{3+}, 4 mg/l FA

For a mean concentration of flocculation aid of

$$\bar{c}(FA) = 1.5 \text{ mg/l}$$

the mean concentration of coagulant could be reduced to

$$\bar{c}(Fe^{3+}) = 10 \text{ mg/l}$$

although the inlet turbidity was higher than in the experiment discussed previously.

Summary

Cylindrical stirred tanks achieve a high flocculation efficiency due to their narrow local velocity gradient and residence time distributions. This could be shown by measurements in bench, pilot and full scale tests, which suggested that mean residence times of only one to two minutes are needed to obtain settleable flocs. Particularly, if only a small area is available for the erection of the treatment plant, cylindrical stirred tanks can be used with high efficiency.

As an example, the treatment process of a filtrate from dewatering digested sludge from a lake is discussed. The process consists of a tube reactor for coagulation and precipitation, a cylindrical stirred tank for flocculation and a counter-current lamella separator for floc separation. In order to obtain the water quality required for discharge into the lake (c_{ak} = 2 FTU, $c(P_T)$ = 0.06 mg/l), a total mean residence time of only 14 min is sufficient (0.5 min for coagulation, 2 min for flocculation, 11.5 min for floc separation). By means of a feedback control which uses the continously measured turbidity 0,75 < c_{ak} < 2 FTU as a control value and by dosing coagulants and flocculation aids in variable concentrations, a third of the coagulant $FeClSO_4$ could be saved in comparison to the standard process. The treatment plant has been in operation for nearly two years without any problems.

References

[1] Hässelbarth, U. (1977): Die Phosphorelimination aus dem Zufluß von Seen zur Sanierung stehender Gewässer. Z. Wasser Abwasser Forsch. *10*, 120 - 125

[2] Reiter, M., Schmidt, M., Wiesmann, U. (1980): Der durchströmte Zylinderrührer im Bereich der wirbelbehafteten Strömung — ein Apparat mit interessanten Eigenschaften. Verfahrenstechnik *14*, 577—582

[3] Reiter, M., Wiesmann, U., Grohmann, A. (1985): The Cylinder Stirred Tank — A New High Performance Flocculation Reactor. German Chemical Engineering *5*, 307 - 313

[4] Reiter, M. (1983): Untersuchungen zur Flokkulation im Zylinderrührreaktor und im durchströmten Rohr. Dissertation D 83, TU Berlin

[5] Oldenstein, K., Fechter, L., Wiesmann, U.: Wasseraufbereitung durch Eliminierung von Phosphaten und Trübstoffen in einer neuen leistungsstarken Flockungsanlage. Vortrag auf der Leipziger Frühjahrsmesse am 17.3.1987

U. Wiesmann, K. Oldenstein
Institut für Chemieingenieurtechnik
TU Berlin
Straße des 17. Juni 135
1000 Berlin
Fed. Rep. of Germany

L. Fechter
Ingenieurbüro für Energie- und Umwelttechnik
Halberstädter Str. 2
1000 Berlin 31
Fed. Rep. of Germany

Pretreatment for Wastewater Reclamation and Reuse

T. Asano and R. Mujeriego

Abstract

The recent trend toward the use of reclaimed municipal wastewater for purposes such as landscape and food crop irrigation, groundwater recharge, and recreational impoundment often requires tertiary or advanced wastewater treatment. These water reuse applications result in exposing the public to reclaimed wastewater, thus assurance of microbiological and, particularly, virological safety is of utmost importance. The principal treatment processes and operations for reuse in these situations are similar to surface water treatment for potable water supply; both normally include chemical coagulation followed by flocculation, sedimentation, filtration, and disinfection. Because of the considerably higher and variable concentrations of organics and turbidity in wastewater, optimization of the chemical coagulation-flocculation and filtration of secondary effluent has been difficult to achieve in practice. The high degree of pathogen removal achieved by a properly operated treatment system ensures the safety of the reclaimed wastewater.

To achieve efficient virus removal or inactivation in tertiary wastewater treatment, two major operating criteria must be met: (1) the effluent must be low in suspended solids and turbidity prior to disinfection to reduce shielding of viruses and chlorine demand and (2) sufficient disinfectant dose and contact time must be provided for wastewater. To satisfy the first criterion, tertiary filtration with adequate chemical coagulation normally follows secondary treatment.

The lack of practical information on the optimization of these process trains has, however, hampered the establishment of more cost-effective wastewater reclamation methods for tertiary treatment. The purpose of this paper is, therefore, to review the recent studies conducted in California on the performance of tertiary treatment systems for wastewater reclamation and reuse. Following the Pomona Virus Study [6, 7, 8], three recently completed studies: the Health Effects Study [9], the Las Virgenes Filtration-Disinfection Study [15], and the Monterey Wastewater Reclamation Study for Agriculture [10, 14] are examined.

Introduction

There are a number of factors which affect the implementation of municipal wastewater reclamation and reuse projects. Generally, the impetus for water reuse in industrialized countries has resulted from four motivating factors:

1) Increasing costs of freshwater development.
2) Desirability of establishing comprehensive water resource planning, including water conservation and wastewater reuse.
3) Availability of high quality effluents.
4) Avoidance of more stringent water pollution control requirements such as needs for advanced wastewater treatment facilities.

The general factors affecting wastewater reuse decisions include: (1) local and regional water supply conditions, (2) water quality requirements for intended water reuse applications, (3) existing or proposed wastewater treatment facilities, (4) requirements for degree of treatment and process reliability, (5) potential health risks mitigation, (6) public acceptance, and (7) financing reuse facilities including sale of reclaimed water.

The recent trend toward the use of reclaimed municipal wastewater for purposes such as landscape and food crop irrigation, groundwater recharge, and recreational impoundment often requires tertiary or advanced wastewater treatment. These water reuse applications often result in exposure to reclaimed wastewater; thus assurance of microbiological and, particularly, virological safety is of utmost importance. Principal treatment processes and operations for tertiary treatment of municipal wastewater for reuse and treatment of surface water for drinking water are similar; both normally include chemical coagulation followed by flocculation, sedimentation, filtration, and disinfection. Alternatively, direct filtration with lower chemical doses and without sedimentation is also used. It is known that both bacterial pathogens and viruses are removed in these processes in varying degrees. The high degree of pathogen removal achieved by a properly operated treatment system ensures the safety of the reclaimed wastewater.

In many instances, however, optimization of the chemical coagulation-flocculation and filtration of secondary effluent has been difficult to achieve in practice. Lack of information on the optimization of these processes in wastewater reclamation and reuse has hampered the establishment of more cost-effective wastewater reclamation methods for tertiary treatment. Therefore, the purpose of this paper is to review the recent studies conducted in California on the performance of tertiary treatment systems for producing reclaimed water that has an extremely low probability of bacterial and viral contamination.

Virus Concentrations in Municipal Wastewater

Although virus concentrations have been reported to be lower in municipal wastewater from U.S. sources than from many other countries, over 100 types of enteric viruses can be present in untreated municipal wastewater [1, 2]. Many of the enteroviral diseases reported have not been associated directly with water route transmission. However, Hepatitis type A, the virus causing infectious hepatitis, and documented to be transmitted by contaminated water, is the virus reported most frequently. Diseases associated with Rotavirus and Norwalk agent are becoming of increasing concern to public health officials. There are undoubtedly incidents of waterborne transmission of viruses that are not recognized, investigated, or reported [3].

The actual viral concentration in untreated municipal wastewater varies considerably in different locations and seasons. In the United States, reported virus concentrations range from a low of about 200 plaqueforming units per liter (PFU/l) in cold climate to 7,000 PFU/l in warm months [4, 5]. However, reported virus concentrations must be viewed with care. There is no universal procedure for the cultivation of all viruses. Each procedure is selective with respect to the viruses enumerated and is affected by the method of concentrating the viruses in the sample, selection of a host cell, and the type of culture techniques used [4]. In addition, because low concentrations of naturally occurring viruses are expected in the tertiary treatment effluents (e.g., less than 0.5 PFU/l), overall concentration factors of 15,000 to 75,000 are often necessary when 379 l samples are processed. The average overall recovery efficiency was reported to be about 20 percent in this circumstance [6, 7, 8]. However, in the recently completed Health Effects Study [9], a newly developed portable virus concentrator was capable of concentrating up to 3,790 l of well water and 1,137 l of wastewater with average virus recoveries of 83 percent and 42 percent, respectively, as determined with seeded poliovirus.

In unchlorinated secondary effluent, viruses were detected in 27 samples out of 60 samples in the Pomona Virus Study [6] in California. The geometric mean of the natural virus concentration that yielded plaques was about 5 PFU/l. Polio-, Reo-, Echo-, and Adenoviruses were identified. Animal viruses were also isolated from unchlorinated activated sludge effluent from the Castroville Wastewater Treatment Plant in California where the Monterey Wastewater Reclamation Study for Agriculture (MWRSA) was conducted. The influent to the two pilot tertiary treatment plants (unchlorinated activated sludge effluent) contained measurable viruses 80 percent of the time sampled, averaging 22 PFU/l ranging from 1 to 734 PFU/l and the highest virus concentrations were found in October and November, the warmest months in this region [10].

At the Orange County Water Factory 21 in California, a total of 35 unchlorinated secondary effluent samples were analyzed for enteric viruses and 27 samples were found to be positive during the study period. The geometric means of virus concentration were 0.025 to 0.015 most probable number of cytopathic units (MPNCU) per liter for the Buffalo Green Monkey Kidney (BGM) and RD cell lines, respectively, when activated sludge effluent was tested. It was also reported that the activated sludge effluent contained 1.5 log orders of magnitude less viruses than the trickling filter effluent [11].

Viruses were also isolated routinely from unchlorinated secondary effluent samples collected for the Health Effects Study [9]; they ranged from 0.1 to 17 PFU/l. With improved virus concentration and assay methods, detection of viruses in treated effluents is expected to increase.

Treatment Processes Capable of Producing Essentially Virus-free Effluent

Because viruses have been detected in unchlorinated secondary effluents and even in chlorinated secondary effluents, it is prudent to remove them where a high degree of public exposure to the reclaimed water is expected. Viruses are associated with suspended and colloidal solids and also may be embedded in organic solids and human excrement [12, 13]. Thus, removal or inactivation of viruses from wastewater depends on the level of wastewater treatment, residual solids concentration, and disinfection.

To achieve efficient virus removal or inactivation in tertiary treatment, two major criteria must be met: (1) the effluent must be low in suspended solids and turbidity prior to disinfection to reduce shielding of viruses and chlorine demand, and (2) sufficient disinfectant dose and contact time must be provided for wastewater.

Virus removal or inactivation during municipal wastewater treatment has been studied in field-scale operations and even more intensely in laboratory bench-scale units. There are potential problems in the extrapolation of virus data derived at one treatment plant to other similar treatment plants. The non- specificity of *in situ* virus enumeration is an important cause of the difficulty. Differences in virus removal data from one plant to another may be a reflection of the types of viruses present in a given wastewater, and also a reflection or virus species resistance to the treatment processes. In laboratory or in large scale seeding studies, a specific type of virus (usually vaccine strains of poliovirus) is used. The extrapolation of these data is subject to the aforementioned problem. The most important aspect of virus seeding experiments is that non-zero virus concentrations can be achieved even in highly treated effluents allowing calculations of virus removal efficiency. It should be pointed out, however, that the virus concentrations used were far in excess (usually in a range of 10^4 to 10^{11} PFU/l) of the indigenous virus concentrations common in municipal wastewater, thus the similitude of virus removal and/or inactivation is not exactly known [5, 16].

Importance of High Quality Secondary Effluent

In the evaluation of virus removal capabilities in the coagulation-filtration system, the County Sanitation Districts of Los Angeles County conducted a series of tests related to the determination of optimal alum and polymer dosages and their relation to the headloss buildup in the filter. The investigation known as the Pomona Virus Study [6, 7, 8] provided useful information on the performance of tertiary treatment systems related to virus removal efficiency and reliability. The performance of the chemical coagulation, flocculation, sedimentation, and filtration system (T-22 System) operated at various alum doses ranging from 55 to 255 mg/l was reported. Each series of headloss data for the dual-media filter represented the observed headloss at the end of the first 7.5 hours of a filter run. A filter effluent turbidity of 0.2 – 0.4 FTU was achieved at an alum dosage of about 155 mg/l and a polymer (Calgon WT-3000 anionic polymer) dose of 0.2 mg/l.

The performance of the dual-media filter with direct filtration mode (coagulation and filtration without sedimentation, Fe Process) indicated that the filter effluent turbidity of 0.2 FTU or less could not be achieved at any of the alum and polymer doses evaluated. In the absence of alum, however, filter effluent turbidities ranging from 0.2 to 0.8 FTU were observed [6]. In other words, the optimization of coagulation-flocculation and filtration processes in a direct filtration system was not readily established. It appears that flocculation time prior to the dual-media filtration was not long enough for effective turbidity removal, although short flocculation time-filtration processes have been designed and operated successfully in water treatment.

A similar difficulty in process optimization was experienced in the Monterey Wastewater Reclamation Study for Agriculture (MWRSA). To improve turbidity and virus removal efficiency with direct filtration, it was necessary to add a mechanical turbine rapid mixer and flocculation chamber prior to the dual-media filter. A series of tests were conducted to determine an optimal combination of operational parameters including alum/polymer dosages, energy inputs, and flocculation time [10, 14].

To produce essentially virus-free effluent using the direct filtration system, the secondary effluent must be of a high quality. To meet a filtered effluent turbidity of less than 1 NTU, the water quality of secondary effluent, based on the available data, must be in the neighborhood of: suspended solids 10 – 15 mg/l, turbidity 3 – 6 NTU, and total COD 40 – 80 mg/l. In full - scale operating plants, a secondary effluent turbidity of about 5 NTU or less is recommended to meet consistently an average operating turbidity of 2 NTU in the filtered effluent using the direct filtration system. It appears that the secondary effluent turbidity of 10 NTU is an economic dividing line in this case above which improved operation of secondary treatment is more cost-effective and warranted [16].

If secondary effluents do not meet the water quality ranges cited above, more costly complete treatment, the T-22 system, must be employed. The tertiary treatment system required would be high- dose chemical coagulation, flocculation, sedimentation, and filtration followed by chlorination, and costs can be excessive for most reuse applications.

Table 1 summarizes the concentrations of BOD, TSS, and turbidity of the secondary effluent, filtered effluent (FE system), and T-22 effluent in the MWRSA. A consistent ratio of about 2 to 1 can be observed in the performance of the two tertiary treatment sequences (FE and T-22) in terms of their respective total suspended solids and turbidity.

With regard to the 2:1 ratio observed with turbidity and TSS between the two treatment trains, it is believed that this difference is due to the high dose of chemicals and the sedimentation process in the T-22 scheme, and the resultant difference in filterability.

Virus Removal in Tertiary Treatment and Disinfection

In the Pomona Virus Study [6, 7, 8], a series of experiments were conducted to determine the virus removal efficiency of the tertiary treatment systems. The overall virus removal efficiencies for both the complete treatment (T-22) and the direct filtration (FE-process) were virtually the same when residual chlorine doses of about 10 mg/l and a two-hour chlorine contact time were used. Approximately, 5.0 to 5.2 log removal of seeded poliovirus was observed in these pilot-scale experiments. When an average combined chlorine residual of 5.0 mg/l and a two-

Table 1. Log-normal probability distribution of treatment data in the Monterey Wastewater Reclamation Study for Agriculture [10, 14]

Parameter	No. samples	Percent chance of parameter value being less than or equal to that listed below						Maximum value
		50	80	90	96	98	99	
BOD_5 [mg/l]								
SE	74	14.3	22.3	28.0	35.9	42.1	48.6	53
Total suspended solids [mg/l]								
SE	302	13.4	19.5	23.7	29.2	33.4	37.7	38
FE	286	1.6	3.1	4.3	6.3	8.0	10.0	17
FC	275	4.4	7.6	10.1	13.8	16.8	20.2	59
T-22	273	0.8	1.5	2.1	3.0	3.8	4.7	12
Turbidity [NTU]								
SE	286	3.8	5.5	6.7	8.2	9.3	10.5	12.0
FE	282	1.1	1.7	2.2	2.9	3.4	4.0	9.4
T-22	262	0.6	0.9	1.1	1.5	1.7	2.0	3.4

Key: NTU Nephelometric Turbidity Units
 SE secondary effluent
 FE filtered effluent with flocculator installed
 FC flocculator-clarifier effluent
 T-22 Title-22 effluent (with 50-200 mg/l alum + 0.2 mg/l polymer followed by flocculation, sedimentation, and filtration)

hour contact time were used, differences in the virus removal efficiencies between two different treatment systems became apparent.

The average removal of poliovirus 1 in the coagulation-flocculation and sedimentation system (complete treatment) was in the range of 1.3 to 1.5 log removal or about 95 percent. The direct filtration removed about 1 log of virus removal or 90 percent. The seeded secondary effluent (unchlorinated) had viral counts in the range of 1.3×10^5 PFU/l whereas the geometric mean of the background level of naturally occuring viruses was 5 PFU/l [6].

The Monterey Wastewater Reclamation Study for Agriculture (MWRSA) conducted extensive virus monitoring with respect to: (1) pilot plant influent and effluent for the presence of animal enteric viruses, (2) poliovirus seeding studies designed to estimate virus removal efficiency in the tertiary treatment systems including complete treatment and direct filtration, (3) a study of poliovirus survival on *in situ* crops at Castroville, California [10].

Both the complete treatment system (T-22 System) and the direct filtration system (FE System) were seeded with vaccine-strain poliovirus to determine their virus removal and inactivation efficiencies. It was found that the direct filtration system was somewhat less efficient with an average of 6.1 log removal compared to the complete treatment train which removed about 7.5 logs. It was suspected that the difference in virus removal efficiency was due to the improper flocculation time inherent to the direct filtration design used in MWRSA. In addition, the low chlorine residuals (about 5 mg/l) used in the disinfection step with 90 minute theoretical contact time appear, on the average, to be less effective (6.1 to 3.2 log reduction). With optimization of the flocculation process and/or higher chlorine residuals (about 10 mg/l), it is expected that the virus removal capabilities of the tertiary treatment system at MWRSA will improve and become comparable to those of the Pomona Virus Study.

Ancillary to the performance and reliability study of full-scale advanced wastewater treatment system at the Orange County Water Factory 21, monitoring for viruses and parasites was conducted and the effectiveness of their removal by the advanced wastewater treatment processes was determined. From the data obtained in this study, it appears that a very low virus level (approximately 2 viruses per day in 57,000 m^3 of water) may pass through the last treatment barriers, the reverse osmosis unit and chlorination process. Enteric viruses were detected in the advanced waste treatment effluents only once during the three- year study [11].

The Las Virgenes Filtration-Disinfection Study is being conducted to determine the optimum filtration and disinfection criteria for the Tapia Water Reclamation Facility in California that would meet the State of California Wastewater Reclamation Criteria shown in Table 2. The most restrictive wastewater reclamation criteria specify a tertiary process train that results in an effluent that is essentially free of viruses. At the Tapia Facility, the tertiary process consisted of coagulation with 4 mg/l chlorine and 1 mg/l alum, filtration with single-medium deep bed filters, and disinfection [15]. The results of the microbiological investigations will be used for determining the filtration and disinfection requirements which involve intensive virus monitoring. The target operating conditions for the tertiary processes are: filtration rate $167-224$ $l/min/m^2$; alum addition 1 mg/l; chlorine residual $5.5-11.0$ mg/l.

For all combinations of conditions studied, the data, so far, indicated that at least 99.9% of the total coliforms entering the tertiary process were removed during coagulation, pre-disinfection, and filtration. After additional chlorine was added, following filtration, to maintain a target residual chlorine of 11.0 mg/l in the contact basin, an additional 94.2 to 95.9% removal was achieved after one hour in the contact basin. When the target residual chlorine level was 5.5 mg/l, 89.4% of the total coliforms were removed after one hour of contact time.

Enteric viruses were detected in all 12 unchlorinated secondary effluent samples analyzed. All samples were positive for enteric viruses, with a geometric mean concentration of 0.03 MPNCU/l [15].

Water Quality Criteria for Irrigation

The health significance of an extremely small probability of virus contamination and few orders of magnitude difference in virus removal efficiency has not been established. It is highly uncertain, however, that additional epidemiological and laboratory studies would yield significant new information on the relationship between viruses in reclaimed municipal wastewater and disease. The long-term health effects of ingesting or being exposed to such reclaimed wastewater are unknown, but will always be of concern regardless of the final water quality.

A basic objective of the State of California regulations, entitled "Wastewater Reclamation Criteria" (issued by the Department of Health Services, 1978), is to assure health protection without unnecessarily discouraging wastewater reclamation. The regulations specify wastewater reuse standards for uses involving agricultural and landscape irrigation, impoundments, and groundwater recharge. The regulations include water quality standards, treatment process requirements, sampling and analysis requirements, operational requirements, and treatment reliability requirements. The required degree of treatment increases as the likelihood of human exposure to the wastewater increases. The treatment and quality requirements for the irrigation uses covered by the Wastewater Reclamation Criteria are summarized in Table 2. The reclamation criteria are intended to assure an adequate degree of health protection from disease transmission and do not specifically address the potential effects of reclaimed water on the crops or soil in agricultural and landscape irrigation.

Table 2. Wastewater treatment and water quality criteria for irrigation *

Treatment level	Total Coliform (Median MPN/100 ml)	Type of use
Primary		Surface irrigation of orchards and vineyards, fodder, fiber, and seed crops
Oxidation and disinfection	≤23/100 ml	Pasture for milking animals
		Landscape impoundments
		Landscape irrigation (golf courses, cementeries, etc.)
	≤2.2/100 ml	Surface irrigation of food crops (no contact between water and edible portion of crop)
Oxidation, coagulation, clarification, filtration**, and disinfection	≤2.2/100 ml max. = 23/100 ml	Spray irrgation of food crops
		Landscape irrigation (parks, playgrounds, etc.)

* Excerpted from "Wastewater Reclamation Criteria", California Administrative Code, Title 22, Division 4, Environmental Health (1978)
** The turbidity of filtered effluent cannot exceed an average of 2 turbidity units (NTU) during any 24-hour period.

In the most stringent requirements, such as for spray irrigation of food crops and landscape irrigation of parks and playgrounds, tertiary effluent that is "pathogen-free" is required (as shown in less than 2.2 total coliform, MPN/100 ml on seven-day average and the number of coliforms does not exceed 23 MPN/100 ml, in any sample). It is assumed, in this case, that essentially virus-free effluent is obtained through various combinations of tertiary treatment and disinfection.

The unit cost of tertiary treatment trains which produced reclaimed water of essentially virus-free quality was estimated in the MWRSA. The full 1.3 m^3/s (30 mgd) capacity of the wastewater reclamation facilities is expected to be required for irrigation during an average of 250 days per year. The costs of producing filtered effluent (FE system) is estimated to be $0.06/m^3, and $0.09/m^3 for the T-22 System [10].

Summary and Conclusions

Because of uncertainties associated with risk assessment of viruses in reclaimed wastewater, improvements in wastewater treatment technology and operation of both conventional and tertiary wastewater treatment plants are warranted. By optimizing wastewater treatment and the coagulation-filtration processes coupled with effective chlorination, it is reasonable to expect that essentially virus-free reclaimed water can be produced in a cost-effective manner. The preferred methods of achieving essentially virus-free reclaimed wastewater are (1) to focus more attention on improving the quality of secondary effluent and operational reliability and (2) to optimize chemical coagulation-flocculation in direct filtration systems. Both of these measures would allow for the use of lower chlorine dosages in disinfection without jeopardizing virus removal and/or inactivation efficiency.

References

[1] Akin, E.W. et al. (1978): Health Hazards Associated with Wastewater Effluents and Sludge: Microbiological Considerations. In: Sagik, B.P., Sorber, C.A. (eds.), Risk Assessment and Health Effects of Land Application of Municipal Wastewater and Sludges, Center for Applied Research and Technology, the University of Texas, San Antonio, Texas

[2] Melnick, J.L.: Are Conventional Methods of Epidemiology Appropriate for Risk Assessment of Virus Contamination of Water? Op cit.

[3] Gerba, C.P., Goyal, S.M. (1985): Pathogen Removal from Wastewater during Groundwater Recharge. In: Asano, T. (ed.), Artificial Recharge of Groundwater, Butterworth Publishers, Boston

[4] Malina, J.E. Jr. (1977): The Effect of Unit Processes of Water and Wastewater Treatment on Virus Removal. In: Borchardt, J.A. et al. (eds.), Virus and Trace Contaminants in Water and Wastewater, Ann Arbor Science Publishers Inc., Ann Arbor, Michigan

[5] Shaffer, P.T.B.: Virus Detection Methods — Comparison and Evaluation. Op cit.

[6] California State Water Resources Control Board, Sacramento (1977): Sanitation Districts of Los Angeles County, Pomona Virus Study — Final Report.

[7] Dryden, F.D., Chen, C.L., Selna, M.W. (1979): Virus Removal in Advanced Wastewater Treatment Systems. Journal WPCF 51, 8, 2098

[8] Miele, R.P., Selna, M.W. (1977): Virus Sampling in Wastewater - Field Experiences. J. Environ. Eng. Div., ASCE 103, EE4, 693

[9] Nellor, M.H., Baird, R.B., Smyth, J.R. (1984): Health Effects Study — Final Report, prepared for the Orange and Los Angeles Counties Water Reuse Study, County Sanitation Districts of Los Angeles County. Whittier, California, March 1984

[10] Final Report, Monterey Wastewater Reclamation Study for Agriculture, Prepared for Monterey Regional Water Pollution Control Agency by Engineering - Science, Inc., April 1987

[11] McCarty, P.L. et al. (1984): Advanced Treatment for Wastewater Reclamation at Water Factory 21, prepared for Municipal Environmental Research Laboratory, US Environmental Protection Agency, Cincinnati, Ohio, EPA-600/2-84-031, January 1984

[12] Moore, B.E., Sagik, B.P., Malina, J.F. Jr. (1975): Viral Association with Suspended Solids. Water Research 9, 197, Pergamon Press, Oxford, England

[13] Sproul, O.J. (1980): Critical Review of Virus Removal by Coagulation Processes and pH Modifications. Prepared for Municipal Environmental Research Laboratory, US Environmental Protection Agency, Cincinnati, Ohio, EPA-600/2-80-004, June 1980

[14] Kirkpatrick, W.R., Asano, T. (1986): Evaluation of Tertiary Treatment Systems for Wastewater Reclamation and Reuse. Water Science and Technology *10*, Vol. 18, 83
[15] Colbaugh, J.E. et al. (1987): Fate of Microorganisms in a Tertiary Plant for Meeting California Primary Contact Standards. Presented at the Water Reuse Symposium IV, Denver, Colorado, August 2-7, 1987
[16] Asano, R., Tchobanoglous G., Cooper, R.C. (1984): Significance of Coagulation-Flocculation and Filtration Operations in Wastewater Reclamation and Reuse. Proceedings of the Water Reuse III, San Diego, California, August 26-31, 1984

T. Asano
California State Water Resources Control Board
P. O. Box 100
Sacramento, California 95801
and
Department of Civil Engineering
University of California
USA

R. Mujeriego
Escuela Técnica Superior
de Ingenieros de Caminos
Universidad Politécnica
de Cataluña
Jorge Girona Salgado, 31
08034 Barcelona
Spain